SECOND EDITION

PHYSICS MATTERS

NICK ENGLAND

Hodder & Stoughton

A MEMBER OF THE HODDER HEADLINE GROUP

Contents

Preface

The first edition of *Physics Matters* was written in 1989 to meet the demands of the newly introduced GCSE courses. The purpose then was to make the book lively, interesting and relevant to the world around us. The second edition continues with these aims, but has been updated to take account of the changes introduced by the National Curriculum. This edition is still aimed at those people who wish to take Physics as a single subject. A parallel edition *Physical Processes*, together with *Materials*, and *Life and Living* forms a series of books for a co–ordinated Science Course.

The new major features of the book are chapters on Astronomy and Earth Sciences. The chapter on Astronomy covers the Solar System, Stars, Galaxies and Cosmological Theories. The Earth Science chapter, written by Rosalind Jones, deals with Meteorology, Soils, Rocks, Seismology and Plate Tectonics. The chapter on Waves now includes additional material on Communication Systems. You will also find a variety of new questions.

I am particularly grateful to the following people who have helped me prepare this edition: Rosalind Jones for writing the chapter on Meteorology and Geology; David Mackin and Jim Bennetts who have offered invaluable advice about content and style; Sue England for her forbearance, encouragement and secretarial skills; Gina Walker and Julia Cousins for their editorial work. In addition I would like to thank my colleagues in the Physics Department at Wellington College, particularly David Harrison and Paul Barratt, for their continuing support and advice.

Nick England
Crowthorne

⚠ Safety Note

Although this is essentially a theoretical book, it has been checked carefully in the context of health and safety. I am grateful to Mr L Jefferies who has carried out a full safety review. In particular, we have tried to ensure that:

- all recognised hazards have been identified
- suitable precautions are suggested
- proposed procedures are in accordance with commonly adopted general risk assessments.

However, teachers should be aware that it is impossible to cater for every situation. Therefore, before any practical work is carried out, teachers should assess the risks involved. In particular, teachers must adhere to local rules issued by their employer.

The following references offer detailed advice, for general risk assessments:

- *Hazcards* (CLEAPSS, 1989)
- *Topics in Safety* (ASE, second edition, 1988)
- *Safeguards in the School Laboratory* (ASE, ninth edition, 1988)
- *Hazardous Chemicals. A Manual for Schools* (SSSERC/Oliver & Boyd, 1979)

SECTION A
Forces

If this spider could make her web out of steel of a similar thickness it would probably break when a fly hit it

1 Density

Concorde is made from aluminium to give it a low density and high strength

Material	Density (kg/m³)
gold	19 300
mercury	13 600
lead	11 400
steel	8 000
titanium	4 500
aluminium	2 700
glass	2 500
water	1 000
cork	200
air	1.3
hydrogen	0.09

Table 1

A tree obviously weighs more than a nail. Sometimes, you hear people say 'steel is heavier than wood'. What they mean is this. A piece of steel is heavier than a piece of wood with the same volume.

To compare the heaviness of materials we use the idea of **density**. Density can be calculated using this equation:

$$\text{density} = \frac{\text{mass}}{\text{volume}} \quad or \quad d = \frac{m}{V}$$

Density is usually measured in units of kg/m³.

Using density in engineering

Knowing the density of materials is very important to an engineer. This allows her to calculate the mass of building materials.

Example. What is the mass of a steel girder which is 10 m long, 0.1 m high and 0.1 m wide?

$$\text{The volume of the girder} = 10\,\text{m} \times 0.1\,\text{m} \times 0.1\,\text{m}$$
$$= 0.1\,\text{m}^3$$
$$d = \frac{m}{V}$$
$$\text{So } m = d \times V$$
$$= 8000\,\text{kg/m}^3 \times 0.1\,\text{m}^3$$
$$= 800\,\text{kg}$$

Steel is a very common building material because it is so strong. Despite this, in aeroplanes aluminium and titanium are used because they have low densities. It is important to make an aeroplane as light as possible.

Glass fibre is one of the most important modern building materials. It is made by strengthening plastic with glass fibres. Table 2 allows us to compare steel and glass fibre. Glass fibre is actually a little stronger than mild steel. This means a larger force is needed to break it. Glass fibre has a much lower density than steel. This makes it ideal for building small boats. Unfortunately glass fibre cannot be used for very large boats because it bends too much.

	Steel	Glass fibre
relative strength	40 000	50 000
density (kg/m³)	8000	2000
$\frac{\text{strength}}{\text{density}}$	5	25

Table 2

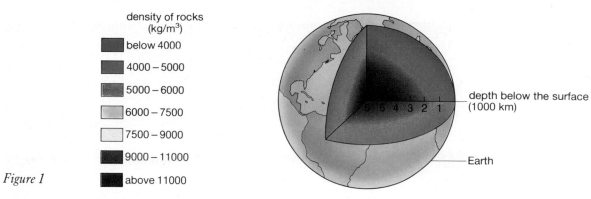

Figure 1

Density of rocks

Geologists are interested in the density of rocks. Figure 1 shows the density of different rocks in the Earth. Blue shows rocks of low density and red shows rocks of higher density. The Earth gets denser towards the centre.

Rocks near the surface of the Earth, such as granite, have densities of about 2700 kg/m³. Volcanic rocks have higher densities. This is because the lava that is thrown out of volcanoes comes from deep below the Earth's surface, where the density is higher.

Will it float or sink?

The density of an object determines whether it sinks or floats. Cork floats on water, steel sinks. Cork is less dense than water, steel is more dense. An object only floats on a liquid if it is less dense than the liquid.

Geologists use this idea to separate out minerals. Pitchblende, a valuable material because it contains uranium, is usually found in granite rocks. Granite and pitch-blende can be separated because of their different densities. The mixture of the two is crushed up. Then it is all put into bromoform (tribromomethane), a dense, toxic and carcinogenic liquid. The pitchblende sinks but the lighter granite floats.

⚠ Tribromomethane is toxic and carcinogenic

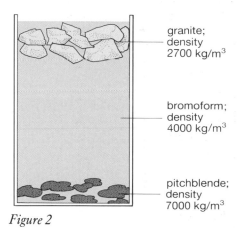

Figure 2

Questions

1 (a) Explain why aluminium and titanium are used to build aeroplanes.
(b) In Table 2, the last row is headed 'strength/density'. Explain why this is an important ratio.
2 Look carefully at Figure 1.
(a) Sketch a graph to show the density of rocks in the Earth (*y*-axis) against depth below the surface (*x*-axis).
(b) The outer part of the Earth is solid; this is called the mantle. The inner part of the Earth is liquid; this is called the core. Use your graph to guess where the mantle meets the core.
3 Carole is a geologist. She wants to work out the density of a rock. First she weighs the rock, then she puts it into a beaker of water to work out its volume.

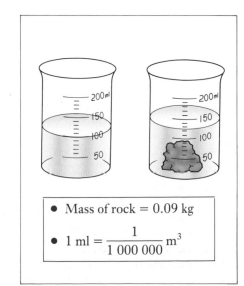

- Mass of rock = 0.09 kg
- $1 \text{ ml} = \dfrac{1}{1\,000\,000} \text{ m}^3$

(a) Use the diagrams to calculate the rock's volume. Give your answer in m³. (The markings on the beaker are all in ml.)
(b) Now calculate the rock's density in kg/m³.
4 What is the volume of:
(a) 1000 kg of aluminium?
(b) 100 kg of cork?
5 White dwarf stars are extremely dense. They have a density of about 100 million kg/m³. If you had a matchbox full of material from a white dwarf, what would its mass be? (Hint: A matchbox has a volume of about 0.000 05 m³.)

2 Forces Near and Far

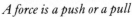

A force is a push or a pull

What is a force?

A force is a push or a pull. Whenever you push or pull something you are exerting a force on it. The forces that you exert can cause three things:
- You can change the shape of an object. You can stretch or squash a spring. You can bend or break a ruler.
- You can change the speed of an object. You increase the speed of a ball when you throw it. You decrease its speed when you catch it.
- A force can also change the direction in which something is travelling. We use a steering wheel to turn a car.

The forces described so far are what we call **contact forces**. Your hand touches something to exert a force. There are also **non-contact forces**. Gravitational, magnetic and electric forces are non-contact forces. These forces can act over large distances without two objects touching. The Earth pulls you down whether or not your feet are on the ground. Although the Earth is 150 million km away from the Sun, the Sun's gravitational pull keeps us in orbit around it. Magnets also exert forces on each other without coming into contact.

The size of forces

The unit we use to measure force is the **newton**. A force of 1 newton (1 N) is defined in terms of how quickly that force can change the speed of a 1 kg mass (see page 53). The box below will help you to get the feel of the size of several forces.

- The pull of gravity on a fly = 0.001 N
- The pull of gravity on an apple = 1 N
- The frictional force slowing a rolling football = 2 N
- The force required to squash an egg = 50 N
- The pull of gravity on you = 500 N
- The frictional force exerted by the brakes of a car = 5000 N
- The push from the engines of a rocket = 1 000 000 N

Two important forces

The pull of gravity and friction are two forces that we notice every day of our lives.

Weight is the name that we give to the pull of gravity on an object. Near the Earth's surface the pull of gravity is 10 N on each kilogram. We say that the Earth's gravitational field strength is 10 N/kg.

Example. What is your weight if your body has a mass of 50 kg?

$$\text{weight} = \text{pull of gravity}$$
$$= 50 \text{ kg} \times 10 \text{ N/kg}$$
$$= 500 \text{ N}$$

Friction is the contact force that slows down moving things. Friction can also prevent stationary things from starting to move when other forces act on them. Figure 1 helps you to understand why frictional forces occur. Any surface is not perfectly smooth. If you look at a surface through a powerful microscope you will be able to see that it has many rough spikes and edges. When two surfaces move past each other these rough spikes catch onto each other and slow down the motion.

Friction is often a nuisance because the rubbing between two surfaces turns kinetic (motion) energy into heat. Some ways of reducing friction are shown in Figure 2. Sometimes, though, friction is useful. Brakes work by friction to slow down cars. Also, when you walk, the frictional forces between your foot and the floor push you forward.

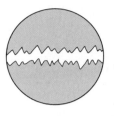

Figure 1
How two surfaces appear when seen through a powerful microscope.

Figure 2
Reducing friction

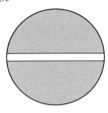

(a) If the surfaces are highly polished, friction is less.

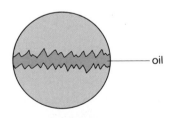

oil

(b) A layer of oil between two surfaces acts as a cushion to stop the edges catching.

The frictional drag from the sea on this hovercraft is reduced by a cushion of compressed air.

Questions

1 In the diagrams (right) some forces are acting on various objects. An arrow such as this: → 5 N means that a force of 5 N acts to the right. For each diagram state the effect that the forces produce.

2 What is the weight of a 2 kg bag of sugar?

3 One of the world's most important ways of reducing friction was invented by the ancient Egyptians in 3000 BC. What is it called?

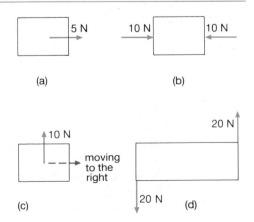

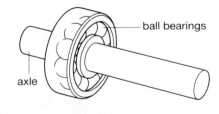

(c) Steel balls reduce friction by allowing surfaces to roll over each other.

3 Pressure

You always choose a sharp knife when you want to chop up meat or vegetables ready for cooking. A sharp knife has a very thin edge to the blade. This means that the force which you apply is concentrated into a very small area. We say that the pressure under the blade is large.

$$\text{pressure} = \frac{\text{force}}{\text{area}} \qquad P = \frac{F}{A}$$

The unit of pressure is $\mathbf{N/m^2}$ or **pascal (Pa)**; $1\,\text{N/m}^2 = 1\,\text{Pa}$.

Pressure points

The photograph shows a physics teacher lying on a bed of nails. How can he lie there without hurting himself? You all know that nails are sharp and will make a hole in you if you tread on one. The teacher has spread his body out though, so that it is supported by a lot of nails. The area of nails supporting him is large enough for it not to hurt (too much!).

Stiletto heels are pressure points. They concentrate the wearer's weight into a very small area. This can make it difficult to walk across soft surfaces (such as grass), and stiletto heels often damage polished floors

⚠ You need to take care getting on and off a nail bed!

What point is he trying to make here?

When Gillian, an engineer, designs the foundations of a bridge she must think about pressure. In the example shown in Figure 1, the bridge will sink into the soil if it causes a pressure greater than $80\,\text{kN/m}^2$. What area must the foundations have to stop this happening?

The bridge has a weight of 1.2 MN, so each pillar will support 0.6 MN.

$$P = \frac{F}{A}$$

$$\text{So} \quad A = \frac{F}{P}$$

$$= \frac{0.6\,\text{MN}}{80\,\text{kN/m}^2} = \frac{600\,000\,\text{N}}{80\,000\,\text{N/m}^2} = 7.5\,\text{m}^2$$

Figure 1

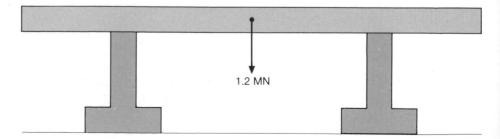

1.2 MN

What would happen if these foundations were poor?

Crushing pressure and breaking stress

Engineers also have to think about the pressure that a material can withstand without breaking; this is called the **crushing pressure**. A lot of modern buildings are made out of concrete, which is cheap and easy to produce. Concrete is very strong when it is compressed (squashed) but weak when it is stretched. When materials are stretched we say that they are **under stress**. We define stress like this:

$$\text{stress} = \frac{\text{stretching force}}{\text{area}}$$

material	crushing pressure MN/m^2	breaking stress MN/m^2
concete	70	1
cast iron	70	10
steel	400	200

Table 1

So stress is rather like pressure, but it stretches something rather than squashing it.

If a concrete beam is going to be stretched (under tension) in a building it is strengthened with steel bars. As you can see from Table 1, steel is far stronger than concrete under tension. A breaking stress of 200 MN/m^2 means that a steel bar of area 1 m^2 could support 200 MN before breaking.

Questions

1 (a) When a pressure of 4 N/mm^2 is applied to your skin it hurts. In the nail bed shown in the photograph each nail has an area of 1 mm^2. The teacher admits to having a weight of 650 N. What is the smallest number of nails that he must lie on, if he is not to be hurt?

(b) Explain why he has to be very careful getting on and off the nail bed.

2 Look carefully at the bridge in Figure 1.

(a) Although Gillian got her calculation right, the chief engineer, Mr Kruszewki, thinks that each foundation ought to have a base area of 30 m^2. Why might he think this?

(b) Use the information in Table 1 and Figure 1 to work out the smallest area that the concrete pillars can have if they are to hold up the bridge safely. What do you think Mr Kruszewki would say about your answer?

3 The diagram in the next column shows a concrete beam.

(a) Make a copy of the diagram and show which parts of the beam are stretched and which are compressed.

(b) Explain the positioning of the steel bars.

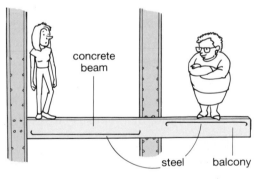

concrete beam

steel   balcony

4 Opposite you can see Boris, Vladimir and Leonid. Boris is twice as tall, twice as long and twice as wide as Vladimir, but they have the same density.

(a) How may times more massive than Vladimir is Boris?

(b) The area of Boris's paw is greater than the area of Vladimir's paw. By how many times is it bigger?

(c) Is the pressure bigger under Boris's paws or under Vladimir's paws? By what factor does the pressure differ under the two cats' paws?

(d) Leonid the lion is a distant cousin to Boris. Lions, as you know, are much bigger than domestic cats. Can you use the result of part (c) to explain why a lion's legs are proportionately thicker than a cat's?

Boris (cat)

Vladimir (kitten)

Leonid (lion)

4 Pressure in Liquids

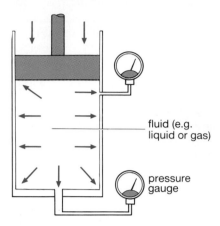

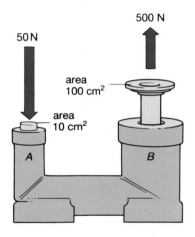

Figure 1
The pressure in a fluid acts equally in all directions

Figure 2
The principle of a hydraulic jack

Transmitting pressures

When you hit a nail with a hammer the pressure is transmitted downwards to the point. This happens only because the nail is rigid.

When a ruler is used to push a lot of marbles lying on a table, they do not all move along the direction of the push. Some of the marbles give others a sideways push. The marbles are behaving like a fluid.

In Figure 1 you can see a cylinder of fluid which has been squashed by pushing a piston down. The pressure increases everywhere in the fluid, not just by the piston. The fluid is made up of lots of tiny particles, which act like marbles, to transmit the pressure to all points.

Hydraulic machines

We often use liquids to transmit pressures. Liquids can change shape, but they hardly change their volume when compressed. Figure 2 shows how a hydraulic jack works.

A force of 50 N presses down on the surface above A. The extra pressure that this force produces in the oil is:

$$P = \frac{F}{A}$$

$$= \frac{50 \text{ N}}{10 \text{ cm}^2} = 5 \text{ N/cm}^2$$

The same pressure is passed through the liquid to B. So the upwards force that the surface above B can provide is:

$$F = P \times A$$

$$= 5 \text{ N/cm}^2 \times 100 \text{ cm}^2 = 500 \text{ N}$$

With this machine you can lift a load of 500 N, by applying a force of only 50 N. Figure 3 shows another use of hydraulics.

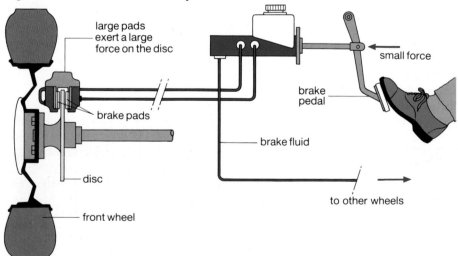

Figure 3
Cars use a hydraulic braking system. The foot exerts a small force on the brake pedal. The pressure created by this force is transmitted by the brake fluid to the brake pads. The brake pads have a large area and exert a large force on the wheel disc. The same pressure can be transmitted to all four wheels.

Increase of pressure with depth

The pressure below the surface of a liquid depends on three things:
- **depth, h**
- **density, d**
- the **pull of gravity** per kilogram, g ($g = 10$ N/kg)

The pressure that acts on a diver depends on the weight of water above him. As he goes deeper the weight of water on him increases, so the pressure also increases.

The diver feels a bigger pressure under 10 m of sea water than under 10 m of fresh water. Sea water has a higher density than fresh water.

The strength of the gravitational pull on the diver affects the force that the water exerts on the diver. The Earth gives a downwards pull of 10 N per kilogram.

To calculate the pressure P under a liquid you can use this formula:

$$P = g \times h \times d$$

The manometer

Figure 4 shows how you can use a **manometer** to measure the pressure of a gas supply. The two points X and Y are at the same level in the liquid. This means that the pressure at X and Y is the same.

Pressure at X = gas supply pressure.
Pressure at Y = atmospheric pressure + pressure due to 27.2 cm of water.

So the gas supply pressure is greater than atmospheric pressure, by an amount equal to the pressure due to a column of water 27.2 cm high. We say that the extra pressure (above atmospheric pressure) of the gas supply is '27.2 cm of water'.

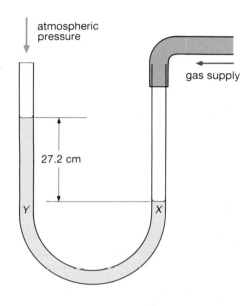

Figure 4
A manometer

Questions

1 The pressure of the gas supply in Figure 4 is 27.2 cm of water. Say what this pressure is in: (i) cm of mercury, (ii) cm of oil, (iii) pascal.

Liquid	Density in kg/m^3
water	1 000
mercury	13 600
oil	800

2 The diagram opposite shows the principle of a hydraulic jack. Two cylinders are connected by a reservoir of oil. A 200 N weight resting on a piston at A, can be used to lift a larger weight sitting on a piston at B.
(a) How big is the pressure in the oil at X?
(b) Explain why the pressure at Y is about the same as it is at X.
(c) Calculate the load, W, which can be lifted at Y.
(d) If you want to lift W by 10 cm, by how much do you have to move piston A?

3 What is the pressure on a diver 50 m below the surface of the sea? (The density of sea water is 1030 kg/m^3.)

4 Sometimes after a road accident the Fire Brigade uses inflated air bags to lift up a vehicle, to free trapped passengers. Explain how such bags can lift a large load.

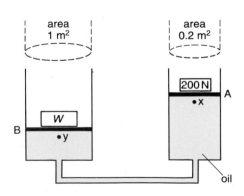

5 Atmospheric Pressure

Large and fragile objects like sheets of glass can be lifted with atmospheric pressure by using suction pads

⚠ Keep hands and feet away from the weights in a box

A simple experiment can show you the action of atmospheric pressure. A vacuum pump is used to remove the air from a metal can. As soon as the pump is switched on, the can collapses. This shows that:
- Atmospheric pressure acts in all directions.
- Atmospheric pressure is very large.

The size of atmospheric pressure

Figure 1 shows a hollow metal cylinder, which comes apart into two pieces. When the air is pumped out of it, a large force is needed to pull the two halves apart. You can measure the size of this force, by attaching weights to the bottom half of the cylinder.

Example. You might find that a total mass of 50 kg is needed to pull the halves apart. When this happens, the pressure from the weight of the 50 kg balances atmospheric pressure.

$$\text{So atmospheric pressure, } P = \frac{F}{A}$$

$$= \frac{500 \text{ N}}{50 \text{ cm}^2}$$

$$= 10 \text{ N/cm}^2 \; or \; 100\,000 \text{ N/m}^2$$
$$(\text{since } 1 \text{ m}^2 = 10\,000 \text{ cm}^2)$$

This is a very large pressure. The atmosphere exerts a force of about 100 000 N (the weight of 100 large men) on the outside of a window (area about 1 m²). The window does not break, because air inside the house pushes back with a force the same size.

Using atmospheric pressure

Piles which anchor the foundations of buildings are usually knocked into the ground by a large pile driver. However, circular or ring piles can be driven into the ground by atmospheric pressure (Figure 2). The idea is to place a dome over the top of the piles. Then a vacuum pump takes the air out of the space under the dome. There is now a large pressure difference between the inside and outside, which pushes the piles into the ground.

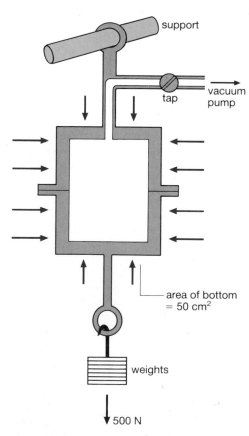

Figure 1
You can measure atmospheric pressure using this apparatus

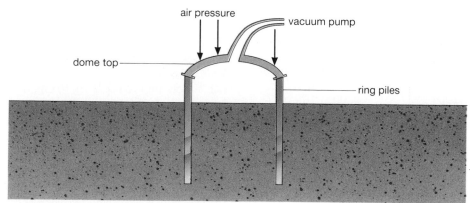

Figure 2
Driving in piles with air pressure

Measuring atmospheric pressure

Instruments that measure atmospheric pressure are called **barometers**. The most accurate type is the mercury barometer (Figure 3). The pressure at the points x and y is the same. So the pressure due to 760 mm of mercury, Hg, is equal to the atmospheric pressure. This is the average pressure of the atmosphere, and we sa   it is 760 mm Hg.

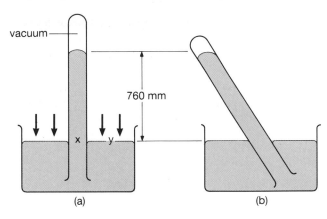

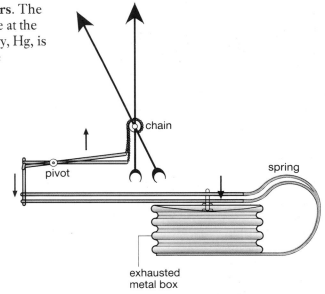

Figure 3
A mercury barometer. The average
atmospheric pressure is 760 mm of
mercury. It does not matter if the barometer
is tilted, as the vertical height of the
mercury column stays the same.

Figure 4
An aneroid barometer. The expansion of the
metal box, when the pressure drops, causes
the movement of a pointer.

⚠ Mercury presents a toxic vapour hazard. Wear protective gloves.

Questions

1 This question is about driving in piles with atmospheric pressure.
(a) The diagram shows the top of the circular dome seen in Figure 2. Use the scale to work out the area.

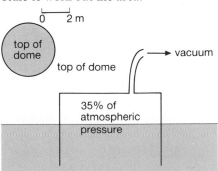

(b) Use your answer to work out the force with which the atmosphere pushes down on the dome top.
(c) The pile sinks into the ground when the air pressure is reduced to 35% of atmospheric pressure. What force then pushes upwards on the dome?
(d) Suppose you are the chief engineer on a building site that uses these piles. What is the maximum load that you can allow to rest on these piles?
(e) The atmosphere also acts on the sides of the dome. Why do you not need to include this effect in your calculations?

2 You take the apparatus shown in Figure 1 to the planet Zeta where the pull of gravity is 25 N on each kilogram. When you evacuate the apparatus, as described opposite, you find that a mass of 40 kg is needed to pull apart the two halves of the cylinder.
(a) How big is the pull of gravity on the 40 kg mass?   (b) Calculate the atmospheric pressure on Zeta .

Strong winds can produce spectacular
events like this waterspout

6 Stretching

Figure 1 shows a simple experiment to investigate the behaviour of a spring. The spring stretches when a load is hung on the end of it. The increase in length of the spring is called its *extension*. You can plot a graph of load against extension.

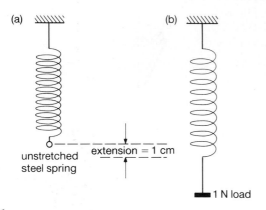

Figure 1

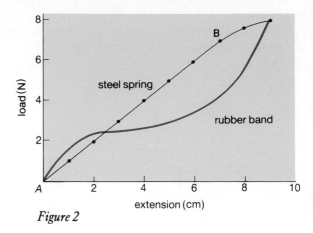

Figure 2

You will find that it produces a straight line such as *AB* in Figure 2. This shows that the extension is proportional to the load. A material that behaves in this way is said to obey Hooke's Law:

$\boxed{\text{Extension is proportional to the load}}$

At point *B* the spring has reached its *elastic limit*. Hooke's Law is no longer obeyed. Over the region *AB* of the graph the spring shows elastic behaviour. This means that when the load is removed from the spring, it returns to its original length and shape.

However, if a load of more than 7 N (beyond point *B*) is applied to this spring, it changes its shape permanently. When the load is removed it does not return to its original shape. This is called *plastic deformation*. When you bounce a hard rubber ball it deforms *elastically*. How do you think Plasticene deforms?

Only some materials obey Hooke's Law. You can see from Figure 2 that a rubber band certainly does not.

Figure 3 shows what happens when a load is put onto two springs. When you put two springs in series each one is pulled by the force of 1 N. Each spring is extended by 1 cm, and the total extension is now 2 cm (Figure 3(a)). When springs are put in parallel (side by side) each one supports half of the load. Each spring only extends by 0.5 cm (Figure 3(b)).

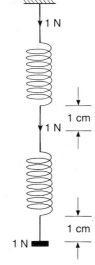

(a) The total extension is 2 cm

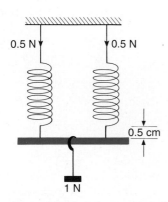

(b) Each spring extends 0.5 cm

Figure 3

Building with steel

This is the sort of problem that a construction engineer might have to solve. Tests in the laboratory on a sample of steel have produced the load/extension graph shown in Figure 4. Gillian, the engineer, needs to know if the girder can support a load of 18 000 N.

Gillian knows from the graph that the steel sample starts to deform plastically when it supports a load of 2000 N. The steel girder in the building must not deform in this way.

When working out what the girder can support, the length does not matter. The same force acts through every part of the girder. However, the area of the girder is important. The girder has an area ten times as big as the sample (Figure 4(a)). It can support ten times the load. So the maximum load that the girder can support is 20 000 N.

However, Gillian would not want to put 18 000 N on to this girder, because it does not leave much room for error. Engineers like to have a built-in safety factor.

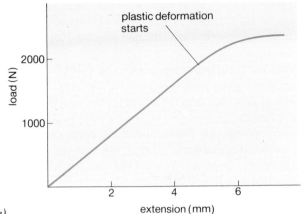

Figure 4(a)
The load/extension graph for a sample of steel in the laboratory can be used to predict how a steel girder will behave.

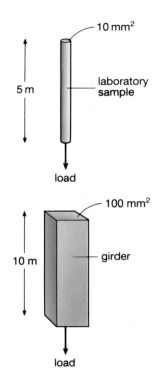

Figure 4(b)

Questions

1 In diagram (i) below: the 2 N weight extends the spring by 4 cm. Diagram (ii) shows an arrangement with three more of the same type of spring. How far will the point X move upwards when the 4 N weight is removed?

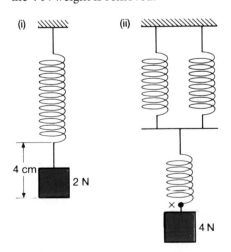

2 Figure 2, in the text, shows the force extension graphs for a rubber band and a spring. Use this graph to work out:

(i) the extension caused by a force when the band and spring are in series

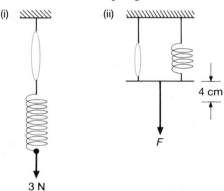

(ii) the force needed to produce a 4 cm extension when they are in parallel.

3 The diagram to the right shows a steel girder under test. Gillian needs to know how far the beam bends when it is loaded. The table shows the sag, h, in the middle, for different loads.

(a) Plot a graph of load (y-axis) against h (x-axis).
(b) Gillian made a mistake in one of her measurements of h. Which measurement is wrong and what should it have been?
(c) Using the same axes sketch a graph to show how a longer beam will sag under the same loads.
(d) Which side of the beam is being stretched and which side is being squashed?

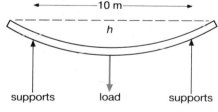

sag, h (cm)	1.3	3.2	5.2	7.0	9.1
load (N)	1000	2500	4000	5500	7000

7 Turning Forces

Figure 1

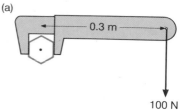

(a)

(a) *Turning moment*
   = 100 N × 0.3 m
   = 30 Nm

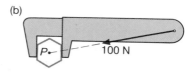

(b)

(b) *A force acting through the nut has no turning moment.*

If you have ever tried changing the wheel on a car, you will know that you are not strong enough to undo the nuts with your fingers. You need a spanner to get a larger turning effect. A tight nut will need a long spanner and a large force.

The size of the turning effect of a force about a point is called a **turning moment**.

> Turning moment = force × perpendicular distance
>   of the force from the point

Perpendicular means 'at right angles'. Figure 1 shows you why this distance is important. If you push the spanner towards the nut (Figure 1(b)) you get no turning force at all.

The same idea applies to lifting heavy loads with a mobile crane (Figure 2). If the turning effect of the load is too large the crane will tip over. So, inside his cab, the crane operator has a table to tell him the greatest load that the crane can lift for a particular working radius.

Table 1 shows you how this works. For example, the crane can lift a load of 60 tonnes safely with a **working radius** of 16 m. If the crane is working at a radius of 32 m, it can only lift 30 tonnes. You get the same turning effect by doubling the working radius and lifting half the load.

Working radius (m)	Maximum safe load (tonnes)	load × radius (tonne × m)
12	80	960
16	60	960
20	48	960
24	40	960
28	34	960
32	30	960
36	27	960

Table 1
A load table for a crane operator

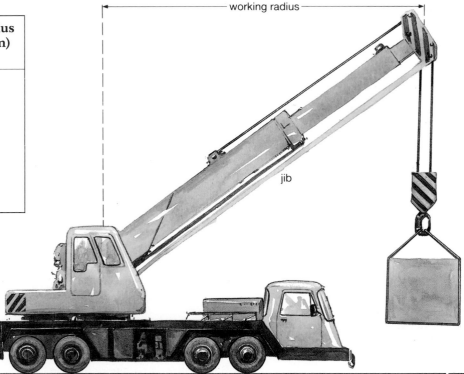

0   10 m

Figure 2
A mobile crane

Balancing

In Figure 3 you can see three children playing on a see-saw. Jaipal and Dominic are sitting at the ends, 2 m from the pivot. Where would Mandy sit to balance the see-saw?

When the anticlockwise turning effect of Jaipal is equal to the clockwise turning effect of Mandy and Dominic, then the see-saw balances.

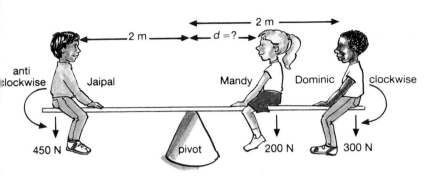

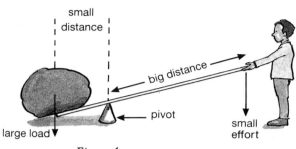

Figure 4
A simple lever

Figure 3
How is the see-saw balanced?

Jaipal's anticlockwise turning moment = 450 N × 2 m

= 900 Nm

Dominic's and Mandy's turning moment = 300 N × 2 m + 200 N × d

So if the see-saw is to balance:

900 Nm = 600 Nm + 200 d

So 200 d = 300 Nm

d = 1.5 m

Mandy must sit 1.5 m away from the pivot.
When the see-saw is balanced we say it is in **equilibrium**.
    In equilibrium:

The sum of the anticlockwise turning moments	=	The sum of the clockwise turning moments

We use this idea of balancing moments to our advantage to increase leverage.
    In Figure 4 the rock can be lifted when its turning effect is balanced by the turning effect of the man's push. With a long lever a small effort can lift a large load.

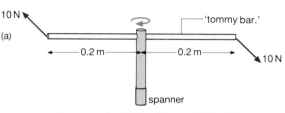

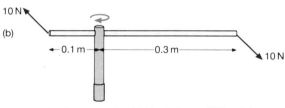

Figure 5
When two equal forces act to produce a turning effect, they are called a couple. A couple causes a rotational effect, but the resultant force on the bar is zero. Notice also that whatever the position of the bar relative to the spanner, the same turning effect is produced.

Questions

1 (a) When you cannot undo a tight screw, you use a screwdriver with a large handle. Explain why.
(b) Explain why door handles are not put near hinges.
(c) Where do you have to put the spare penny to balance the ruler?

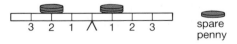

2 Look at the crane in Figure 2.
(a) What is meant by working radius?

(b) Use the scale in Figure 2 to calculate the working radius of the crane.
(c) Use the data in Table 1 to calculate the greatest load that the crane can lift safely in this position.
(d) Laura, a student engineer, makes this comment: 'You can see from the driver's table that the crane can lift 960 tonnes, when the working radius is 1m.' Do you agree with her?
3 Alice decides she would like to join the other children on the see-saw in

Figure 3. She has a weight of 170 N. How far will Mandy have to move to balance the see-saw if Alice sits (i) above the pivot (ii) on Dominic's lap?
4 What turning moment, about the nut, does this 200 N force have?

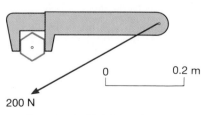

8 Stability

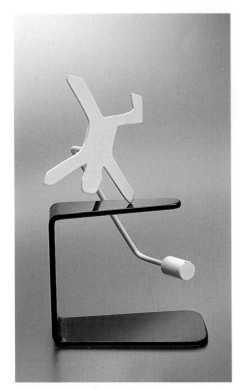

This rocking toy is stable because its centre of gravity is below the pivot

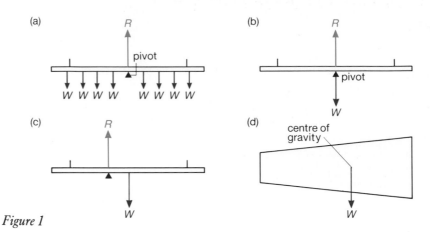

Figure 1

Centre of gravity

Figure 1(a) shows a see-saw that is balanced about its midpoint. Gravity has the same turning effect on the right-hand side of the see-saw, as is has on the left-hand side. The resultant turning effect is zero.

The action of the weight of the see-saw is the same as a single force, W, that acts downwards through the pivot (Figure 1(b)). This force has no turning moment about the pivot. As the see-saw is stationary, the pivot must exert an upwards force R on it, which is equal to W.

When the see-saw is not pivoted about its midpoint, the weight will act to turn it (Figure 1(c)).

The point that the weight acts through is called the **centre of gravity**. The centre of gravity of the see-saw lies at its midpoint because it has a regular shape. In Figure 1(d) the centre of gravity lies nearer the thick end of the shape.

Jumping higher

When you stand up straight your centre of gravity lies inside your body. But in some positions you can get your centre of gravity outside your body.

This idea is most important for pole vaulters and high jumpers (Figure 2(a)). Suppose a high jumper's legs can provide enough energy to lift her centre of gravity to a height 2 m above the ground. If she can make her centre of gravity pass under the bar, she can clear a bar higher than 2 m (Figure 2(b)).

Why do Fosbury-floppers usually win?

Figure 2
(a) Centre of gravity inside the body

(b) Centre of gravity outside the body

Equilibrium, stability and toppling

Something is in equilibrium when both the resultant force and resultant turning moment on it are zero. We talk about three different kinds of equilibrium, depending on what happens to the object when it is given a small push.

- A football on a flat piece of ground is in **neutral equilibrium**. When given a gentle kick, the ball rolls, keeping its centre of gravity at the same height.
- A tall thin radio mast is in **unstable equilibrium**. It is balanced with its centre of gravity above its base. However, a small push (from the wind) will move its centre of gravity downwards. To prevent toppling the mast is stabilised with support wires, as shown in the photograph.
- A car is in **stable equilibrium** (Figure 3(a)). When the car is tilted a little the centre of gravity is lifted, (b). In this position the action of the weight keeps the car on the road. In (c) the centre of gravity lies above the wheels; the car is in a position of unstable equilibrium. If the car tips further (d), the weight now provides a turning moment to topple the car. Cars are more stable if they have a low centre of gravity and a wide wheel base.

The support wires on this radio mast stop it from falling

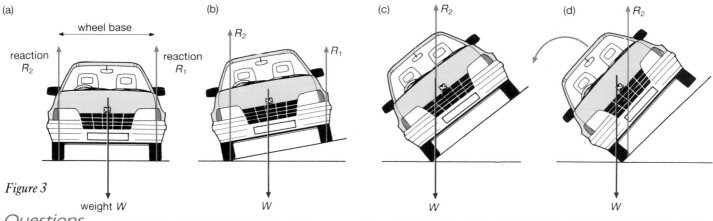

Figure 3

Questions

1 You are in the business of manufacturing lager glasses. You want to make your profits as big as possible. So you decide to produce a very unstable glass, which will get knocked over very easily. Then people will have to buy more glasses. Discuss which shape of glass will be most suitable for your purpose.

(a)　　　(b)　　　(c)

2 The diagram below shows a cantilever bridge which is constructed from three spans of concrete beams. The centre span has a weight of 6 MN, and each of the other arms has a weight of 10 MN. The arrows show the forces on the right-hand arm, which is 20 m long.

(a) Explain why B is the centre of gravity of the right-hand arm. Why is the centre of gravity not in the middle of this span?

(b) The pillar at C holds this span by exerting a force of 10 MN. How big must the force R be?

(c) By calculating turning moments about A, show that the downwards turning moments of the forces acting through D and B are balanced by the upwards turning moment of the force acting through C.

(d) Which pillars on the bridge would you make the strongest?

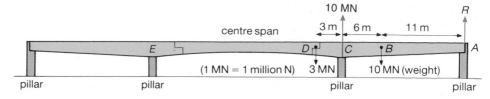

9 More about Forces

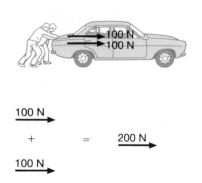

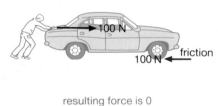

Figure 1

Figure 2

Vectors and scalars

If you want to move a chair you have to give it a push. The direction in which the chair moves depends on the direction of your push. A hard push will move the chair quickly. So both the *direction* and the *size* of a force are important.

Force is a **vector**. Vector quantities have both size and direction. Other examples of vector quantities are: velocity (wind 50 km/h north), displacement (2 paces backwards), acceleration and momentum (see section B).

A quantity that has only a size is called a **scalar**. Some examples of scalar quantities are: mass (2 kg of oranges), temperature (20°C), and energy (100 J).

Adding forces

Adding scalars is always easy: 3 kg of apples + 3 kg of apples = 6 kg of apples. Adding vectors can be a little harder. Here are three examples to show you how to do it:

- When two forces act in the same direction they add up to give a larger *resultant* or *net force*. In this case 100 N + 100 N = 200 N (Figure 1).
- In the second example the driver has left the hand brake on (Figure 2). Your push of 100 N to the right is cancelled out by a frictional force acting to the left. The resultant force is zero.
- In Figure 3 you can see two tugs pulling a passenger liner into port. To work out the resultant force in this case we need to make a scale drawing. The first step is to choose a scale. We will make a length of 1 cm represent a force of 100 000 N. The lines representing the forces are then drawn to scale. (These are the blue lines in Figure 3(b).) Then two more lines are drawn to complete a parallelogram (the black lines). The resultant force can now be worked out by drawing a line across the diagonal of the parallelogram (the thick blue line). In this example, the line is 6 cm long, so the resultant force on the liner is 600 000 N. You can see that the direction of this resultant force pulls the liner straight ahead.

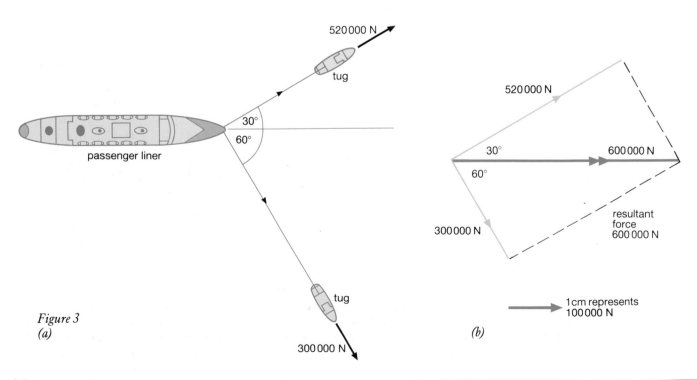

Figure 3
(a)

(b)

Newton's third law

Newton's third law states that 'to every force there is an equal and opposite force'. So forces always occur in pairs. This sounds an easy law to apply, but there are a few tricks. Here are some examples:

- When you walk, the Earth pushes your foot forwards; your foot pushes back on the Earth (Figure 4).

- A spacecraft returning to Earth is pulled downwards (Figure 5). The spacecraft pulls the Earth back. This means that as the spacecraft moves, the Earth moves too. But the Earth is so big that it moves only a tiny amount—far too little for us to notice.

- Figure 6 shows a man sitting on a chair. There are two pairs of forces to think about. The Earth's gravitational pull on the man is W. The man's gravitational pull on the Earth is W^1. The chair exerts a reaction force R on the man. He exerts a force R^1 on the chair.

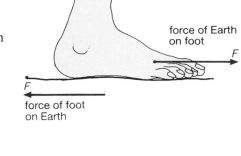

Figure 4

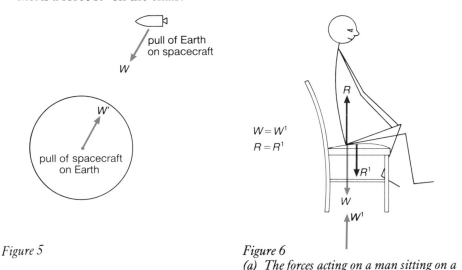

Figure 5

Figure 6
(a) The forces acting on a man sitting on a chair

$$W = W^1$$
$$R = R^1$$

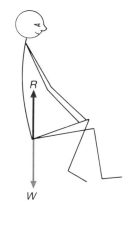

(b) The man stays where he is because R = W.

Questions

1 What is the resultant force on this rocket?

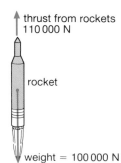

thrust from rockets 110 000 N

rocket

weight = 100 000 N

2 The diagram shows a large ship being pulled by two tugs. As the ship moves through the water there is a frictional force that acts on it. Work out the resultant force on the ship.

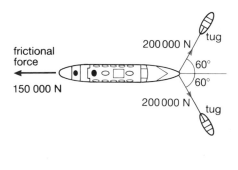

frictional force

150 000 N

200 000 N   tug

60°
60°

200 000 N   tug

3 (a) The diagram shows two sumo wrestlers, Taiho and Toshimitsu, locked in combat. Make a sketch of the diagram and mark in as many pairs of forces as you can.

Taiho   Toshimitsu

(b) Taiho wins the bout by pushing Toshimitsu down. Koki, who was watching the fight says: 'Taiho is the stronger so Taiho's push on Toshimitsu was bigger than Toshimitsu's push on Taiho'. Explain what is wrong with this comment.

SECTION A: QUESTIONS

1 The diagram shows two forcemeters, A and B. Each meter has a weight of 1 N. What is the reading on A?

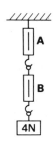

2 Make a copy of each of these drawings. Mark in the centre of gravity of each object.

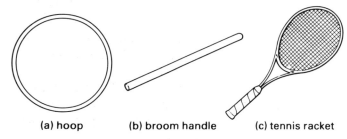

(a) hoop   (b) broom handle   (c) tennis racket

3 In the diagram at the top of the next column, each of the springs is identical. A load of 3 N stretches the springs 3 cm, until the top plate rests firmly on two fixed plates.
(a) Explain how this device can be used to measure both small and large loads accurately. Where in the home might you use such a device?
(b) Sketch a graph to show how the movement of the bottom plate changes for loads from 0–12 N. Label the axes carefully.

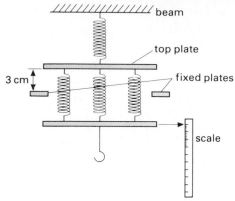

4 The table below shows the properties of some materials used in the manufacturing and building industries. When choosing a material for a particular job, you need to consider the stiffness, strength and density of the material. You need to be aware of the cost too.
(a) Explain why concrete, brick and wood are often used for building houses.
(b) Explain what is meant by **stiffness**.

Why is the $\frac{stiffness}{density}$ ratio important?

(c) **Strength** is a measure of what load can be placed on 1 m² of a material before it crumbles or breaks.

Why is the $\frac{strength}{density}$ ratio important?

(d) Explain why steel is used for building bridges, but aluminium is used to build aircraft.
(e) GRP (fibre glass) can be used to build small boats, and even small minesweepers for the Navy.
(i) Why is it useful to make minesweepers from GRP?
(ii) Why are large ships not made from GRP?
(f) Turbine blades for aircraft engines need to be strong, stiff and light. Which material in the table would you choose for making the blades?

Material	Relative stiffness	Relative strength*	Density (kg/m³)	Stiffness/Density	Strength/Density	Cost per tonne (pounds)
Steel	21 000	40 000	7800	2.7	5.1	100
Aluminium	7300	27 000	2700	2.7	10	400
Wood	1400	2700	500	2.8	5.4	30
Concrete	1500	4000	2500	0.6	1.6	6
Brick	2100	5500	3000	0.7	1.8	10
GRP	2000	50 000	2000	1	25	70
CFRP	20 000	100 000	2000	10	50	1700

*These strengths are for when the materials are under compression.
GRP: Glass Reinforced Plastic (fibre glass).   CFRP: Carbon Fibre Reinforced Plastic.

5 Popeye is blowing into the sail in his boat to push it forwards. Is this possible?

6 Below, you can see two wells. Oil, salt water and natural gas are trapped by oil-proof layers of rock.

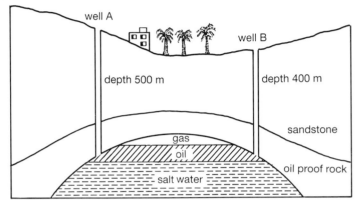

(a) (i) Why is the salt water lying below the oil?
(ii) Why is the natural gas lying above the oil?
(b) When well A is drilled, oil shoots 10 m above the ground. When well B is drilled, oil shoots much higher above the ground. Explain why oil shoots much higher from well B.
(c) As the oil escapes from the wells, what happens to (i) the volume of the natural gas; (ii) the pressure of the natural gas?
(d) After a few months, oil can still be obtained from well A but it has to be pumped out. Explain why the oil does not shoot out by itself.

ULEAC

7 (a) State clearly *two* conditions necessary for a body to be in equilibrium.
(b) The diagram shows a mousetrap. The idea is that as the mouse walks towards the cheese he tips the plank up and

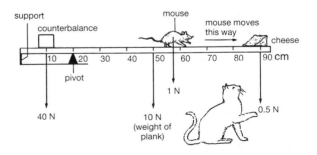

then falls conveniently near to Boris the cat. Go through the following calculations to see how far along the plank the mouse can go before it tips up.
(i) Calculate the turning moment of the counter balance about the pivot.
(ii) The weight of the plank may be taken to act through the 50 cm mark. Calculate the turning moment of the cheese and the plank together.
(iii) Use your answers to (i) and (ii) to find where the mouse will overbalance the trap.
(c) Use the information in the diagram to calculate the upwards force that the pivot exerts on the plank, just as the 'trap' overbalances.

8 Bristol airport lies about 150 km due west of Heathrow airport.

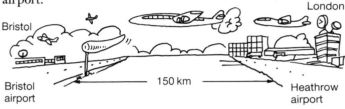

(a) On a windless day a light aircraft sets off from Heathrow to Bristol. The plane flies at 250 km/h. How long does it take to get there?
(b) On the return journey, a strong wind blows due north. The wind speed is 50 km/h. If the pilot sets off to fly east to Heathrow, in which direction will he end up going? Draw a diagram to illustrate your answer.
(c) He corrects his course so that he actually flies eastwards. In which direction must he point the plane? How long does his journey take? (You might find it useful to make a scale drawing.)

9 Mike is a physicist who likes to weigh himself in unusual ways. He knows that atmospheric pressure (100 000 N/m²) supports a column of mercury 75 cm high. He has attached a hot water bottle to a mercury manometer. When he stands on it the mercury moves as shown.

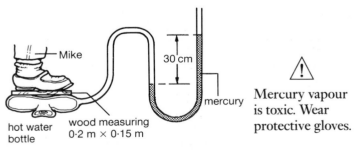

(a) What extra pressure does Mike cause?
(b) What is the area of the wood that Mike stands on?
(c) Now work out Mike's weight. What is his mass in kg?

QUESTIONS

Unfortunately the bottle burst just as Mike was reading the height of the mercury. So, he has devised another method of weighing himself. He has borrowed a Newton balance that can measure weights up to 500 N. The diagram shows his next weighing experiment.

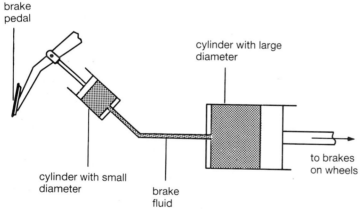

(d) Use the diagram to work out Mike's weight.
(e) How big is the force *F*?

10 (a) (i) State what is meant by **pressure**.
(ii) A suitable unit for pressure is the pascal (N/m²). State another unit used to measure pressure.
(b) Explain briefly, in terms of pressure, why:
(i) the sharp edge of a knife, and not the blunt edge, is used for cutting.
(ii) the thickness of the wall of a dam, used to store water, is greater at the base than at the top.
(iii) a tin can collapses when the air is removed from it.
(iv) aneroid barometers may be used as altimeters in aircraft.
(c) Here is a simple type of hydraulic braking system:

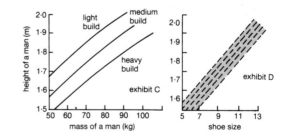

The areas of cross-section of the small cylinder and large cylinder are 0.0004 m² and 0.0024 m² respectively. The brake pedal is pushed against the piston in the small cylinder with a force of 90 N.
(i) Determine the pressure exerted on the brake fluid.
(ii) Determine the force exerted by the brake fluid on the piston in the large cylinder.

11 Colonel Carruthers has been murdered at Campbell Castle. Inspector Grappler of the Yard has been sent to investigate. He finds old Carruthers dead in the library. Outside the library window there are some footprints in the

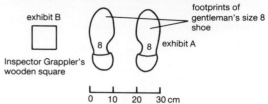

flower bed (exhibit A). This is an important clue; the inspector realises that he can estimate the suspect's height. Inspector Grappler carries out an experiment in the flower bed. He uses a wooden square (exhibit B) and some weights. He piles the weights onto the square and measures how far the square sinks into the flower bed. His results are in the table.

Mass of the inspector's weights (kg)	Depth of the hole made by the square (mm)
5	7
10	14
15	20
20	25
25	30

(a) Work out the area of the suspect's shoes.
(b) Inspector Grappler measured that the shoes had sunk 23 mm into the flower bed. Use the data in the table and your answer to part (a) to make an estimate of the suspect's mass.
(c) Exhibit C shows roughly how the height of a man depends on his mass. Exhibit D shows roughly how a man's shoe size depends on his height. Use these graphs to make an informed guess of the suspect's height.
(d) Of Colonel Carruthers' servants, which one do you suspect?

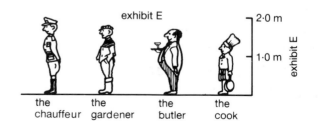

SECTION B

Forces in Motion

Surfers make use of the energy provided by the waves to travel at high speed

1 How Fast Do Things Move?

Linford Christie won the gold medal for the men's 100 metres at the Barcelona Olympics in 1992. He covered the distance in 9.96 seconds; what was his average speed?

Moving object	Speed (m/s)
glacier (Rhonegletsher)	0.000001
snail	0.0005
human walking	2
human sprinter	10
express train	60
Concorde	600
Earth moving round the Sun	30 000
light and radio waves	300 000 000

This snail would take nearly two and a half days to go 100 metres!

When you travel in a fast car you finish your journey in a short time. When you travel in a slow car your journey takes longer. We might say that the speed of a car is 100 kilometres per hour (100 km/h) which means that the car will travel a distance of 100 kilometres in one hour. We can write a formula connecting distance, speed and time:

$$\text{average speed} = \frac{\text{distance travelled}}{\text{time taken}} = \frac{d}{t}$$

We write average speed because the speed of the car may change a little during the journey. When you travel along a motorway your speed does not remain exactly the same. You slow down when you get stuck behind a lorry and speed up when you pull out to overtake a car.

Example. At top speed a guided-missile destroyer travels 170 km in three hours. What is its average speed in m/s?

$$\text{average speed} = \frac{d}{t}$$

$$= \frac{170 \times 1000 \, \text{m}}{3 \times 3600 \, \text{s}}$$

$$= 16 \, \text{m/s}$$

remember 1 km = 1000 m;
1 hour = (60 × 60 s) = 3600 s

Velocity

It is not just the speed that is important when you go on a journey. The direction matters as well. Figure 1 shows three possible routes taken by a helicopter leaving London. The helicopter travels at 300 km/h. So in one hour the helicopter can reach Liverpool, Paris or Brussels depending on which direction it travels in.

When we want to talk about a direction as well as a speed we use the word **velocity**. For example, when the helicopter flies towards Liverpool, we can say that its velocity is 300 km/h, on a compass bearing of 330°. Velocity is a vector, speed is a scalar (see page 18).

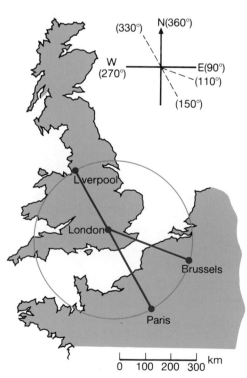

Figure 1

Time-tabling railway trains

You have probably sat on a slow train in a railway station wondering why the train was taking so long to start. As likely as not you would have been waiting for an express train to overtake you. There is often only one line between railway stations. The trains must be time-tabled so that the faster express trains can overtake a slower train that has stopped at a station.

Suppose you are a passenger train planning officer working for British Rail. You have to make sure that a slow train going from Reading to Chippenham does not get in the way of the London to Cardiff express train. The slow train travels at an average speed of 70 km/h and the express train at an average speed of 120 km/h.

Plotting a graph of the distance travelled by these trains against time helps you to solve the problem. You can see from Figure 2 what happens if both trains leave their stations at 0900 h. The slow train travels the 70 km to Swindon, and the faster train travels from London to Swindon, in the same time. You can also see from the graph that there is a slow train to Newport leaving Bristol Parkway at 1015 h that is going to get in the way of the express train.

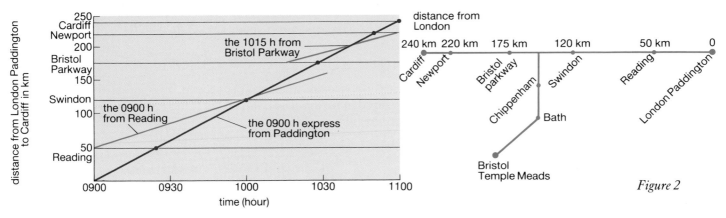

Figure 2

Questions

1 The world record for running 1500 m is held by Saïd Aouita. He ran that distance in 3 minutes and 29 seconds. What was his average speed?

2 When Richard Noble broke the land speed record in *Thrust II* in October 1983 he travelled at an average speed of about 1000 km/h. How long did it take him to travel 1 km?

3 Sketch a graph of distance travelled (*y*-axis) against time (*x*-axis) for a train coming into a station. The train stops for a while at the station and then starts again.

4 This question refers to the train time-tabling shown in Figure 2.
(a) How can you tell from the graph that the express train travels faster than the other two?

(b) At what time does the express train reach Bristol Parkway?
(c) At what time should the 1015 h from Bristol leave so that it meets the express train at Newport?

5 Ravi, Paul and Tina enter a 30 km road race. The graph shows Ravi's and Paul's progress through the race.
(a) Which runner ran at a constant speed? Explain your answer.
(b) What was Paul's average speed for the 30 km run?
Tina was one hour late starting the race. During the race she ran at a constant speed of 15 km/h.
(c) Copy the graph and add to it a line to show how Tina ran.
(d) How far had Tina run when she overtook Paul?

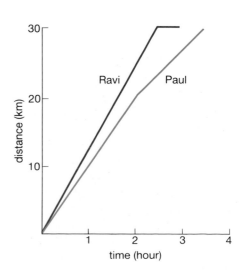

2 Acceleration

Drag cars accelerate to a high speed in a short time

When a car is speeding up, we say it is *accelerating*. When it is slowing down we say it is *decelerating*. A deceleration can be thought of as a negative acceleration.

You may be interested in driving a sports car which can accelerate quickly away from the traffic lights when they turn green. A car that accelerates rapidly reaches a high speed in a short time. For example, a sports car speeds up to 45 km/h in 5 seconds. A truck speeds up to 45 km/h in 10 seconds. We say the acceleration of the car is twice as big as the truck's acceleration.

You can work out the acceleration of the car or truck using the formula:

$$\text{acceleration} = \frac{\text{change of velocity}}{\text{time}}$$

$$\text{acceleration of sports car} = \frac{45 \text{ km/h}}{5 \text{ s}}$$

$$= 9 \text{ km/h per second}$$

$$\text{acceleration of the truck} = \frac{45 \text{ km/h}}{10 \text{ s}}$$

$$= 4.5 \text{ km/h per second}$$

Some more about acceleration

We usually measure velocity in metres per second. This means that acceleration is usually measured in **m/s per second**. What is the acceleration of our sports car in m/s per second?

$$45 \text{ km/h} = \frac{45\,000 \text{ m}}{3600 \text{ s}} = 12.5 \text{ m/s}$$

$$\text{So acceleration} = \frac{\text{change of velocity}}{\text{time}}$$

$$= \frac{12.5 \text{ m/s}}{5 \text{ s}} = 2.5 \text{ m/s per second}$$

or

$$2.5 \text{ m/s}^2 \text{ (metres per second squared)}$$

Average Speed

When a cyclist accelerates from rest at a constant rate up to a speed of 10 m/s, her average speed over that period is 5 m/s – half way between zero and 10 m/s. In general, for constant acceleration,

$$\text{Average speed} = \frac{u + v}{2}$$

u = initial speed, v = final speed

Velocity/time graphs

You have already seen that distance/time graphs can be helpful when looking at problems involving moving cars or trains. It can also be helpful to plot graphs of velocity against time.

Figure 1 shows the velocity of a cyclist as she cycled through a town. The following things happened in her journey.
(1) For the first 20 seconds, as she came into the town, she travelled at a constant velocity.
(2) Then she started to cycle up a hill, so she slowed down.
(3) At the top of the hill she came to some traffic lights. She had to wait for them to change.

(4) When the traffic lights changed to green she accelerated away.
The slope of a velocity/time graph is useful, because it shows how quickly the cyclist changes velocity. If the slope is steep, she is changing velocity (accelerating or decelerating) quickly.

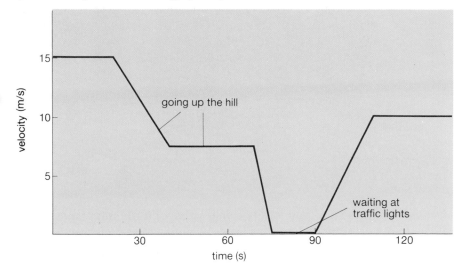

This car has been in a serious crash. When a fast-moving, heavy object like a car decelerates rapidly, it exerts a large force on anything that tries to stop it. It is the reaction to this force that damages the car (and its passengers)

Figure 1

Questions

1 This question refers to the graph in Figure 1.
(a) What was the cyclist's velocity after 60 s?
(b) How long did she have to wait at the traffic lights?
(c) Which was larger, her deceleration as she stopped at the traffic lights, or her acceleration when she started again? Explain your answer.

2 Below is a table showing how the speed, in km/h, of Damon Hill's Williams racing car changes, as he accelerates away from the starting grid at the beginning of the Brazilian grand prix.
(a) Plot of graph of speed (y-axis) against time (x-axis).
(b) Use your graph to estimate the acceleration of the car in km/h per second at: (i) 16 s, (ii) 1 s.

	Starting speed (m/s)	Final speed (m/s)	Time taken (s)	Acceleration (m/s^2)
Cheetah	0	30	5	6
Second stage of a rocket	450	750	100	
Aircraft taking off	0		30	2
Car crash	30	0		−150

3 Copy the table above and fill in the missing values.

4 In America drag cars are designed to cover distances of 400 m in about 6 seconds. During this time the cars accelerate very rapidly from a standing start. At the end of 6 seconds, a drag car reaches a speed of 150 m/s.
(a) What is the drag car's average speed?
(b) What is its average acceleration?

5 Describe the motion in the following two cases:

speed (km/h)	0	35	70	130	175	205	230	250	260	260
time (s)	0	1	2	4	6	8	10	12	14	16

3 Observing Motion

Athletes use photographs like this one to analyse their motion so that they can improve their performance

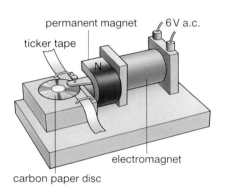

Figure 1

A trainer who wants to know how well one of his athletes is running stands at the side of the track with a stop watch in his hand. By timing the athlete over a set distance, the trainer knows how fast he is running. But this only tells the trainer an average speed. If he wants to learn more, he needs to look at the athlete's movements over very short time intervals. The best way to look at something that is moving is to film it. This is what sports scientists do to check on an athlete. By looking at each frame of film the scientists can see if the athlete is wasting any effort in his running action.

Another way to look at motion is to use multiflash photography. On the next page you can see a photograph of a golfer hitting a ball. In this technique the golfer swings his club in a darkened room, while a lamp flashes on and off at regular intervals. When the images of the club are close together, it is moving slowly. When the images are far apart, the club is moving quickly.

Ticker timer

For studying motion in the lab we use a **ticker timer** (Figure 1). A ticker timer has a small hammer that vibrates up and down 50 times per second. The hammer hits a piece of carbon paper which leaves a mark on a length of tape.

Figure 2 shows you two tapes that have been pulled through the timer. You can see that the dots are close together over the region *PQ*. Then the dots get further apart, so the object moved faster over *QR*. Then the movement slowed down again over the last part of the tape, *RS*. Since the timer produces 50 dots per second, the time between dots is $\frac{1}{50}$ s or 0.02 s. So we can work out the speed:

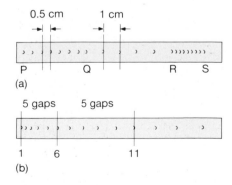

Figure 2

$$\text{speed} = \frac{\text{distance between dots}}{\text{time between dots}}$$

Between *P* and *Q*, speed $= \dfrac{0.5 \text{ cm}}{0.02 \text{ s}} = 25 \text{ cm/s}$ *or* 0.25 m/s

In tape B the dots get further and further apart, so the object attached to this tape was accelerating all the time.

Measuring acceleration

Figure 3 shows how you might measure the acceleration of a trolley moving down a slope. When you have let the trolley go down the slope the tape stuck to it will look a bit like tape B in Figure 2. It is helpful to cut up the tape into 5-tick lengths. You do this by cutting through the first dot, then the sixth and eleventh and so on. Each length of tape is the distance travelled by the trolley in $\frac{1}{10}$ s (5 dots means 5 x $\frac{1}{50}$ seconds). You can then use your pieces of tape to make a graph and see how the trolley moved. If your pieces of tape form a straight line then the acceleration was constant (see Figure 4).

Cutting up your tape like this is like plotting a graph of speed against time; the steeper the slope the greater the acceleration.

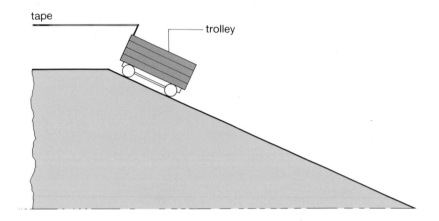

Figure 3

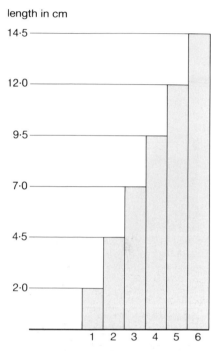

length in cm

Figure 4
Each length of tape shows the movement of the trolley over 0.1 s

Questions

1 Work out the speed of the tape in Figure 2(a) in: (i) the region *QR*, (ii) the region *RS*.
2 (a) Why can you tell from Figure 4 that the trolley accelerated down the slope at a constant rate?
(b) Work out the average speed of the trolley in: (i) interval 4, (ii) interval 6.
(c) What is the time between the middle of interval 4 and the middle of interval 6?
(d) Now use your results from parts (b) and (c) to calculate the acceleration of the trolley.
3 Examine the photograph of the golfer. Where is his club moving fastest? Explain your answer.
4 In aircraft the airspeed, *v*, is usually measured using an airspeed indicator. This indicator makes use of the difference in air pressure between still

and fast-moving air. The pressure of fast-moving air is less than the pressure of still air.
(a) Use the data in the table to plot a graph of pressure difference (*y*-axis) against v^2 (*x*-axis). Explain why the graph shows that the pressure difference is proportional to the square of the air speed.

Difference between still and moving air pressures ($N m^2$)	Airspeed, *v* (m/s)
400	40
1600	80
3600	120
6400	160
10000	200

(b) Use the graph to work out the airspeed when the pressure difference is: (i) 2000 N/m^2, (ii) 12000 N/m^2.

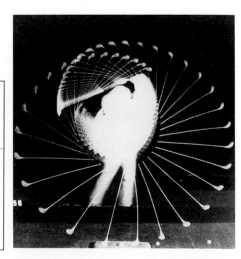

4 Forces and Motion

This unit is about **Newton's first law of motion**. This law says that when a force pushes or pulls an object, its velocity will change. When there is no force, the object remains stationary, or carries on moving in a straight line with a constant speed.

(a)

In Figure 1(a) you can see the Russian spacecraft *Vostock*. It is a very great distance from the Earth, so we can forget about any gravitational pull. At the moment *Vostock* is not moving and its rockets are turned off.

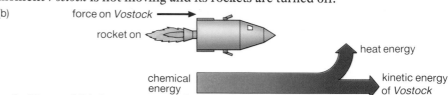
(b)

In Figure 1(b) the cosmonaut has turned on a rocket. There is now a force pushing *Vostock* forwards and its speed increases. The stored chemical energy in the fuel is turned into kinetic energy of the moving spacecraft, and heat.

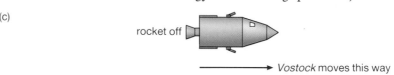
(c)

In Figure 1(c) the rocket has been turned off. Again no forces act on *Vostock* and it carries on moving at a constant speed. In space there is no air so nothing gets in the way to slow *Vostock* down. The only way to slow it down is to fire rockets that are pointing forwards. The blast from these produces a backwards force on the spacecraft and its speed decreases (Figure 1(d)).

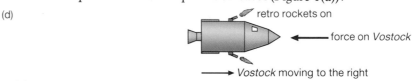
(d)

If we look at a similar example of motion on the Earth, things are not quite so simple. The problem is that on Earth there are frictional forces. When a car wheel turns on its axle, the moving surfaces rub together and slow the car down. The car also has to push its way through the air. The air gives rise to another force that slows the car; we call it wind resistance. If you put your hand out of the window of a fast-moving car, you can feel how strong this force is.

When a car is parked on a level road it stays where it is. This is because there is no force to push it along the road. In the top photograph the engine has been started and the car accelerates. But wind resistance begins to act. Wind resistance gets larger as the speed increases.

In the second photograph the car is going along the road at constant speed. The forwards push from the wheels is equal to the backwards push from the wind resistance.

In the third photograph the engine has been switched off. Now only frictional and wind resistance forces push on the car. These slow down the car. In space it is possible to move at a constant speed without using fuel because there are no frictional forces. On the road a car needs energy from fuel to work against wind and frictional resistance.

The resultant force on the car is forwards, so the car accelerates

There is no resultant force, so the car moves at a constant speed

Wind resistance now makes the car decelerate

Streamlining

If you want to go fast you need to make the effect of wind resistance as small as possible. There is a bigger wind resistance on objects which have a large area. Pointed objects can pass easily through air in the same way that a knife will cut through something. Fast cars are carefully designed so that air can flow easily past them; this is called **streamlining**.

You see the same idea in ships. Those that are built for speed are very thin and have pointed bows. Skiers and cyclists crouch so that they go head first into the wind. This makes the area that goes into the wind as small as possible. They also wear very smooth, skin-tight clothing which reduces the effect of the wind.

Here are some examples of streamlining in technology and nature. The smoke streams directed across the car in a wind tunnel help designers improve the car's streamlining. The solid rear wheel of the bicycle is designed to reduce air resistance. The dolphin's streamlined shape allows it to move very quickly through the water

Questions

1 (a) The diagrams below show the forces that act on an aeroplane at different times during a flight. The longer an arrow, the larger the force. For each of the cases below describe how the plane is moving.
(b) Draw another diagram to show the forces acting on the plane just after it has taken off.

2 A student wrote this in an exam: 'When something is moving forwards, there must be a force acting on it. As soon as you turn the engine off in a car or the rocket off in a spaceship, they stop.' The student has made some mistakes; rewrite this paragraph explaining where he went wrong.

3 The data in the table show how the wind resistance on a car varies as the speed increases.
(a) Use the data to work out the fastest the car can travel when the wheels produce a forwards push of 2600 N.
(b) When the car is made more streamlined it travels faster. Explain why.

Speed (m/s)	Wind resistance (N)
0	0
5	600
10	1200
15	1800
20	2400
25	3000

lift
thrust
drag
weight

5 Force, Mass and Acceleration

The large force produced by a powerful engine acts on a small mass to give this racing car a large acceleration

When a force acts on a car it will accelerate. What affects how big the acceleration is? If you have ever had a car with a flat battery then you will have given the car a push to start. When one person tries to push a car alone then the acceleration of the car is very slow. It takes a long time for the car to increase its speed. When three people give the car a push, it accelerates much more rapidly.

The acceleration is proportional to the force:

$$\text{acceleration} \propto \text{force}$$

You also know from everyday experience that large massive objects are difficult to set in motion. When you throw a ball you can accelerate your arm more quickly if the ball has a small mass. You can throw a cricket ball much faster than you can put a shot. A shot has a mass of about 7 kg and the force that your arm can apply cannot accelerate it as rapidly as a cricket ball.

The acceleration is inversely proportional to the mass:

$$\text{acceleration} \propto \frac{1}{\text{mass}}$$

It is difficult to get large things moving, but it is also very difficult to stop large things when they are moving. The world's largest ship is the *Seawise Giant*; it is 460 m long and has a mass of about 560 000 tonnes. Once a ship like this is moving the captain has to allow about 10 km for the ship to slow down.

The frictional force of water on a large mass like a ship can produce only a small deceleration, so it takes a long time for a ship like this one to slow down

Newton's second law of motion

With some careful experiments in the laboratory, you can see how the size of the acceleration is connected to the size of the force. Figure 1 shows the idea. A trolley placed on a table is accelerated by pulling it with an elastic cord. This is stretched so that it always remains the length of the trolley. The acceleration of the trolley is measured by putting a piece of ticker tape on the back of the trolley.

The first experiment is to increase the force acting on one trolley. You can see how our ticker tape graphs get steeper as the force increases. This means the acceleration is getting bigger. A 1 kg trolley accelerates at a rate of 1 m/s^2 for a force of 1 N; 2 m/s^2 for a force of 2 N; and 3 m/s^2 for a force of 3 N.

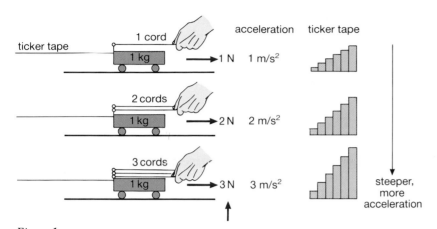

Figure 1
(a) Experiment 1
Keep the mass constant and change the force

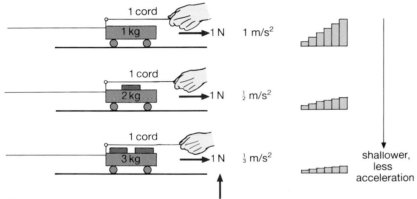

Figure 1
(b) Experiment 2
Keep the force constant and change the mass

In the second experiment (Figure 1(b)), you change the mass of the trolley but you pull with the same force. As the mass increases the ticker tape graphs become less steep. This means the acceleration is getting less. For a force of 1 N acting on a 1 kg trolley the acceleration is 1 m/s^2, for a 2 kg trolley the acceleration is $\frac{1}{2} \text{ m/s}^2$, for a 3 kg trolley the acceleration is $\frac{1}{3} \text{ m/s}^2$. For each of the experiments in Figure 1, you can see that the following equation is true:

$$\boxed{\text{Force} = \text{mass} \times \text{acceleration}}$$

This is **Newton's second law of motion**. This equation lets us define the Newton in this way: *A force of 1 N will accelerate a mass of 1 kg at a rate of 1 m/s².*

Example. A sledge has a mass of 10 kg. A boy pulls it with a force of 20 N. What is its acceleration?

$$F = m \times a$$
So
$$a = \frac{F}{m} = \frac{20 \text{ N}}{10 \text{ kg}} = 2 \text{ m/s}^2$$

Questions

1 Trains accelerate very slowly out of stations. Why is their acceleration very much slower than that of a car?
2 The manufacturers of Formula 1 racing cars try to make them as light as possible. Why do you think they do that?
3 This question refers to the experiments described in Figure 1.
(a) How many cords are needed to accelerate a trolley of mass 4 kg at a rate of 0.5 m/s^2?
(b) What acceleration is produced by four cords acting on a trolley of mass 3 kg?
(c) The trolley in this experiment has frictional forces acting on it. What effect does friction have on your results?
(d) What can you do to help compensate for friction?
4 A driver travelling on the motorway at 30 m/s takes her foot off the accelerator and slows down to 27 m/s in 6 s.
(a) Calculate the size of her deceleration.
(b) The mass of the car is 1000 kg. Calculate the size of the wind resistance and frictional forces that are acting on the car.

6 Free Fall

The diver will enter the water head first. She is streamlined, so the forces acting on her are small. If she belly flops a large force will slow her down, which will hurt

When you drop something it accelerates downwards, moving faster and faster, until it hits the ground. There is a story that Galileo used the leaning tower of Pisa to demonstrate the effect of gravitational acceleration. He dropped a large iron cannon ball and a small one. Both balls reached the ground at the same time. They accelerated at the same rate of about $10\,\text{m/s}^2$.

Weight and mass

The size of the Earth's gravitational pull on an object is proportional to its mass. The Earth pulls a 1 kg mass with a force of 10 N and a 2 kg mass with a force of 20 N. We say that the strength of the **Earth's gravitational field**, g, is 10 N/kg.

The **weight**, W, of an object is the force that gravity exerts on it, this is equal to the object's mass × the pull of gravity on each kilogram.

$$W = mg$$

The value of g is roughly the same everywhere on the Earth, but away from the Earth it has different values. The Moon is smaller than the Earth and pulls things towards it less strongly. On the Moon's surface the value of g is 1.6 N/kg. In space, far away from all planets, there are no gravitational pulls, so g is zero, and therefore everything is weightless.

The size of g also gives us the **gravitational acceleration**, because:

$$\text{acceleration} = \frac{\text{force}}{\text{mass}} \qquad or \qquad g = \frac{W}{m}$$

Example. What is the weight of a 70 kg man on the Moon?

$$W = mg$$
$$= 70\,\text{kg} \times 1.6\,\text{N/kg}$$
$$= 112\,\text{N}$$

Parachuting

You have read earlier that everything accelerates towards the ground at the same rate. But that is only true if the effects of air resistance are small. If you drop a feather you know that it will flutter slowly towards the ground. That is because the size of the air resistance on the feather is only slightly less than the downwards pull of gravity.

The size of the air resistance on an object depends on the area of the object and its speed:
- the larger the area, the larger the air resistance.
- the larger the speed, the larger the air resistance.

Figure 1 shows the effect of air resistance on two balls, which are the same size and shape, but the red ball has a mass of 0.1 kg and the blue ball a mass of 1 kg. The balls are moving at the same speed, the air resistance is the same, 1 N, on each. The pull of gravity on the red ball is balanced by air resistance, so it now moves at a constant speed. The red ball will not go any faster and we say it has reached **terminal velocity**. For the blue ball, however, the pull of gravity is greater than air resistance so it continues to accelerate.

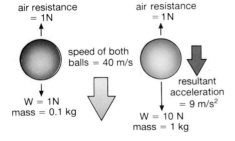

air resistance = 1N   air resistance = 1N

speed of both balls = 40 m/s

resultant acceleration = 9 m/s²

W = 1N   mass = 0.1 kg   W = 10 N   mass = 1 kg

Figure 1
At this instant both balls have a speed of 40m/s. At this speed the weight of the red ball is balanced by air resistance, but the heavier blue ball is still accelerating

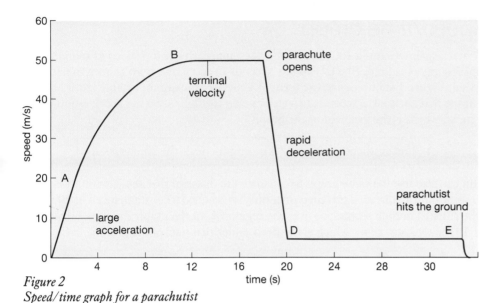

Figure 2
Speed/time graph for a parachutist

The sky diver has reached her terminal velocity. She is not in a streamlined position. How could she go faster?

Figure 2 shows how the speed of a sky diver changes as she falls towards the ground. The graph has five distinct parts:
(1) *OA*. She accelerates at about $10 \, \text{m/s}^2$ just after leaving the aeroplane.
(2) *AB*. The effects of air resistance mean that her acceleration gets less as there is now a force acting in the opposite direction to her weight.
(3) *BC*. The air resistance force is the same as her weight. She now moves at a constant speed because the resultant force acting on her is zero.
(4) *CD*. She opens her parachute at *C*. There is now a very large air resistance force so she decelerates rapidly.
(5) *DE*. The air resistance force on her parachute is the same size as her weight, so she moves with a constant speed until she hits the ground at *E*.

Questions

1 A student wrote 'my weight is 67 kg'. What is wrong with this statement, and what do you think his weight really is?
2 A hammer has a mass of 1 kg. What is its weight (i) on Earth (ii) on the Moon (iii) in outer space?
3 Explain this observation: 'when a sheet of paper is dropped it flutters down to the ground, but when the same sheet of paper is screwed up into a ball it accelerates rapidly downwards when dropped'.
4 This question refers to the speed/time graph in Figure 2.
(a) What was the speed of the sky diver when she hit the ground?
(b) Why is her acceleration over the part *AB* less than it was at the beginning of her fall?

(c) Use the graph to estimate roughly how far she fell during her dive. Was it nearer 100 m, 1000 m, or 10 000 m?
5 The graph (right) shows how the air resistance force on our sky diver's parachute changes with her speed of fall.
(a) What is the resistive force acting on her if she is travelling at a constant speed of 5 m/s?
(b) Explain why your answer to part (a) must be the same size as her weight.
(c) Use the graph to predict how fast these people would fall using the same parachute: (i) a boy of weight 400 N, (ii) a man of weight 1000 N.
(d) Make a copy of the graph and add to it a sketch to show how you think the air resistance force would vary on a

parachute of twice the area of the one used by our sky diver.

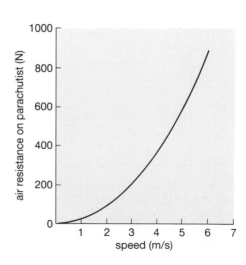

7 Projectiles

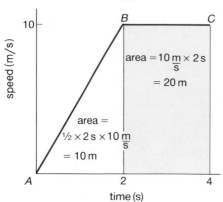

Figure 1
The speed/time graph of a sprinter

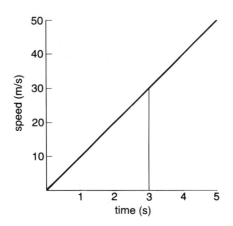

Figure 2
The speed/time graph for a falling ball

Time (s)	Speed (m/s)	Distance fallen = ½ speed × time (m)
0	0	0
1	10	5
2	20	20
3	30	45
4	40	80
5	50	125

Table 1 Distance fallen as a function of time. Check that the distance fallen $= \frac{1}{2} \times 10 \times t^2$ (t = time). Can you explain why this formula works?

Speed/time graphs

When a sprinter runs a 100 m race, she does not reach her top speed as soon as the starting pistol is fired. She takes a few seconds to accelerate from rest up to her top speed. Figure 1 shows this in the form of a speed/time graph. From Figure 1 you can see that she took 2 seconds to accelerate up to 10 m/s. So her acceleration is 5 m/s², which is the gradient of the graph.

> The gradient of a speed/time graph equals the acceleration

You can also use the same graph to calculate the distance that she has run. It is easy to work out the distance travelled once she has reached a constant speed of 10 m/s. She runs 10 m each second, so over the region BC of the graph she runs a distance of 10 m/s × 2 s = 20 m, which is the **area** under that part of the graph.

> The area under a speed/time graph equals the distance travelled

We can check this formula for the first 2 seconds of her race. The distance travelled in that time is equal to the area of the triangle under the line AB.

$$\text{Area of triangle} = \tfrac{1}{2}\,\text{base} \times \text{height} = \tfrac{1}{2} \times 2\,\text{s} \times 10\,\text{m/s}$$
$$= 10\,\text{m}$$

We would have found the same answer if we had used the formula:

> Distance = average speed × time.

The average speed over the first 2 seconds is 5 m/s, halfway between 0 and 10 m/s

$$\text{So } d = \text{average speed} \times \text{time} = 5\,\text{m/s} \times 2\,\text{s}$$
$$= 10\,\text{m}$$

Falling

When an object is allowed to fall freely towards the ground, it accelerates at a rate of 10 m/s². Provided wind resistance is very small in comparison with the object's weight, this acceleration remains constant. Figure 2 shows the speed/time graph for a falling ball; each second its speed increases by 10 m/s. The distance the ball has fallen after 3 seconds, for example, may be calculated from the area under the graph.

$$d = \tfrac{1}{2} \times 30\,\text{m/s} \times 3\,\text{s}$$
$$= 45\,\text{m}$$

The table shows the distance travelled in intervals of 1 second. Check the calculations for yourself. Figure 3 shows the position of the ball during the first 3 seconds, after it is released.

Falling sideways

What happens to the ball if it is thrown sideways off the edge of a cliff with a speed of 10 m/s? Provided air resistance can again be ignored, the problem is not too difficult. We can treat the downwards and sideways motions of the ball separately.

- Downwards motion: the ball is acted on by the pull of gravity, so it accelerates as before, falling 5 m after 1 second, 20 m after 2 seconds, etc.
- Sideways motion: since air resistance is very small, no force acts sideways on the ball. Therefore it moves sideways with a constant speed of 10 m/s (Newton's first law). After 1 second it has travelled sideways 10 m, after 2 seconds 20 m, etc.

Figure 4 shows the positions, in intervals of 1 second, of a ball (A) dropped off the edge of the cliff, and a ball (B) thrown sideways with a speed of 10 m/s. The balls are released at the same instant. B falls in a **parabolic path** – it moves sideways at a constant speed, while accelerating downwards at a constant rate.

Any object that is hit, launched or thrown will follow a parabolic path. Figure 5 is a multiflash photograph of two balls launched in the laboratory. The balls are illuminated by a stroboscope, so that their positions are shown every 0.1 seconds. The balls were launched at different angles and speeds, and the two paths were superimposed on the same film.

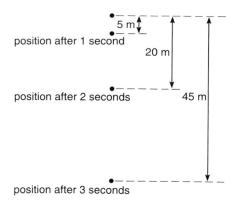

Figure 3
The ball travels further as it travels faster, so the gaps between the ball's position grow each second

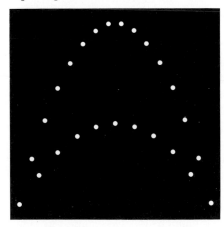

Figure 5

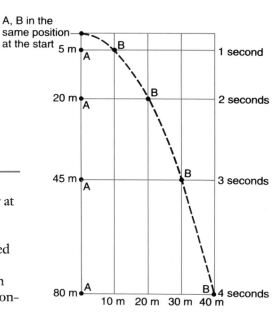

Figure 4
Note that both balls fall the same distance in the same time. B moves at a constant speed sideways. B falls in a parabolic path

Questions

1 Copy Figure 4 carefully onto graph paper.
(a) Mark carefully where ball B will be after 2.5 seconds.
(b) Draw the path B would follow if thrown sideways at only 5 m/s.
2 On the moon the gravitational acceleration is 1.6 m/s². A stone is dropped from a cliff; how far does it fall in 6 seconds?
3 This questions is about the photograph in Figure 5. Explain your answers carefully.
(a) Which ball was travelling faster just after it was released?

(b) Which ball was travelling faster at the top of its path?
(c) Each ball travelled the same horizontal distance, 3 m, this is called the ball's range.
Using the fact that the time between flashes was 0.1 s, calculate the horizontal speed of each ball.
(d) Make a sketch of the upper path from Figure 5. Add to it a second path of a ball, which is launched with the same upwards speed but twice the sideways speed. What is the range of this ball?

8 Applying Forces

Most of the energy of an impact in a car crash is absorbed by the crumple zones. This photograph shows the crumple zones on a Volvo

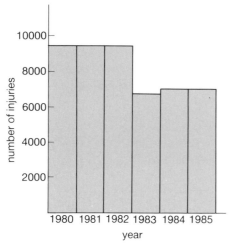

Figure 1
The number of serious injuries to front-seat passengers and drivers in cars and light vans in Great Britain. In January 1983 it became compulsory to wear seat belts in the front seats. The number of passengers and drivers wearing seat belts rose form 45% to 95%. The statistics speak for themselves. Will you drive a car without a seat belt?

Car crashes

In the last unit you read about a parachutist who landed on the ground travelling at 5 m/s. This may not sound like a very high speed but it is quite fast enough for a parachutist to suffer an injury such as a broken ankle. You hit the ground at the same speed when you jump down from a garden wall. To avoid injuring themselves, parachutists are taught to bend their knees on landing. This helps them to decelerate more slowly, because their decrease in speed takes longer. When the deceleration is less, the force acting on the parachutist's legs is less. Now there is a better chance of avoiding injury.

So, when a rapid deceleration occurs a large force acts to cause it. This idea is most important when designing cars with safety in mind. The photograph above shows a typical family saloon car. In the centre of the car is a rigid passenger cell which is designed not to buckle in a crash. However, the front and back of the car are designed to crumple on impact. These structures are called the **crumple zones**. In an accident, the deceleration that the passengers feel will be reduced. This is because of the time taken for the front of the car to crumple. This works in the same way as bending your legs on landing on the ground.

Another factor that has helped road safety is the wearing of seat belts (Figure 1). The photograph below shows what happens to a passenger if he is not wearing a seat belt. When the car stops suddenly he carries on moving because there is no force acting on him. The first thing he hits is the windscreen. He now experiences a large decelerating force which could fracture his skull. Glass from the window will also cut his face very badly.

This is what can happen if you don't wear a seat belt. The large force produced by rapid deceleration can throw passengers out of a car. Only dummies forget to put their seat belts on

Figure 2 shows the deceleration of two passengers in a car crash; Helen was wearing a seat belt and John was not. Helen decelerated over a period of 0.1 s. During this time the front end of the car was crumpling up. John was not wearing a seat belt, so he did not take advantage of the time provided by the crumple zone. Instead he carried on moving until he hit the windscreen; his deceleration was then much faster and the force acting on his head was very large indeed. John was killed; Helen survived.

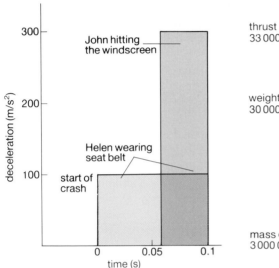

Figure 2
Graph to show approximate decelerations in a car crash

Figure 3
Acceleration of the Saturn V rocket

A Saturn V rocket takes off, accelerated by 33 million newtons of thrust

Overcoming gravity

You have just read that large forces will cause large accelerations or decelerations. Large forces are also used to accelerate large masses. One of our greatest technological achievements has been to land men on the moon. At launch the mass of a Saturn V rocket was 3 million kg and the thrust from its engines was about 33 million N. What sort of acceleration did this force produce? (see Figure 3).

Questions

1 Explain why it is a good idea to wear your safety belt in a car. Why is it safer for the driver if the back seat passengers are also strapped in?

2 What is a crumple zone in a car?

3 This question refers to John and Helen's car crash. John's mass was 80 kg and Helen's was 60 kg.
(a) Use Figure 2 to calculate the force that was acting on each of them to slow them down during the crash.

(b) How can you tell from the graphs that before the collision they were travelling at 10 m/s?
(c) Use the information to draw speed/time graphs for each of Helen and John during the crash.
(d) Helen was driving and was responsible for the crash. The police found that she had a large amount of alcohol in her blood. What measures should be taken to stop people from drinking and driving?

4 (a) After take off, a *Saturn V* rocket burns 14 000 kg of fuel per second. How much mass has the rocket lost after 2 minutes?
(b) What is the mass of the rocket after 2 minutes? (The rocket started with a mass of 3 million kg.)
(c) Now calculate the acceleration of the rocket after 2 minutes. The thrust from the engines is 33 million N.

9 Moving in Circles

Two Scottish players combine to force an England winger into touch

In the photograph you can see a rugby international between Scotland and England in progress. England's right wing has the ball and is running to score a try by the corner flag. However, the Scottish winger is running across to stop him, and he succeeds by giving the English winger a push. The sideways push does not slow him down but it changes the direction that he is running in. He is forced out of play, and England fails to score a try.

This account of rugby players demonstrates an important idea. So far, when you have seen forces acting on moving objects, the forces have caused something to speed up or slow down. However, if a force is applied at right angles to the direction of motion then there is no change of speed, but the direction of the motion is changed (Figure 1).

Earlier you met acceleration and you used the equation:

$$\text{acceleration} = \frac{\text{change of velocity}}{\text{time}}$$

When we use the word **velocity** we need to state the direction of motion as well as the speed. The rugby player changes his velocity but does not alter his speed. So it is possible to have an acceleration without changing speed. This is a very difficult idea to understand, but makes sense if you think about this: the player was pushed, and you know that pushes cause acceleration ($F = ma$).

Figure 1

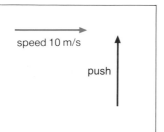

speed 10 m/s

push

(a) Before the push

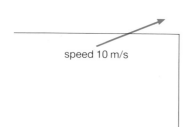

speed 10 m/s

(b) After the push the direction in which the man runs is changed. There is no change of speed, but there is a change of velocity

The friction between the tyres and the track provides the force needed to change direction

Circular motion

You can fasten a conker to a string and whirl it around your head in a horizontal circle. The conker moves at a constant speed, but its direction is always changing. The velocity of the conker is changing so it must be accelerating. The pull of the string is the force which accelerates the conker. The direction of this force is always at right angles to the motion of the conker. It is towards the centre of the circular path. It is called **centripetal force** (Figure 2).

The photographs on this page show some examples of centripetal forces making things move in a circular path. When an athlete throws a hammer he turns around two or three times before letting go of the hammer. While he turns round, his arms are providing a centripetal force to keep the hammer moving in a circle. As soon as he lets go, the hammer flies off along the direction it was travelling in.

A satellite in orbit around the Earth provides an interesting example of a centripetal force in action. The Earth's gravitational pull acts on the satellite to keep it in its circular path.

There are three factors that affect the size of the centripetal force acting on an object moving in a circular path. More force is needed if:
- the mass is increased
- the speed is increased
- the radius of the circular path gets less.

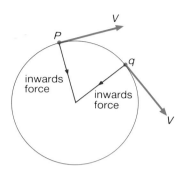

Figure 2
Conker on a string moving around a horizontal circle at constant speed

Igor uses his arms to provide the centripetal force

Questions

1 Explain carefully how it is possible for something to accelerate but not change its speed.

2 The diagram shows a ball attached to a string going around a circle.

(a) Copy the diagram and mark the direction of the force on the ball at *A*.

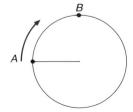

(b) When it gets to *B* the string breaks. In which direction does the ball move?

3 You can read below part of a discussion that Sundip and Jane had about how a spin drier works. Which one of them gives the better explanation? Correct any errors that they have made during their discussion.

Jane: A spin drier turns round very quickly to get the water out of the clothes. It works because when something moves round in a circle there is a force pushing things outwards, so the water is forced out through the holes in the spin drier.

Sundip: When something moves in a circle there is a force towards the centre. The drier makes the washing go in a circle but the water keeps moving in a straight line; that is why the water gets out.

Jane: What gives the washing a force towards the centre? Why doesn't it give the water the same force?

10 Collisions

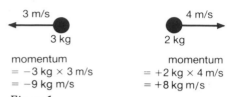

3 m/s ← ● 3 kg   4 m/s ● → 2 kg

momentum
$= -3\,kg \times 3\,m/s$
$= -9\,kg\,m/s$

momentum
$= +2\,kg \times 4\,m/s$
$= +8\,kg\,m/s$

Figure 1
In this diagram we have given anything moving to the right positive momentum, and anything moving to the left negative momentum

Why does a gun recoil when it is fired?

Momentum

Momentum is defined as the product of mass × velocity. We measure it in units of kilogram metres per second (kg m/s).

$$\text{momentum} = m \times v$$

Velocity is a vector quantity, so momentum is too. We must give a direction when we talk about momentum (see Figure 1). Momentum is very useful when we meet problems involving collisions or explosions. There is always as much momentum after a collision as there is before it.

Impulse

When a force, F, pushes a mass, m, we can work out the acceleration using:

$$F = ma$$

But the acceleration, $a = \dfrac{v - u}{t}$

where v is the final velocity, u is the starting velocity and t is the time taken for the velocity to change. The second equation can be substituted into the first to give:

$$F = \frac{m(v - u)}{t}$$
$$\text{or } Ft = mv - mu$$

Ft is called an **impulse** and has units of newton seconds. In words the equation can be expressed as:

$$\text{impulse} = \text{change of momentum}$$

So to cause a change of momentum of 100 kg m/s, for example, we must apply an impulse of 100 Ns. But we could apply a force of 100 N for 1 s, or 1 N for 100 s. Each causes the same change of momentum.

Conservation of momentum

In Section A you met Newton's third law. This law says that when one body pushes against a second body, the second body pushes back with an equal and opposite force. Using this idea we can work out what happens in collisions.

Figure 2 shows two ice hockey players chasing after the puck. The blue player pushes the red player with his stick. The size of the push is 400 N and it lasts for 0.4 s. How fast are the players moving after the push? Since $Ft =$ change of momentum, the red player's momentum increases by 400 N × 0.4 s = 160 Ns. We can calculate his increase in velocity using:

$$\text{increase in momentum} = \text{mass} \times \text{increase in velocity}$$

$$\text{So } 160\,Ns = 80\,kg \times \text{increase in velocity}$$

$$\text{increase in velocity} = 2\,m/s$$

An understanding of forces and momentum can make you into a world champion at the snooker table!

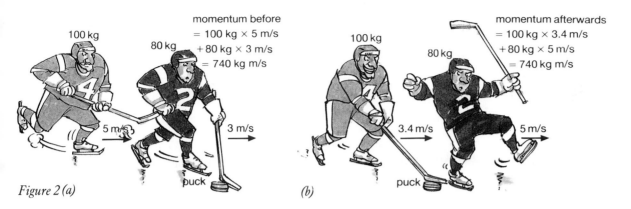

Figure 2 (a)   (b)

This means the red player moves with a velocity of 5 m/s after the push. While the stick was touching the red player, there was also a push back on the blue player. The same force of 400 N acted on the blue player for 0.4 s. So his momentum decreased by 160 Ns.

$$\text{His decrease in velocity} = \frac{\text{decrease in momentum}}{\text{mass}}$$

$$= \frac{160\,\text{Ns}}{100\,\text{kg}}$$

$$= 1.6\,\text{m/s}.$$

His final velocity = 5 m/s − 1.6 m/s = 3.4 m/s.

The momentum of one player increases by 160 Ns, the momentum of the second player decreases by 160 Ns. So the total change in momentum was zero. This leads to an important law:

> The momentum before a collision *always* equals the momentum after the collision.

This is known as the **principle of conservation of momentum**. Figure 3 shows another example: you can see that momentum is conserved.

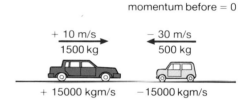

momentum before = 0

+ 15000 kgm/s   − 15000 kgm/s

momentum after = 0

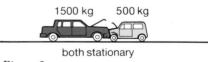

both stationary

Figure 3

Questions

1 (a) Explain why momentum can be measured in units of kg m/s or Ns.
(b) What change of momentum is caused by an impulse of 2 Ns?
2 What is the momentum of a runner of mass 70 kg, running at 10 m/s?
3 In each of the following experiments carried out in a laboratory, the two

trolleys collide and stick together. Work out the speeds of the trolleys after their collisions.
4 In Figure 3 two cars collided head on; each driver had a mass of 70 kg.
(a) Calculate the change of momentum for each driver.
(b) The cars stopped in 0.25 s. Calculate the average force that acted on each driver.

(c) Explain which driver is likely to be more seriously injured.
5 A field gun of mass 1000 kg, which is free to move, fires a shell of mass 10 kg at a speed of 200 m/s.
(a) What is the momentum of the shell after firing?
(b) What is the momentum of the gun just after firing?
(c) Calculate the recoil velocity of the gun.
(d) Why do you think very large guns are mounted on railway trucks?

11 Rockets and Jets

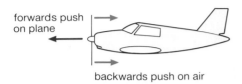

Every time you go swimming you demonstrate Newton's third law. As you move your arms and legs through the water, you push the water backwards. But the water pushes you forwards.

The same idea applies to an aeroplane in flight powered by a propellor (Figure 1). The propellor accelerates air backwards, but the air exerts an equal forwards force on the aircraft. As you will see below, rockets and jet engines use the same principle.

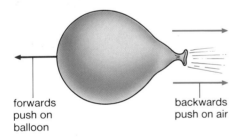

Figure 1
The forces acting on an aeroplane powered by a propellor

Rockets

A propellor takes advantage of air resistance. If air did not exert a force on us as we move through it, a propellor would not work. This is why an ordinary aeroplane cannot fly in space, where there is no air for its propellors to push against. The simplest way to demonstrate the action of a rocket is with an air-filled balloon. If you blow up a balloon and then let it go, it whizzes around the room. It would also whizz around in space where there is no air, because the balloon gets its forwards push from the escaping air. You can explain this by using the principle of conservation of momentum. Before you let go of the balloon, the momentum of the balloon and air is nothing. Once the air is allowed to escape from the balloon the *total* momentum is still zero, but the balloon has forwards momentum and the air has backwards momentum (Figure 2).

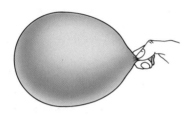

Figure 2
(a) No momentum

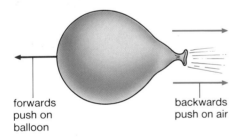

forwards push on balloon

backwards push on air

(b) Balloon has momentum to the left
Air has momentum to the right
Total momentum is still zero

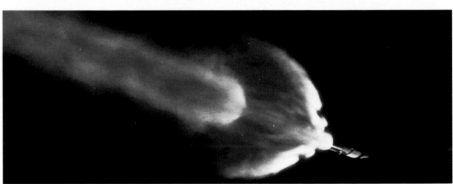

Saturn 1B at a height of 20 kilometres. The burnt fuel is driven away from the rocket with great force. In turn, the fuel gives an equal and opposite force to the rocket, which drives it into space

Figure 3 shows a simple design for a rocket. The rocket carries with it fuel, such as kerosene, and liquid oxygen (oxygen is needed for the fuel to burn). These are mixed and burnt in the combustion chamber. High pressure gases are then forced out backwards through the nozzle at speeds of about 2000 m/s.

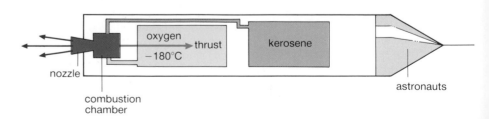

oxygen
−180°C
thrust
kerosene
nozzle
combustion chamber
astronauts

Figure 3
A simple rocket

44

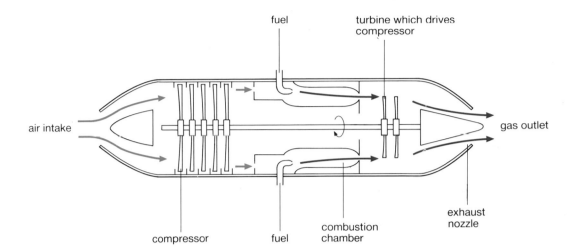

fuel

turbine which drives
compressor

air intake

gas outlet

compressor

fuel

combustion
chamber

exhaust
nozzle

Figure 4
The jet turbine engine

Jets

Figure 4 shows a simplified diagram of the sort of **jet turbine engine** that is
found on an aeroplane. At the front of the engine a compressor sucks in air,
acting rather like a propellor, but the air is compressed and as a result it becomes
very hot. Some of the heated air then goes through into the combustion chamber
and is mixed with the fuel (kerosene). The fuel burns and causes a great increase
in the pressure of the gases. The hot gases are then forced out at high speed
through the exhaust nozzle. The escaping gases provide a forward thrust on the
engine.

The engine needs to be started with an electric motor to rotate the compressor,
but once the engine is running some of the energy from the exhaust gases is used
to drive a turbine. The turbine is mounted on the same shaft as the compressor so
air can now be sucked in without the help of an electric motor.

*This is a large jet engine. You can clearly
see the turbines*

Questions

1 (a) Explain carefully how a rocket
works.
(b) An aeroplane powered by jet
engines cannot fly in space, but a rocket
can. Why?
2 On the right you can see a velocity/
time graph for a firework rocket. The
graph stops at the moment the
fireworks stop burning. A positive
velocity on this graph means the rocket
is moving upwards.
(a) How can you tell from the graph
that the acceleration of the rocket is
increasing?
(b) Why does the acceleration rise?

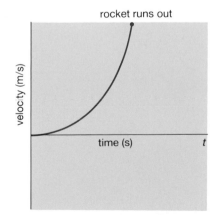

rocket runs out

velocity (m/s)

time (s)

t

(Hint: What happens to the mass of the
rocket as it burns?)
(c) Copy the graph and sketch how the
velocity changes until the rocket hits
the ground.
3 In the last unit you met the equation

$$F = \frac{mv - mu}{t}.$$

(In words: force equals the rate of
change of momentum). Use this
equation to calculate the thrust from a
rocket that uses 2000 kg of fuel and
oxygen each second, and ejects its
exhaust gases at a speed of 1000 m/s.

SECTION B: QUESTIONS

1 This graph of velocity against time represents the motion of a cyclist.

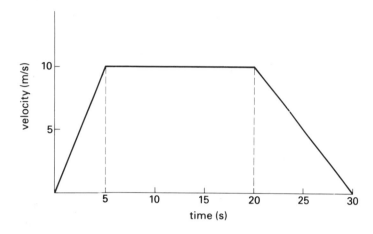

velocity (m/s)

time (s)

(a) Calculate her deceleration in the last 10 seconds of the journey.
(b) Calculate the distance that she travelled during the journey.

2 A sky diver, whose weight is 800 N, jumps out of an aeroplane. After he has been falling for some time he is travelling at a constant speed. How big is the force of air resistance acting on him: is it (a) 800 N, (b) greater than 800 N, or (c) less than 800 N?

3 Sara drops a stone off the edge of a cliff. It takes 2 s to reach the bottom.
(a) Calculate the speed of the stone just before it reaches the ground.
(b) Calculate the stone's average speed.
(c) Calculate the height of the cliff.

300 N

Mike

4 Mike has worked out a new way to measure his mass. He is in a trolley that accelerates down the slope when released. Use the data below to answer these questions.
● It takes 300 N to hold Mike stationary on the slope.
● When released Mike travels 36 m in 6 s.
● Mass of the trolley = 32 kg.
(a) What is Mike's average speed over the 36 m?
(b) Assuming he accelerates at a steady rate, what is his final speed after travelling 36 m?
(c) Now work out his acceleration.
(d) Calculate Mike's mass.

5 The diagram below shows an alpha particle colliding head on with a stationary proton in a cloud chamber. Before the collision the alpha particle has a speed of 10^7 m/s. After the collision its speed is 0.6×10^7 m/s. The mass of the alpha particle is 4 times that of the proton. Calculate the speed of the proton.

alpha   stationary
particle   proton

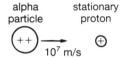

10^7 m/s

Before the collision

alpha
particle   proton

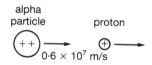

0.6×10^7 m/s

After the collision

6 Archimedes' Principle states that 'if a body is wholly or partially immersed in a fluid then the upthrust is equal to the weight of fluid displaced'.

● Total mass of sandbags + Boris + basket = 24 kg
● Volume of balloon = 30 m³
● Density of air at sea level ≈ 1.2 kg/m³
● Density of helium in balloon ≈ 0.2 kg/m³

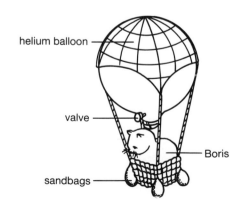

helium balloon

valve

Boris

sandbags

(a) What is the force due to gravity on the balloon and its contents? (Include the helium.)
(b) What upthrust is there on the balloon?
(c) What is Boris's initial acceleration upwards?
(d) Assuming this acceleration is constant how far will he have travelled after 2 seconds?

The graph shows how the density of air varies with height above sea level.

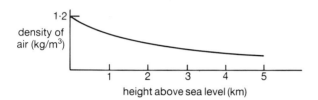

(e) Deduce from the graph how high Boris will rise.
(f) How could he go higher?
(g) How could he get down again?

7 Ravi and Sarah are very good friends. But unfortunately they live 10 km apart. However, at 7 pm every evening they set out to meet each other; Ravi walks at 6 km/h and Sarah walks at 4 km/h.

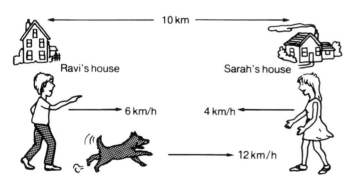

Ravi has a dog called Rover who always goes out with him. Rover thinks they must be mad to walk so far every night. However, he likes a good run. So as soon as he leaves Ravi's house he runs at 12 km/h until he meets Sarah. As soon as he meets Sarah he turns round until he meets Ravi. Rover runs backwards and forwards between Ravi and Sarah until they meet. How far has Rover run when they meet?

8 The diagram shows an accelerometer, which can be used to test the performance of a train. Springs can exert a force on a mass, M. The scale shows the size and direction of the resultant force. When the train is at rest the reading on the scale is zero.

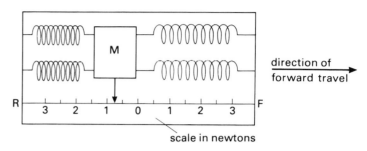

(a) What will the reading on the scale be when the train has been travelling at a steady speed for some time? Explain your answer.
(b) When the train accelerates forwards, the mass moves towards the end marked R. Explain why.
(c) Calculate the acceleration shown in the diagram (the mass of M is 0.5 kg).
(d) The train brakes suddenly and the deceleration is 3 m/s². What is the reading on the scale now?
(e) How would you adapt this accelerometer to measure greater accelerations?

9 When a bomber releases a load of bombs they fall freely under the action of gravity. The table below gives positions of the bombs at 2 s intervals.
(a) Plot a graph of the height of the bombs above the ground (y-axis) against the distance travelled horizontally (x-axis).
(b) How fast is the plane travelling?
(c) Use the graph to predict the position of the bombs after: (i) 6 s, (ii) 20 s.
(d) The bomber now flies on a mission at a height of 1500 m at the same speed. How far away from the target must the bombs be released if they are going to hit? Use your graph to help you answer this.
(e) Another bomber flies at a speed of 200 m/s. Draw a second graph, using the same axes as before, to show the position of bombs that fall from this plane.

Positions of bombs	Time (s)										
(see question 9)	0	2	4	6	8	10	12	14	16	18	20
height above the ground (m)	2000	1980	1920		1680	1500	1280	1020	720	380	
distance travelled horizontally (m)	0	600	1200		2400	3000	3600	4200	4800	5400	

QUESTIONS

10 The diagram shows an experiment which is designed to calculate the acceleration due to gravity, *g*.

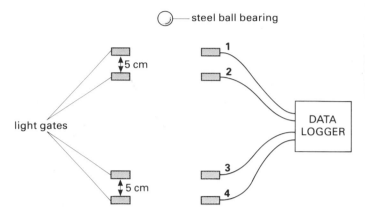

steel ball bearing

5 cm

5 cm

light gates

1
2
3
4

DATA LOGGER

A small ball bearing is allowed to fall past four 'light gates'. The time at which the ball bearing passes through each light gate is recorded by a data logger. The times recorded for such an experiment are shown in the table; the times are in milliseconds (ms). The gates are arranged in two pairs. The separation of the gates in each pair is 5 cm, as shown in the diagram.

Light gate	Time (ms)
1	0
2	25
3	310
4	320

(a) Work out the average speed of the ball bearing as it fell past gates 1 and 2.
(b) Work out the average speed of the ball bearing as it fell past gates 3 and 4.
(c) Work out the acceleration of the ball bearing as it fell. Explain your working carefully.
(d) Explain how a light gate works.
(e) The same experiment can also be done by attaching a ticker tape to a heavy ball. Explain two advantages that the light gates have over the ticker tape method.

11 This question is about the performance of the Montego 2.0 HL, manufactured by the Rover company. Examine the data shown provided by the manufacturers. Graph A is a velocity /time graph, showing how the velocity of the car changes as it accelerates through the gears. This was obtained by attaching a data logger to the speedometer. Graph B shows the available force at the wheels, for different gears and speeds. The total resistive force acting on the car is also shown.

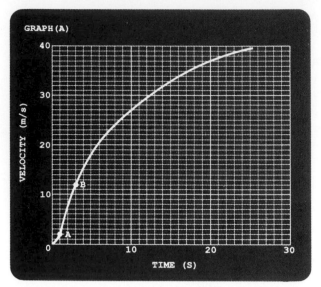

(a) The manufacturers claim that the Montego can cover 400 m, starting from rest, in under 20 seconds. Use the velocity/time graph to show this is correct.
(b) Use the velocity/time graph to calculate the acceleration between times of 1 s and 3 s. (The region *AB* of the graph.)
(c) Use graph (b) to work out the maximum possible acceleration of the car, in first gear. The mass of the Montego is 1000 kg.
(d) Explain why the car's acceleration is less in the higher gears.
(e) Which gear would you drive in to obtain the maximum acceleration from a speed of 20 m/s?
(f) Use graph (b) to show that the car's maximum speed is 48 m/s. In which gear is this speed reached?

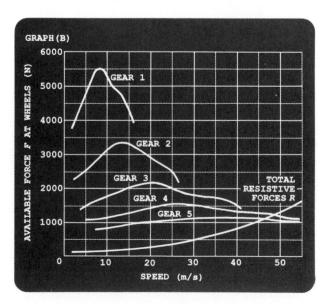

SECTION C
The Earth in Space

Physicists try to explore and explain the universe in which we live

1 The Solar System

The brightest thing that you can see in the sky is our **Sun**. The Sun is a star, which provides all the heat and light that we need to live. The Earth is a **planet** which moves around the Sun. Altogether, there are nine planets which move round the Sun in curved paths. These paths are called **orbits**. The Earth is the third planet out from the Sun. Planets do not produce their own light like the Sun; we can see planets at night because they *reflect* the Sun's light. The Sun and its nine planets are known as the **solar system**. The word 'solar' means 'belonging to the sun'. Some of the planets have moons that move around them. The Earth has only one moon.

The four planets nearest to the Sun, including the Earth, have hard, solid and rocky surfaces. The planet closest to the sun is **Mercury**. This has a cratered surface that looks like our Moon's surface. During the day, Mercury is baking hot. Even lead would melt on its surface. Mercury travels quickly round the Sun, taking 88 days to complete one orbit.

Venus lies between Mercury and the Earth. It has a thick atmosphere; its clouds are made from burning hot sulphuric acid. Nothing could survive on its hot, poisonous surface.

Mars is the last of the rocky planets. It is a lot colder than the Earth because it is further away from the Sun. Like the Earth, it has large polar icecaps, which can be seen through a telescope. At some stage in the past, Mars had active volcanoes. The largest is called Olympus Mons. It towers 25 km above the surrounding land, this is more than $2\frac{1}{2}$ times the height of Mount Everest above sea level. As continents move on the Earth, mountains grow. The Himalayas are growing now at about the same rate as your fingernails. Mountains may still be growing on Mars.

Beyond the orbit of Mars, are four giant planets: **Jupiter, Saturn, Uranus** and **Neptune**. Each of these is much larger than the four inner planets (Figure 1). In Figure 2, you can see that these four giant planets move in orbits that are far away from the Sun. Each of these planets is made from gas. If you landed on one of these planets, you would sink into it.

Illustration of the planet Mars to show the large polar ice cap. A system of canyons can be seen at the equator. At the top left of the picture you can see the Olympus Mons volcano, with three smaller volcanoes below it

Pluto Neptune Uranus Saturn Jupiter Mars Earth Venus Mercury Sun

Figure 1
The relative sizes of the Sun and its planets. (Figure 2 shows the positions of the planets relative to one another)

The largest planet is Jupiter, which has a diameter 11 times bigger than that of the Earth. More than 1000 Earths would be needed to fill Jupiter's enormous volume. The swirling clouds in Jupiter's atmosphere blow around at hurricane wind speeds of 200 miles per hour. Jupiter has 14 moons which orbit around it. Four of these are about the same size as our Moon. The innermost of these four large moons is called Io; when Voyager 1 flew by Io in 1979, it photographed active volcanoes. Saturn is best known for its beautiful rings.

The planet furthest from the Sun is Pluto; it is smaller than our Moon. It is a frozen and dead world. Its temperature is 240 degrees below freezing. From its cold surface, the Sun would look like a bright star, not the provider of life giving light and warmth that we know.

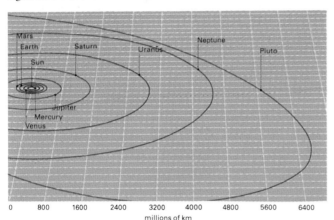

Facts about the Sun:
diameter 1 400 000 km;
surface temperature 6000°C.

Figure 2
The orbits of the planets around the Sun. The four inner planets are very close to the Sun; the gaps between the outer planets are very large

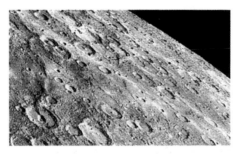

Mercury is only 58 million km from the Sun. It has a hot surface with no atmosphere to protect it from meteors. This photograph shows how the surface has been heavily cratered by meteors

Planet	Diameter of planet	Average distance of planet from the Sun	Time taken to go round the Sun	Number of moons	Average temperature on sunny side
Mercury	4900 km	58 million km	88 days	0	350°C
Venus	12 000 km	108 million km	225 days	0	480°C
Earth	12 800 km	150 million km	$365\frac{1}{4}$ days	1	20°C
Mars	6800 km	228 million km	687 days	2	0°C
Jupiter	143 000 km	780 million km	12 years	14	–150°C
Saturn	120 000 km	1430 million km	29 years	24	–190°C
Uranus	52 000 km	2800 million km	84 years	15	–220°C
Neptune	49 000 km	4500 million km	165 years	3	–240°C
Pluto	3000 km	5900 million km	248 years	1	–240°C

Table 1
Facts about the planets

Questions

1 Using Table 1:
(a) Which is the largest planet?
(b) Which planet has a temperature closest to that of our Earth?
(c) Which planet takes just under two years to go round the Sun?
(d) Are the temperatures of the planets related to their distance from the Sun? Why is Venus hotter than Mercury?
(e) Is there any pattern in the number of moons that planets have? Try and explain any pattern that you find.
(f) Which planets will go round the

Sun more than once in your lifetime?
2 (a) Here is something active to do. You can make a model of the solar system. The scale of your solar system should be 1 m for 10 million kilometres. Use Table 1 to calculate how far each planet needs to be placed from the Sun. Now go outside and make your model – you will need a lot of space!
(b) Now work out where the nearest star, Alpha Centauri, should be placed on your model. Should it be in the next street, the next town or where?

(Alpha Centauri is 6000 times further away from the Sun than Pluto is.)
3 Try and answer these questions about the solar system. You may need to go to a library to find the answers.
(a) Where is the Sea of Tranquillity?
(b) Which planet has a red spot?
(c) What are Oberon and Titania?
(d) What are 'asteroids'?
(e) Which planet has the shortest day?
(f) How many planets have rings around them?

2 The Four Seasons

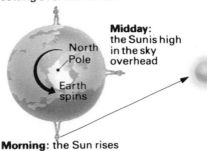

Evening: the Sun is setting over the horizon

Midday: the Sun is high in the sky overhead

North Pole

Earth spins

Sun

Morning: the Sun rises and is seen low down close to the ground

Figure 1

The Earth spins around on its axis. It rotates once in about 24 hours. This is why we have days and nights. When we face the Sun it is day; and when we cannot see the Sun it is night. At night we can see stars. The stars are there all the time, but the sky is too bright for us to see them during the day.

Because the Earth spins, the Sun, Moon and stars seem to move across the sky. Figure 1 shows the Earth as you would see it from above the North Pole. Professor Chandrasekhar lives on the Equator. You can see from Figure 1 that the Professor sees the Sun close to the ground in the morning and evening, but that it is over-head at midday.

The Earth moves around the Sun in an approximately circular orbit. The closest the Earth gets to the Sun is 147 million kilometres; the furthest is 152 million kilometres. The Earth takes $365\frac{1}{4}$ days to complete its path round the Sun. This is why we have 365 days in our calendar. Every four years, we have a **leap year** with an extra day to make up for the $\frac{1}{4}$ day lost every year.

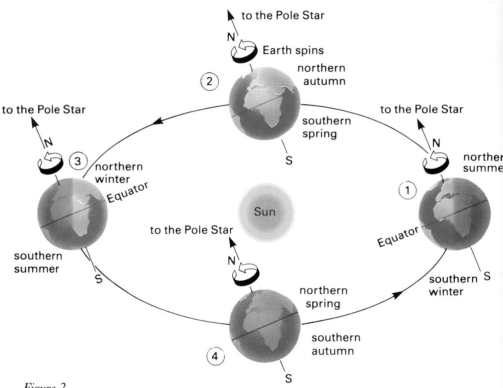

Figure 2
Earth's path around the Sun

Animals like the hedgehog in the photo hibernate over the winter. This reduces their energy demands to a low level so they can survive the cold months when food is scarce

Figure 2 shows the Earth's yearly path. The Earth spins around an axis that goes through the North and South Poles. All the time, the North Pole points towards the pole star (Polaris). The Earth's axis is tilted as shown in Figure 3. This tilt gives the Earth its seasons. In the middle of our summer the northern half of the Earth is tipped towards the Sun by an angle of $23\frac{1}{2}°$ (position (1) in Figure 2). At this time, the south of the Earth is tipped away from the Sun. Because the north is tipped towards the Sun it is hotter, so it is summer. But the south is tipped away, so it is cold and winter. Six months later, the Earth has reached position (3). The north is now tipped away from the Sun, and the south is tipped towards the Sun. It is now winter for us, and summer in the south. In the spring and autumn, the northern and southern halves of the Earth get equal amounts of sunshine.

The tilt of the Earth's axis also causes the length of our days to change. In Figure 3, you can see that it is summer in the north. Look at the half of the Earth north of the equator; more of it is in sunlight than in darkness. This makes our summer days long. More of the south is in darkness, so it has long winter nights. In the northern summer, the North Pole always sees the Sun, so it has 24 hours of daylight each day. But in the winter, the North Pole has no daylight at all. The Equator has days that are 12 hours long all through the year. If you live in Kenya, the Sun rises at 6 am every morning and sets at 6 pm; there is very little twilight. If you live in Oslo, you will get 20 hours of sunlight in the summer, but only four hours of sunlight in winter.

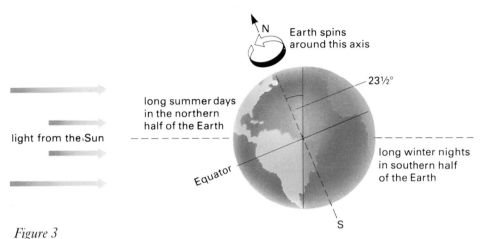

Figure 3

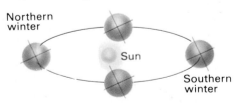

Figure 4
In winter, the Sun is low in the sky and shadows are long In summer, the Sun is high in the sky and so shadows are shorter

Questions

1 Jo knows it is cold at the North Pole. But she thinks that, over a year, someone living at the North Pole would see the Sun for as long as someone living at the Equator. Is she right?

2 Imagine what it would be like on the Earth if it was tilted at a different angle. In the diagram on the left below, the Earth does not tilt at all.

What happens then? Do we have different seasons? Do our days change in length? In the diagram on the right below, the North Pole is facing the Sun. Can you imagine what life would be like now?

3 Mars is quite similar to the Earth its axis of rotation is tilted towards the sun by an angle of 24°, and its day is $24\frac{1}{2}$ hours long.

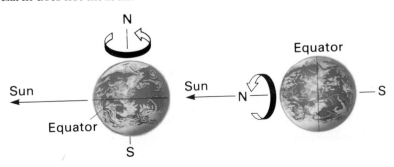

The diagram above shows Mars' orbit round the Sun. In southern winter it is 250 million km away from the Sun, but only 200 million km from the Sun in northern winter. Describe how the seasons on Mars differ from ours. Is a northern winter on Mars colder than a southern winter?

3 Moving Planets

Figure 1
The Earth moves in an elliptical orbit round the Sun. The Moon moves around the Earth

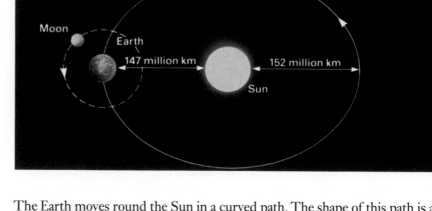

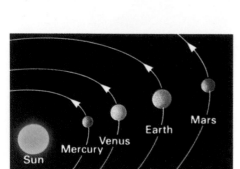

Figure 2
All planets rotate around the Sun in the same direction

The Earth moves round the Sun in a curved path. The shape of this path is an **ellipse.** The ellipse is a bit like a squashed circle. Figure 1 shows how the Earth and Moon move together round the Sun. All the planets move in elliptical paths around the Sun. They all move in the same direction, as shown in Figure 2. The planets near the Sun move faster than those further away. In Figure 2 you are looking down on the solar system from on top. If you could look at it from the side (Figure 3), you would see all the planets in nearly the same plane. This is called **the plane of the ecliptic.** Our Moon also lies in the same plane. This means that you can quite often see the Moon close to some planets in the sky.

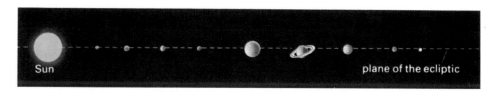

Figure 3
All planets lie very close to the same plane. Will all planets always be on the same side of the Sun?

In ancient Rome, the word 'planet' meant 'wanderer'. The planets were given their name because they appear to wander around the sky. The Earth moves round the Sun once a year. In July, we are always in the same position so we see the same stars at night each year. The stars are so far away that they do not appear to move at all, even over thousands of years. Each July we see the bright star Antares low in the sky. From 1984 to 1988, Saturn could be seen clearly on July evenings close to Antares. But Saturn moves slowly round the Sun taking about 30 years to go once round (Figure 4). During 1986 and 1987 Saturn moved slowly past Antares (Figure 5).

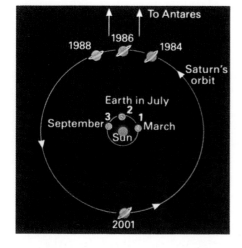

Figure 4
As Saturn moves slowly round the Sun, we see it move against the stars

Yearly motion

Figure 5 shows Saturn's passage through the stars over a period of several years. However, by looking at Saturn's position at the same time each year, the effect of the Earth's motion has been ignored. When you look at a planet every month or so, you will see a more complicated motion.

Figure 6 shows how Mars moved past the Hyades and Pleiades in late 1990 and early 1991. Notice that Mars moved to the left for most of the time; but between October and December it moved backwards, through a **retrograde loop**. This loop occurs because the Earth, which orbits the Sun faster than Mars, overtakes Mars at this point.

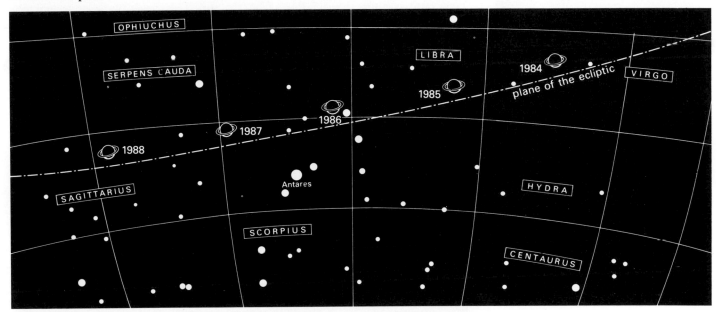

Figure 5
This is how Saturn moved over a period of 5 years

Figure 6
The movement of Mars during 1990 and 1991

Questions

1 (a) Look at the star map in Figure 5. Approximately where will you see Saturn in 2014? (Remember Saturn takes about 30 years to go round the Sun.)
(b) Explain why it will be very difficult to see Saturn in July 2001. In which month will it be best to see Saturn in 2001 ?

2 If you look at Jupiter or Saturn through a telescope, you can see a large planet with clear markings. If you look at Pluto through a telescope it looks like a star – a small point of light. Pluto was discovered in 1930. How did astronomers know that Pluto is a planet and not a fixed star?

3 (a) Make a copy of Figure 4. Draw lines from Earth positions 1, 2 and 3 to Saturn's position in 1986. Use your diagram to explain why Saturn appears to go through a backward loop every year.
(b) Explain why Mars appears to move more rapidly than Saturn.
4 Prepare a five-minute talk for the class on 'planetary motion'.

4 Beyond the Solar System

Look at the sky on a dark night. How many stars do you think you can see? All of these stars are great distances away from us. We measure distances to stars in **light years**. A light year is the distance that light travels in one year. This is about 10 million million kilometres. The brightest star in the sky (after the Sun) is Sirius. It is also one of the nearest, but light takes nine years to reach us from Sirius. Light takes only six hours to reach Pluto from the Sun. This makes our solar system look very small.

Try to look at the sky on a clear night through binoculars or a small telescope. You will be able to see even more stars. You may be able to see the milky way. Many of the stars in the milky way have their own solar systems.

Clusters of millions of stars like the milky way are called galaxies. The Sun is one of the stars in the milky way (Figure 1). The milky way has about 100 000 million stars in it. If you could see our galaxy from the side, it would look like two fried eggs stuck back to back (Figure 1); it is long and thin except for a bulge in the middle. If you could see the galaxy from the top it would look like a giant whirl-pool with great spiral arms. In fact, the galaxy does spin round. Our Sun takes about 220 million years to go once round the centre of the galaxy. In your lifetime, the pattern of stars that you see each night will not appear to change. But over thousands of years the pattern will change as our Sun moves through the galaxy. Our ancestors who lived 100 000 years ago would have seen different constellations from those we can see today.

From the Earth, we see the stars of our galaxy as the milky way

This spiral galaxy is in the constellation of Pisces. It is at a distance of 30 million light years and has a diameter of 80 thousand light years

Figure 1
(a) Side view of our galaxy
(b) Top view of our galaxy
(If you look towards the centre of our galaxy, you see the milky way. On dark summer nights, this looks like a milky band overhead, but it is made from billions of stars)

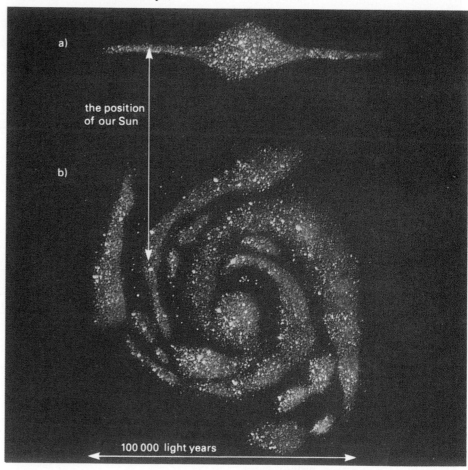

a)

the position of our Sun

b)

100 000 light years

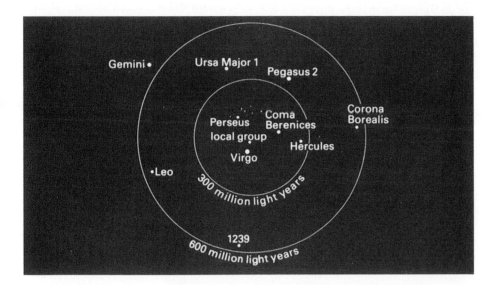

Figure 2
This map shows groups of galaxies close to our local group. The most distant galaxies in the universe are 15 000 million light years away from us. In this map, a large dot means that the group of galaxies contains more than 50 galaxies

About 10 000 million years ago there were no galaxies at all, and the universe was full of hydrogen and helium gas. **Gravity** is a force which acts over enormous distances, even over thousands of millions of light years. Groups of galaxies were formed when gravity gradually pulled large volumes of gas together. After the galaxies were formed, stars were formed inside the galaxies. Look at Figure 2. This shows a map of groups of galaxies. Our local group is in the middle. There are about 20 galaxies in our local group. Some groups of galaxies have as many as 1000 galaxies in them. There is a total of about 10 000 million galaxies in the universe.

This photograph shows part of the Virgo cluster of galaxies. The Virgo cluster is about 50 million light years away. It contains about 1000 galaxies

Questions

1 Explain what is meant by each of the following terms.
(a) moon, (b) planet, (c) star, (d) galaxy, (e) group of galaxies, (f) universe
2 Our Sun is about 4600 million years old. Use the information in the text to calculate the number of times the Sun has rotated round the galaxy. Humans have existed on the earth for about 50 000 years; how many times round the galaxy has the Sun rotated in that time?
3 One of the nearest galaxies to us is the Andromeda galaxy. It is 2 million light years away. The fastest rocket we can make travels at 1/10 000 of the speed of light. Calculate how long it would take us to reach Andromeda. (The speed of light is 300 000 km/s.)

5 The Earth as a Sphere

It is obvious to us that the Earth is a sphere. After all, satellites go round the Earth every day. They send back photographs of a curved surface. The first astronauts who went to the Moon looked back and saw it was like a big ball. But, people have not always thought the Earth was a sphere. In 600 BC, the Greeks thought the Earth was flat. They believed that they lived on a flat Earth, which was surrounded by sea. They thought that the Earth was at the centre of everything. The stars, Sun and Moon went round us each day. People thought that you could fall off the 'underneath' of a sphere. They had no idea about gravity. So they could not understand how gravity would always pull you towards the Earth.

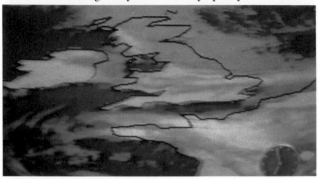

Satellites give us information about weather from space. Maps of this kind are used on television

Modern photographs from space show us that the Earth is spherical

However, careful observations and measurements soon showed the Greeks that the Earth was really a sphere. They watched ships slowly disappear below the horizon. At sea, they knew what it was like to be surrounded by a circular horizon. This suggested that the surface of the Earth was probably curved.

They also noticed that as you go further north, the Pole Star appears higher in the sky at night and the Sun lower at day. In 330 BC, Eratosthenes used this idea to measure the radius of the Earth. Eratosthenes lived in Alexandria which was 800 kilometres north of Syene, now Aswan in Egypt. He knew that in midsummer each year the Sun was directly overhead at noon in Syene. This was because sunbeams falling into a deep well were reflected back up again by the water at the bottom (Figure l(a)). At noon on midsummer's day, he measured the shadow of a tall obelisk in Alexandria. He found that the sunbeams made an angle of 7° with the vertical (Figure 1(b)). This showed Eratosthenes that the Earth had a curved surface. It also allowed him to work out the Earth's radius (Figure l(c)).

Figure 1
(a) At noon in Syene, the Sun is overhead
(b) Midsummer noon in Alexandria
(c) The Sun is not overhead in Alexandria because the Earth is a sphere

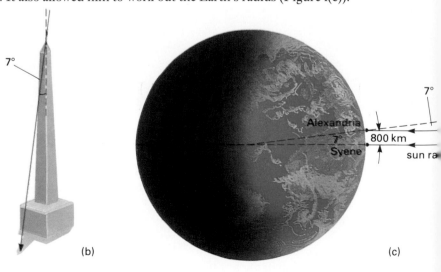

Questions

1 In 1492 Christopher Columbus sailed from Europe and encountered America.

(a) At the time, some people worried that he might fall off the edge of the Earth. Why were they worried?

(b) Suppose you are Christopher Columbus. What would you say to your sailors to persuade them that they would not fall off the edge of the world during the voyage?

2 Draw a diagram to explain why ships disappear over the horizon at sea.

3 At the North Pole, the Pole Star appears overhead. At the Equator, the Pole Star appears close to the northern horizon.

(a) Why does this show the Earth is curved? Explain by drawing a diagram.

(b) What would the directions of the Pole Star be like if the Earth was flat?

4 This question is about Eratosthenes' measurements.

(a) Eratosthenes measured the distance from Syene to Alexandria. How do you think he did it?

(b) Explain why Eratosthenes took more than a year to make his observations. *(Hint:* He could not travel very quickly in those days!)

(c) Look at figure l(c). Explain why you cover a distance of 114 km if you go 1° north from Syene.

(d) There are 360° in a circle. What distance do you go if you walk right round the Earth?

(e) This distance is 6.3 times the radius of the Earth. Calculate the Earth's radius.

5 Aristotle was a Greek philosopher who lived around 340 BC. He believed that the Earth is a sphere. One reason was this: the heavens are a region of unchangeable perfection. The sphere is the perfect, symmetrical solid shape; therefore the Earth, Sun, Moon, planets and stars must all be spheres. Some evidence to support Aristotle was provided from **eclipses**. He noticed that, when the moon was eclipsed by the Earth, the Earth's shadow on the Moon was always circular. The diagram below shows that only a spherical Earth will *always* cast a circular shadow.

(a) Explain carefully why the observation of lunar eclipses supports the idea that the Earth is a sphere.

(b) What other evidence did Aristotle have to support the argument that the Earth is a sphere?

(c) How would you explain to Aristotle that his first reason for a spherical Earth, based on perfection and symmetry, is not based on scientific evidence?

(d) Here are two dictionary definitions.

Philosophy: the study or pursuit of knowledge of things and their causes, either practical or theoretical.
Science: the study or pursuit of systematic and formulated knowledge.

(i) In Newton's time, 'physics' was called natural philosophy. Explain why.

(ii) How well do the words 'scientist' and 'philosopher' apply to Newton and Aristotle?

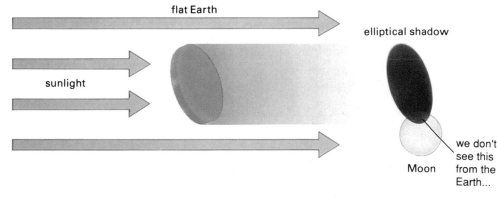

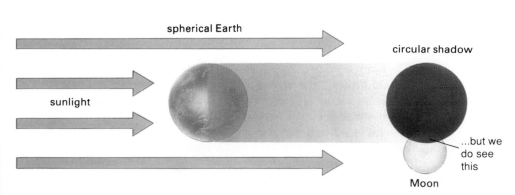

6 Gravitation

Newton's law of gravity

In the 100 years before Isaac Newton was born, astronomers had carefully observed the motion of planets and the positions of stars. From these measurements, the astronomers worked out that the Sun is the centre of our solar system, and that the planets move in orbits around it. But nobody understood what kept the planets moving along their paths.

In 1665, the great plague swept across England. Cambridge University was closed and all the students, including Newton, were sent home. It was then, stranded in Lincolnshire, that Newton worked out his Law of Gravity. He suddenly realised that the Earth's gravitational pull does more than keep our feet on the ground. It reaches out, beyond the highest mountains, and into the depths of space. The Earth's pull stretches 400 000 km across space, and keeps the Moon moving around us.

Newton said that any two masses in the universe attract each other with a gravitational pull. The size of this force is given by:

$$F = \frac{GM_1 M_2}{R^2}$$

F is the force (in newtons) between two masses, M_1 and M_2 (in kg). These masses might be planets for example. R is the distance between the centre of these masses (in m), see Figure 1. G ('big gee') is the universal constant of gravitation; its value is 6.7×10^{-11} Nm²/kg². The pull is too weak to be noticeable between two people. It is only when one of the masses is the size of a planet that we can feel the force of gravity.

Newton's equation shows us that the Earth's pull gets weaker the further out into space we go (Figure 2). What is the Earth's pull on the Moon? The Moon is 60 times further from the centre of the Earth than we are. For us, the pull is 10 N/kg – each kg experiences a pull of 10 N. In the equation, if R gets 60 times bigger, $1/R^2$ gets 60^2, or 3600, times smaller. So the Earth exerts a pull of $10/3600$ N on each kg of the moon; that is about 0.003 N/kg.

Sir Isaac Newton (1642–1727) was the first person to realise that the Earth 's gravitational pull extended beyond the highest mountains. He used his theory to explain the Moon's motion

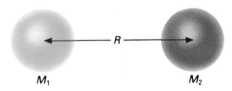

Figure 1

Figure 2
The Earth's pull gets weaker further away, but it is strong enough to keep the Moon in orbit around us

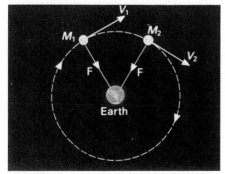

Figure 3
The Earth's pull changes the velocity of the Moon from v_1 to v_2

Orbits

Figure 3 shows the Moon's orbit round the Earth; in position M_1, the moon is moving along the direction v_1. Without any force acting on it, the Moon would continue to move in that direction. But the Earth is always pulling the Moon towards it. This pull is at right-angles to the Moon's motion. The pull does not speed the Moon up; the pull deflects the Moon away from a straight path, into a curved path. At M_2, the Moon has a new velocity v_2. The Earth's pull changes this velocity again. The Earth pulls on the Moon all the way round its orbit.

The Moon has stayed in orbit for billions of years. As it goes round the Earth it does not lose any energy. There are no frictional forces in space to slow it down. The Earth's force on the Moon does not change its speed, because it acts at right-angles to its motion. (See Unit D1; no work is done when the push is at right-angles to the motion.)

Tides

The gravitational pull of the Moon on the ocean causes our **tides**. We get two high tides a day. The Earth-Moon system rotates about a centre of gravity (or **barycentre**) at B (Figure 4). This is inside the Earth but not at its centre. At A, there is a high tide because the Moon pulls more strongly on the water closer to it. At C there is also a high tide. At C the Moon pulls the water less strongly. As the water rotates around B it piles up; this is because the Moon's pull is not strong enough to keep it in a smaller circular path.

This photograph by the Voyager I spacecraft shows Io (left) and Europa above the swirling clouds of Jupiter. Newton also used his theory to explain that all planets (and stars) have their own gravitational pulls. Jupiter pulls its moons to keep them in orbit. The Sun pulls the planets to keep them in orbit too

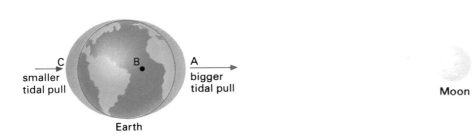

Figure 4
The tidal pulls are exaggerated here; the height of the tides is only a few metres

The Sun also exerts a tidal pull on our seas, but about half as much as the Moon. Twice a month, the Sun and Moon line up to produce a large tidal pull. We then get **spring tides**. When the Sun and Moon pull at right-angles to each other, the high tides are smaller. These are called **neap tides** (Figure 5). Other factors, such as strong winds, also affect the height of tides.

Spring tides occur at full moons (above) and new moons (below)

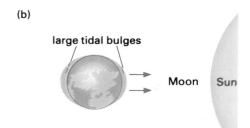

(b)

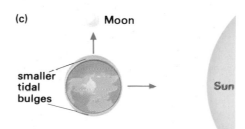

(c)

Figure 5
Neap tides occur when the Sun and the Moon pull at right angles to each other

Questions

1 Jupiter is 325 times as massive as the Earth, and Jupiter is five times further away from the Sun than we are. Calculate the ratio of the Sun's pull on Jupiter to the Sun's pull on the Earth.

2 Explain carefully how Newton's law can account for the orbital motion of satellites around the Earth.

3 On 21 March, the morning high tide at Plymouth was at 0818 hours. On 22 March, the morning high tide was at 0900 hours. Can you explain why? (Hint: the Moon takes about 29 days to go around us.)

4 Use Newton's equation to calculate the gravitational forces on Sarah, who has a mass of 50 kg, in these places: (a) on Earth, (b) on Mars, (c) on the Moon. You will need the data below.

- Mass of Earth 6×10^{24} kg
- Mass of Moon 7×10^{22} kg
- Mass of Mars 6×10^{23} kg
- Radius of Earth 6.4×10^6 m
- Radius of Mars 3.4×10^6 m
- Radius of Moon 1.7×10^6 m
- $G = 6.7 \times 10^{-11}$ Nm^2/kg^2

7 Making a Star

A 'nebula' means a fuzzy or blurred object – nebulas look blurred through a telescope. In the Orion nebula, hydrogen gas condenses to form stars. The middle of the nebula is lit up by bright young stars which are only about 300 000 years old

Like all stars, our Sun was formed from a giant cloud of gas. The photograph opposite shows part of the **orion nebula,** where stars are being formed now. The orion nebula is made mostly from hydrogen gas. The density of the gas is very low – about 100 million million times less dense than water. However, over millions of years, gravity acts to condense the gas into a smaller volume. This warms the gas up. As the gas atoms fall towards each other they speed up; potential energy is turned into kinetic energy. When the atoms collide, their large kinetic energy is turned into heat energy. Eventually, the temperature at the centre of the ball of gas reaches 15 million K and a star is born (Figure 1).

Nuclear fusion

Once the inside of a star reaches a temperature of about 15 million K, nuclear fusion starts. Like the cloud of gas that made it, a star is made mostly of hydrogen. At very high temperatures the hydrogen atoms are ripped apart, leaving only protons and electrons inside the star. At low temperatures two protons cannot collide because their charges repel each other. But at the very high temperatures inside stars, two protons have enough energy to overcome this repulsion. When two protons collide they can join or fuse together. By this process of fusion, protons join to form helium nuclei (Figure 2). The fusion of nuclei releases a lot of energy. This is how stars produce their heat and light.

Our Sun is half way through its life of 10 thousand million years. Throughout its life, the Sun has burnt fairly constantly. This is why life has evolved on our planet, the temperature has been steady. However it seems possible that, from time to time, the Sun's output of energy might change by a very small amount. In the second half of the seventeenth century, the climate was very harsh. The Thames froze over during winter. It is possible that the Ice Ages were caused by a decline in solar power output.

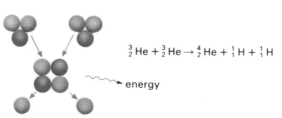

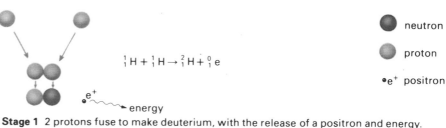

Stage 1 2 protons fuse to make deuterium, with the release of a positron and energy. A positron is a positively charged electron. The energy is carried away by a γ-ray.

$${}^{1}_{1}H + {}^{1}_{1}H \rightarrow {}^{2}_{1}H + {}^{0}_{1}e$$

Stage 2 Deuterium fuses with a proton to form Helium-3, with a further release of energy.

$${}^{2}_{1}H + {}^{1}_{1}H \rightarrow {}^{3}_{2}He$$

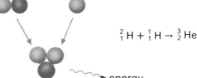

$${}^{3}_{2}He + {}^{3}_{2}He \rightarrow {}^{4}_{2}He + {}^{1}_{1}H + {}^{1}_{1}H$$

Stage 3 The process is completed when 2 Helium-3 nuclei fuse to make Helium-4.

Figure 1
A star is a battle ground. The forces of gravity try to collapse a star, but these inward forces are balanced by the enormous outward pressure exerted by the hot core. The pressure at the centre of the star is about 500 million times bigger than the Earth's atmospheric pressure

Figure 2 Nuclear fusion in the Sun

Eventually the hydrogen in the Sun will run out. Without the high pressure in its centre sustained by nuclear fusion the Sun will collapse. This rapid collapse will warm the core further to 100 million K.

At this temperature the helium nuclei can fuse to make heavier elements. The Sun will swell up and turn into a red giant, with a diameter 100 times bigger than it is now. The Sun will swallow up Mercury, and the Earth will be burnt to a cinder. Later the Sun will collapse again and finish its life as a tiny white dwarf star, before cooling down like a dying fire.

Some stars 'die' in a more dramatic way. Stars that are about four times as massive as our Sun collapse so rapidly at the end of their lives that they turn into **black holes.** The gravitational pull of a black hole is so strong that not even light can escape from it (Figure 3). Stars that are about 10 times as massive as the Sun (blue giants) live brilliant but short lives; they live for about 10 million years only. In that time, they shine more brightly than 10 000 suns. When they die they blow themselves apart in a **supernova explosion.** A supernova is so bright that it can outshine a whole galaxy of stars. Heavy elements are made in supernovas; our own solar system would have been made from the remnants of such an explosion. You are made from a recycled star.

Figure 3 Cygnus X-1 is a massive hot star, which is pulled around by a massive but invisible dark companion. The companion is as heavy as four suns. The companion is thought to be a black hole, which sucks matter out of its neighbouring star. As matter falls into the black hole, X-rays are emitted

We think there are more sun spots when the Sun is hot. At the end of the seventeenth century, no sunspots were seen for about 60 years. This coincided with a 'mini' Ice Age

The bright star Sandulek–69 turned into a supernova in January 1987. 'Supernova' means bright and new. A supernova is a dying star, but it shines brightly where people had not previously seen a star. The star outshone the whole of the Magellan Cloud which is Sandulek's galaxy

Questions

1 Explain carefully how stars are formed. Why do stars heat up when they are formed?

2 (a) Explain what is meant by 'nuclear fusion'.

(b) Why can nuclear fusion only occur at high temperatures?

(c) What is deuterium?

3 (a) How do black holes form?

(b) What is a 'supernova'? How does it get its name?

4 Blue giants live for 10 million years. This is described in the text as a 'short time'; why?

5 Look at the table of data about some stars. Then answer the questions that follow.

Star	Temperature at surface (°C)	Diameter relative to Sun	Mass relative to Sun	Brightness relative to Sun
Antares	3100	300	15	2500
Rigel	30 000	10	30	25 000
B–Centauri	19 000	6	4	4000
Vega	10 600	2.6	3	40
Sun	5800	1	1	1
61 Cygni	3900	0.7	0.5	0.1
Sirius B	9000	0.03	1	0.002

(a) Is there a connection between the brightness of a star and its temperature?

(b) Is there a connection between the brightness of a star and its diameter?

(c) Can you work out which star is a red giant?

(d) Can you work out which star is a white dwarf?

(e) Can you work out which star is a blue giant?

8 The Planets

This photograph shows Saturn with three moons and its rings. Notice the shadows cast by the rings and one of the moons. The gap between A and B rings is known as the Cassini Division

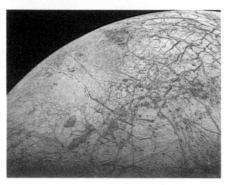

On Europa, one of the moons of Jupiter, ice flows have covered the craters on the surface

The planets: their development and composition

Most scientific theories are developed as a result of experiment or observations. There are a few theories about the development of the solar system. However, they are quite speculative because we cannot observe other solar systems forming. Astronomers think that about one third of all stars have planets, but our telescopes cannot show them. Figure 1 shows a popular theory for the development of the solar system. The planets, Sun and moons are thought to have developed at the same time (about 4500 million years ago) out of a large gaseous cloud.

The Sun is made mostly from hydrogen and helium, but there are traces of other elements too. The composition of the planets can be split into three classes, depending on their volatility: **rocky, icy** and **gaseous.** Rocks are mostly iron, and oxides and silicates of magnesium, calcium and aluminium. There are however many other elements present in all planets too, but in lesser abundance. The gases present in planets are mostly hydrogen and helium, then oxygen, nitrogen, ammonia, carbon dioxide and methane. Water ice is the commonest icy material, but there are also solid and liquid gases to be found. For example, the polar caps on Mars are thought to contain a lot of solid carbon dioxide; the atmosphere of Jupiter contains water and ammonia ice.

1 4500 million years ago a shock wave, in a spiral arm of our galaxy, triggered the collapse of a gas cloud. This developed into a doughnut shape, which flattened out.

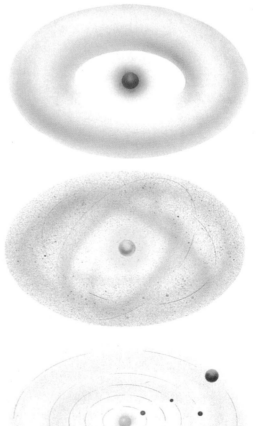

2 Enough hydrogen gathered in the centre for fusion to start in the Sun. Solid particles began to strike each other and stick together.

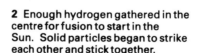

3 Eventually, as the small particles continued to coalesce, just a few large planets and moons were left. Most of the gas and dust in the solar system became attached to a planet, or was removed by a strong solar wind. After millions of years, the gravitational attraction between the planets tended to pull their orbits into the same plane.

Figure 1

The composition of a planet depends very much on its position in the solar system. The planets close to the Sun have a very different composition to those further away. Table 1 summarises the composition of the planets, and Figure 2 their internal structures.

Table 1

Planet	Mass relative to Earth	Radius (Earth= 1)	Relative density (water = 1)	Distance from Sun in A.U.†	% Rocks	% Ice	% Gas	Main gases in atmosphere
Mercury	0.06	0.38	5.4	0.39	nearly all	–	–	none
Venus	0.82	0.95	5.2	0.72	nearly all	–	some in atmosphere	CO_2
Earth	1	1	5.5	1	nearly all	water in oceans, ice at poles	some in atmosphere	N_2 O_2
Mars	0.11	0.53	3.9	1.5	nearly all	ice at poles	some in atmosphere	CO_2
Jupiter	318	11.2	1.3	5.2	10% rock/ice		90%	H_2 He
Saturn	95	9.4	0.7	9.5	30% rock/ice		70%	H_2 He
Uranus	14.6	4.1	1.2	19.1	70% rock/ice		30%	H_2, He CH_4
Neptune	17.2	3.9	1.7	30.1	70% rock/ice		30%	H_2, He CH_4
Pluto	0.1?	0.4?	?	39.4	mostly rock/ice		?	none?

† 1 Astronomical Unit or A.U. is the average Earth–Sun distance.
O_2 oxygen, N_2: nitrogen, CH_4: methane, CO_2: carbon dioxide.

Figure 2

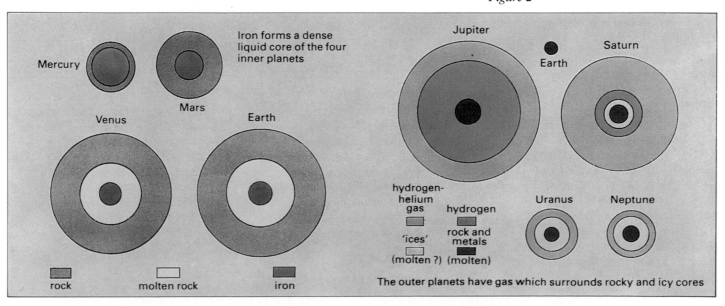

As Voyager flew past Io in March 1979, the cameras caught a volcano as it erupted. This photograph shows a huge plume of material thrown out into space

You should realise that there remains considerable doubt over the planets' exact structure and composition. The four inner planets are rocky. The four large giant planets have rocky and icy cores, but have large amounts of gas in their outer parts. Moons of Earth and Mars are rocky. The moons of the outer planets usually contain a considerable proportion of ice as well as rock. A possible explanation for the composition of the planets is as follows. In its early stages, the Sun was a lot hotter than it is now. The Sun lost a lot of material as a strong solar wind, which removed most of the gas from the inner planets.

Early in the life of the solar system, there were millions of small rocks still whizzing around. Some are still in orbit around the Sun, but most have collided with planets or moons. Mercury, Mars and many moons show the 'impact craters' caused by these rocks. Many of these craters are thought to be 4000 million years old, nearly as old as the solar system. The action of the Earth's atmosphere has eroded away craters. On the Moon and Mars, lava flows from volcanoes have covered over some craters, leaving flat plains. The volcanoes on Mars show that the planet was recently geologically active. Although there are no active volcanoes on the Moon, seismometers left there by the Apollo astronauts have detected small **moonquakes**. Io and Europa, two moons of Jupiter, are very active. Both are squeezed by Jupiter's strong gravitational field as they orbit, which generates a lot of heat. On Io, this heat results in many active volcanoes. On Europa, recent ice flows from beneath the surface have covered up the craters.

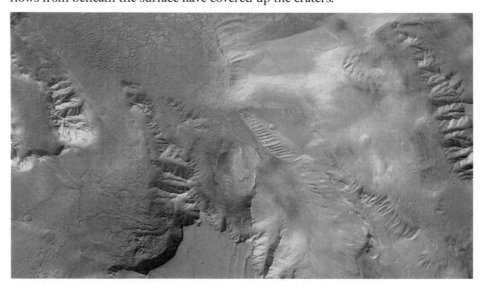

Mosaic of Viking orbiter photos showing the central portion of the giant Martian canyon system, the Valles Marineris. The layered terrain to the left of the image may have been formed by a huge, ancient lake

Questions

1 (a) In the second paragraph, the words *volatility* and *abundance* are used. What do these words mean?
(b) Explain why theories about the formation of the solar system are 'speculative'.
2 Mars has craters on its surface, but few craters are seen near volcanoes. Explain why.
3 Use the information in Table 1 and Figure 2 to answer these questions.
(a) Why are the four inner planets denser than the outer planets?

(b) What makes Neptune more dense than Saturn?
(c) Jupiter contains a larger proportion of gas than Saturn. Yet Jupiter is denser than Saturn. Account for this in terms of the planets' masses.
(d) Mars and Mercury are nearly the same size, yet Mercury is denser. Account for this in terms of their compositions.
(e) Which two planets are (i) the most similar, (ii) the most dissimilar?

4 Go to the library, do some research and write a short article on one of the following:
(a) the moons of Jupiter,
(b) the rings of Saturn,
(c) Uranus and Neptune,
(d) the atmosphere of Venus,
(e) Mars,
(f) rills and craters on the Moon,
(g) Mercury,
(h) the composition of Jupiter and Saturn.

9 Sun, Stand Still!

Earth at the centre

Look at Figure 1. This shows how Pythagoras thought the Earth fitted into space in about 500 BC. He realised (correctly) that the Earth is a sphere. His model (incorrectly) places the Earth at the centre of the universe. The Earth was thought to be surrounded by several **crystal spheres** which carried the heavenly bodies. The outer **celestial** sphere carried the stars. To explain the motion of the stars, Pythagoras said that this sphere rotated once every 24 hours around a stationary Earth. The inner spheres rotated at slightly slower speeds. This allowed Pythagoras to explain the motion of the Moon, Sun and planets reasonably well.

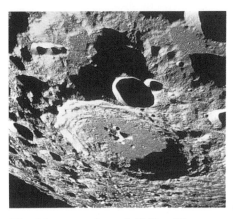

The Moon was formed 4600 million years ago at which time its surface was made of molten rock. There was a large number of meteors whizzing around in space then. The Moon was bombarded by some of these meteors, which caused the cratered surface we see today

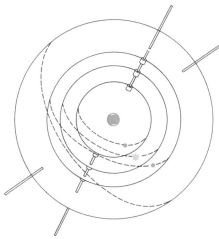

Figure 1
In 500 BC, the Greek philosophers thought heavenly bodies rotated round the Earth on crystal spheres. The stars are on the outer sphere

Pythagoras could not explain why some planets appear to make strange loops. Look at Figure 2. This shows how Jupiter appeared to move past the stars in 1991–2.

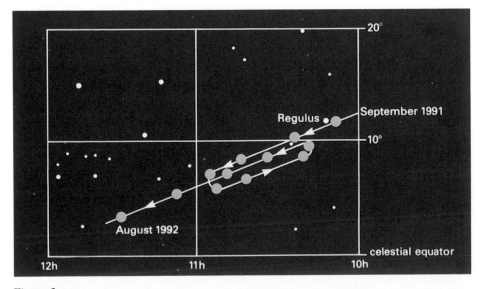

Figure 2
This diagram shows Jupiter's retrograde loop during 1991 and 1992. Jupiter is marked at intervals of one month

Jupiter's red spot has been a feature of its atmosphere for well over a hundred years. The spot is surrounded by swirling clouds blown around by winds of hurricane strength, with speeds in excess of 300 km per hour

Ptolemy (120 AD) produced another model to account for this motion (Figure 3). The arm EA rotates around the Earth every 12 years (in Jupiter's case). The arm AJ rotates around A once a year. The combined motion gives a looping movement. This model was only a calculating machine. It predicted very accurately the positions of planets, but there was no scientific evidence to back up the reality of the model.

Greek astronomers also noticed that Mars and Venus varied in brightness. But they could find no explanation for that.

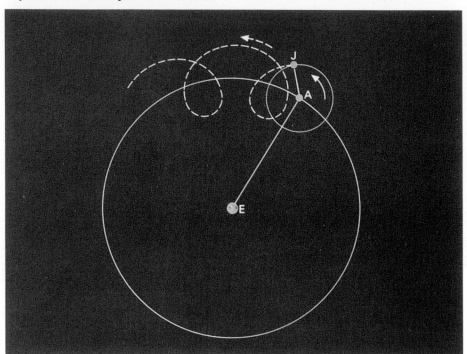

Figure 3
Ptolemy explained the loops of planets with this construction of 'epicycles'

Nicholas Copernicus (1473–1543)

Nicholas Copernicus was a scholarly Polish monk. He divided his time between church duties and studying astronomy. He believed that the universe was a divine creation, but he thought that God's arrangement of the planets would be a simple one. Copernicus took a bold step, against the teaching of the previous 2000 years. He produced a new model, with a stationary Sun placed at the centre. The Earth and other planets rotate around the Sun. Copernicus explained the motion of the Moon by saying it rotates around the Earth. The daily motion of the stars is explained by the Earth rotating on its axis every 24 hours. It is the Earth that spins round, not the celestial sphere of stars. Copernicus also measured carefully the planets' orbits. He worked out that Mercury and Venus move in orbits closer to the Sun than the Earth; Mars, Jupiter and Saturn are farther away from the Sun. This theory was very simple and also produced the following successful explanations.

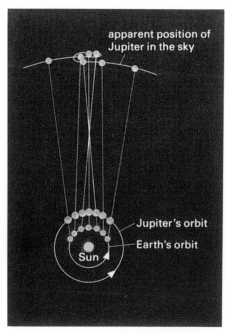

Figure 4
Over a year, Jupiter appears to move through the star pattern. As the Earth overtakes Jupiter, it seems to go backwards

- Copernicus could explain the looping paths of the planets: a loop is caused by a combination of the Earth moving round the Sun, together with the planet moving in a larger orbit round the Sun (Figure 4).

- With the new theory, it was possible to explain why Mars and Venus change in brightness. In Figure 5, Mars appears brighter when it is close to us in position M_1, but duller when it is further away in position M_2.
- Copernicus also predicted that, if our eyesight were better, we should see phases of Venus and Mercury just as we can see the Moon's phases.

Venus has a dense atmosphere of carbon dioxide and sulphuric acid, through which it is impossible to see. However, information about its surface has been gathered using radar

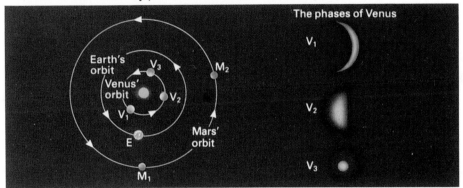

Figure 5
This is how Copernicus thought Mars, Earth and Venus orbited the Sun. The shapes to the right of the diagram show the phases of Venus at positions V_1, V_2 and V_3, when viewed from E

Galileo Galilei

In 1610, Galileo used the new invention of a telescope to look at the skies. By projecting an image of the Sun, he discovered **sunspots**. He also saw craters on the Moon and rings around Saturn. He looked at Venus over a period of time, and saw the phases that Copernicus predicted.

When Galileo turned his telescope to Jupiter, he saw what looked like stars close to it. But each night the number and position of the 'stars' changed (Figure 6).

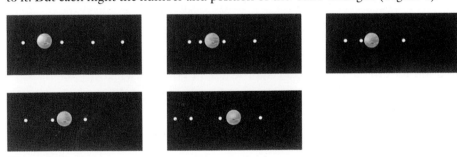

Figure 6
This is how the moons of Jupiter move over a series of nights. On some nights, a moon has disappeared behind Jupiter

He realised that he was seeing four moons in orbit around Jupiter. He could observe their movement over a few hours. This was a miniature solar system that he could see. Jupiter stood in the middle as moons went round it, just as planets rotate about a stationary Sun.

Galileo became a champion of the Copernicus system. He became an outspoken critic of the church's teaching because the church maintained that the Earth and the human race were the centre of God's creation. Eventually, fearing torture or execution, Galileo promised to stop teaching or believing the idea that the Sun stood still.

Questions

1 We now believe Copernicus' model of the solar system and not the early Greek model. What did Copernicus manage to explain that the Greeks did not?

2 How did Galileo's observations support Copernicus?

3 (a) 'Facts come first, then theories.' Discuss how this statement applies to the work of
(i) Pythagoras,
(ii) Copernicus.
(b) Think of another scientific theory. Was the theory based on experimental observations, or did the theory come first?

10 Origins

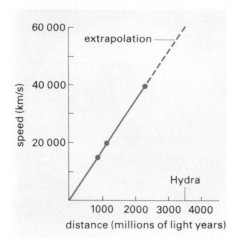

Figure 1
The speed of the galaxies is proportional to their distance away from us, according to Hubble's results

The galaxy is found in this constellation	Distance of galaxy (millions of light years*)	Speed of galaxy (km/s)
Virgo	72	1200
Perseus	400	
Ursa Major	900	15 000
Corona Borealis	1200	20 000
Bootes	2400	40 000
Hydra		60 000

* 1 light year = 10 million million (10^{13}) km. This is the distance light travels in one year.

Table 1

Most religions have similar stories about the creation of our world. At some moment in the past a divine being created the Earth, then the Sun, and Stars. Later animals and finally humans were put on the Earth. According to the account in the Book of Genesis in the Bible, this took six days. Modern science too has a story of creation to tell, but it stretches over 15 000 million years.

The big bang

By 1930, astronomers realised that out Sun is part of an enormous galaxy of stars. They also discovered that there are millions of other galaxies in the universe. By looking carefully at the light sent out by galaxies, Edwin Hubble (1889–1953) noticed that it had been shifted towards the red end of the spectrum. This told him that the galaxies were moving away from us. Table 1 shows the distances of galaxies in various constellations and their calculated speeds.

Hubble's data suggest that the speed of a galaxy is proportional to its distance away from us (Figure 1). The galaxies in Bootes are twice as far away as those in Corona Borealis; the speed of a galaxy in Bootes is twice as fast as the speed of a galaxy in Corona Borealis. From this law, Hubble could make predictions. By extending the graph (extrapolation) he calculated the distance of far off galaxies; for example, Hydra's measured speed is 60 000 km/s, so it must be about 3600 million light years away.

Hubble's work led to an amazing result. In all directions, galaxies are flying away from us; the further away they are, the faster they go. Billions of years ago, the galaxies must have been a lot closer together. Even further back in time, all the galaxies were in the same place. This led to the idea that the universe originated with an enormous cosmic explosion – the **Big Bang**. About 15 000 million years ago, all of the matter in the visible universe exploded out of a point smaller than a pin head. Physicists call this point a singularity. Ever since, matter has been flying outwards away from the explosion.

Evidence for the big bang theory

When a fire engine rushes past you with its siren blaring, you hear an example of the **Doppler effect**; as the engine approaches the note is of a higher pitch than

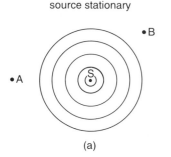

Figure 2

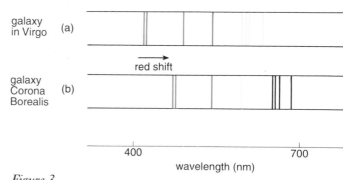

Figure 3
(a) shows a line spectrum emitted by a galaxy in Virgo. Notice how the same pattern of lines is seen in the light emitted from the Corona Borealis galaxy (Figure 3(b)). However, the Corona Borealis galaxy is moving away faster and the pattern has a greater red shift

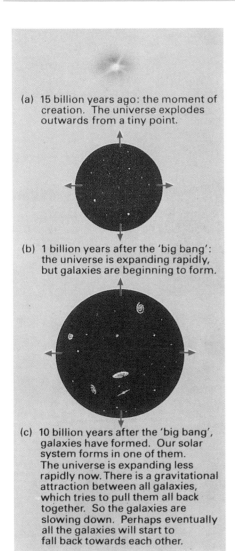

(a) 15 billion years ago: the moment of creation. The universe explodes outwards from a tiny point.

(b) 1 billion years after the 'big bang': the universe is expanding rapidly, but galaxies are beginning to form.

(c) 10 billion years after the 'big bang', galaxies have formed. Our solar system forms in one of them. The universe is expanding less rapidly now. There is a gravitational attraction between all galaxies, which tries to pull them all back together. So the galaxies are slowing down. Perhaps eventually all the galaxies will start to fall back towards each other.

Figure 4

normal, and as it goes away it is lower than normal. Figure 2 helps to explain why. In Figure 2(a) a source of sound, S, is stationary: the waves spread out symmetrically in all directions. People standing at A and B hear the same note. In Figure 2(b) the source is moving to the left. The waves are distorted as a result of this motion, bunching up to the left and spreading out to the right. A person standing at C hears more waves per second than a person at D, because the waves are closer together. So C hears a higher pitch and D hears a lower pitch than normal.

A similar Doppler effect happens with light if a source is moving very quickly; if the source is going away from us the wavelength of the light increases or shifts towards the red end of the spectrum. This is known as the **red shift**. The faster the source moves the greater the shift. Hubble discovered that distant galaxies showed a red shift (Figure 3). The more distant galaxies show a greater red shift and so are travelling faster away from us.

Further evidence for the big bang theory was provided by Penzias and Wilson (1965) who discovered the **cosmic background radiation**. The whole universe emits radiation which is thought to be a distant echo of the big bang.

Cosmology

Cosmology is the science that tries to explain the origins of the Universe. Another theory that briefly held sway was the **steady state theory**, produced by Hoyle 1948. In this theory, it was stated that matter was continuously being created in an expanding universe. A consequence of this theory is that all galaxies would look the same. However, this is now known not to be the case. Quasars which are a very great distance from us are thought to be galaxies at an early stage of that evolution. Now most cosmologists accept the big bang theory.

Minutes after the big bang, the universe was full of only the elementary particles that make up our atoms: electrons, protons and neutrons, together with a lot of radiation. After millions of years (Figure 4) gravity started pulling matter together into large clumps. After a billion (1000 million) years, galaxies were forming. In these galaxies stars began to form. About 10 billion years after the big bang, our Sun and Earth were formed from the remnants of a supernova explosion. In the next $4\frac{1}{2}$ billion years life evolved on the Earth. Eventually, 50 000 years ago, modern man and woman arrived.

Cosmologists have suggested what our universe might have been like right back to the dawn of time. But nobody can explain why there was a big bang. The whole universe was made in an instant, which is even more amazing than other creation stories. Cosmology has neither proved nor disproved the existence of God; the ultimate mystery remains unsolved. Where did the universe come from?

Questions

1 (a) Look at Table 1. Work out the speed of the galaxies in Perseus.
(b) Quasars are some of the brightest, but most distant objects in the universe. They are very small, but hundreds of times brighter than a galaxy. They are thought to be galaxies which are forming. But we see them as they were a long time ago, because they are so distant.

Such a quasar travels away from us at 200 000 km/s. How far away is it?
2 Explain briefly the evidence for the big bang theory.
3 Galaxies are moving away from us, but they are slowing down. Explain why.

4 Hubble's constant H, is defined by the equation:
$v = HR$
v is the speed of recession of a galaxy and R is its distance away from us.
(a) Use Figure 1 to calculate Hubble's constant. Express your answer in
(i) km/s per million light year, (ii) s^{-1}.
(b) Use your answer to (a) part (ii) to show that the age of the universe must be thousands of millions of years.

SECTION C: QUESTIONS

1 As the Moon moves round us, we see it close to different stars each night. The diagrams show the view from Shakila's bedroom window on two consecutive nights in December. On the first night, the moon is near the Pleiades. On the second night, the moon has moved into the constellation of Gemini.

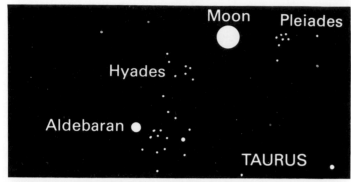

View from Shakila's window on 5 December at 11 pm

View from Shakila's window on 6 December at 11 pm

(a) Draw a diagram to explain why Shakila saw the moon in different positions each night.

(b) Sketch a copy of the stars Shakila saw and show where she saw the moon on 7 December.

(c) Look at the following diagram. Explain where Shakila sees the moon when she gets up on the morning of 12 December.

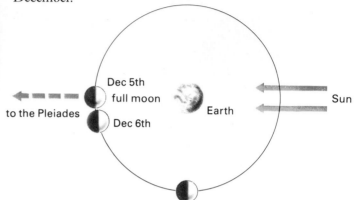

2 Barnard's Star is one of the closest stars to us, at a distance of 6 light years. Careful observation of this star (over some years) shows that it wobbles. The wobble can be explained by the presence of two large planets, which exert a gravitational pull on the star. It is also thought that Vega, a young star, might have planets too. At the moment, very little information has been gathered about other solar systems because Earth-based telescopes have not been able to see planets. It is hoped that, in the future, the Hubble space telescope might show us planets near other stars.

Star name	Star luminosity (Sun = 1)	Star temperature (K)	Star mass (Sun = 1)	Estimated age of star system (millions of years)	PLANET 1		PLANET 2		PLANET 3	
					Mass (Earth = 1)	Distance from star (A.U.)*	Mass (Earth = 1)	Distance from star (A.U.)	Mass (Earth = 1)	Distance from star (A.U.)
Barnard's Star	0.0004	2800	0.1	14 000	80	4	150	7	unknown	unknown
Ophiuchi A	0.5	4900	0.9	10 000	unknown	1	3	2	100	8
Alpha Centauri	1.1	5750	1.1	5000	3	2.5	20	5	70	12
Allenby 247	2.1	5900	1.2	4000	2	1.5	150	7	20	30
Procyon	5.2	6300	1.2	4000	4	2	2	3	15	25
Vega	40	10 600	3	500	1	10	6	200	40	

*A.U. = 1 Astronomical unit = Earth–Sun distance
Table 1

Planet	Distance from Sun (A.U.)	Temperature in day (°C)
Mercury	0.4	350
Earth	1	20
Jupiter	5	−150
Neptune	30	−240

Table 2 The solar system

Table 1 provides some hypothetical information about planets and their stars, gathered by such a telescope. In this question you are asked to speculate about the conditions on these planets. You might find Table 2 and sections C1 and C8 helpful.

(a) How could an alien astronomer, looking at our Sun, detect that it has a solar system. How long would it take this astronomer to be sure?

(b) Why will a telescope on a spacecraft be better than one of the same size on Earth?

(c) Assuming that other solar systems might be similar to ours, suggest which planets in Table 1 might be rocky and which might contain a large proportion of ice and gas (see unit C8).

(d) Describe what conditions might be like on the surfaces of (i) Vega's innermost planet, and (ii) Barnard's Star's outer planet.

(e) The relative heating effects of stars on the surfaces of planets can be compared using the formula: $H = L/R^2$, where H is the average energy arriving each second on 1 m² of a planet's surface, L is the luminosity of the star, and R is the distance between star and planet in A.U. For example, for the Earth, $L = 1$, $R = 1$, and so $H = 1$.

Show that H is 1/8 for the second planet of Ophiuchi A. (This means this planet will be a lot colder than the Earth.)

(f) Use the idea in part (e) to investigate which planet will be nearest to the Earth's temperature and which planet will be nearest to Jupiter's temperature.

(g) Which star system is most likely to support intelligent life? Explain your answer.

3 The (imaginary) planet Zeta Minus rotates around its star Zuben. Zeta Minus is a twin planet of the Earth, and Zuben is a star just like our Sun. Use the data and the diagram to answer these questions.

Data
- 1 A.U. = Earth–Sun distance
- Zyzek–Zuben distance = 23 A.U.
- Zuben–Zeta Minus distance = 1 A.U.
- Day length on Zuben = 24 hrs
- Zeta Minus rotates around Zuben once a year
- Zuben rotates round Zyzek once every 43 years

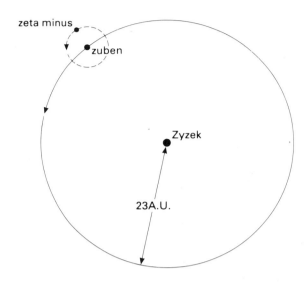

- Zyzek is 25 times brighter than Zuben (as seen from Earth)
- Zeta Minus on average receives 5% of its energy from Zyzek and 95% from Zuben
- Zyzek is 4 times as massive as Zuben

(a) Zyzek is 25 times brighter than Zuben, yet Zyzek provides Zeta Minus with a lot less energy. Explain why.

(b) On average the inhabitants of Zeta Minus live about 120 years. Describe the variations in season and climate that they experience over their lifetime.

(c) In the Zetean language, there are two words for 'night'. The words have a slightly different meaning. Explain what these meanings are.

(d) Zyzek and Zuben are about 50 light years from Earth. Make a sketch to show the observed motion of the two stars over a period of 43 years.

4 Examine carefully the photograph of the moon below. Suggest an explanation, in terms of geological activity, for the features you can see.

QUESTIONS

5 Our Moon orbits around us about once every four weeks. Imagine that we had a second smaller moon, in a closer orbit, which goes round us once in two weeks. Describe what sort of tides we would get. Assume that the second moon has the same tidal effect as the Sun (which is about half that of the Moon).

6 This question sets out to explain geostationary orbit which is where a satellite appears to be stationary above a point on the Earth's equator.

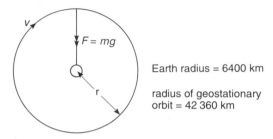

Earth radius = 6400 km

radius of geostationary
orbit = 42 360 km

(a) Explain why a satellite can only be stationary above the equator.
(b) Why does such a satellite take about 24 hours to complete one orbit?
(c) Use the data in the diagram and the equation $v = 2\pi r/T$ to calculate the speed of the satellite (in metres per second). When a satellite moves in a circular path, a force is needed to pull it towards the centre of the Earth; this force is the satellite's weight, $W = mg$.
(d) Show that the radius of the satellite's orbit is 6.6 times bigger than the Earth's radius.
(e) The Earth's gravitational field strength is 9.8 N/kg at its surface. Explain why the field strength at a height of the satellite's orbit is $9.8 \times (1/6.6)^2$ N/kg. Calculate this value.
(f) Calculate the weight of a 100 kg geostationary satellite.
(g) For anything moving in a circular path, there must be a necessary centripetal forces directed towards the centre. For an object of mass m, speed v in a path of radius r, the size of this force is given by: $F = mv^2/r$
Calculate the size of this force for a geostationary satellite of mass 100 kg. Comment on your answer.

7 Over the last million years, it is thought that there have been five Ice Ages. These have been dated by geologists. Diagram (1) shows the calculated variation in the Earth's temperature over this time. This question asks you to consider some theories for the cause of Ice Ages.
Diagram (2) shows the fluctuation in the number of sunspots observed since 1610. Sunspot activity can be seen to vary in a regular cycle. It is possible that the Sun is a little hotter when there are lots of spots. This has little effect on our climate over a short period. However, it is suggested that the absence of spots over a long period might cause the

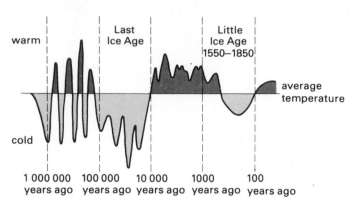

Diagram (1)

Earth to cool. In diagram (2), you can see a period of about 100 years, 'the **Maunder minimum**', when few observations of sunspots were recorded.
(a) Use diagram (2) to calculate the average period of the sunspot cycle.
(b) Is there enough evidence to link the little Ice Age with a decrease in solar activity? (During the little Ice Age the Earth suffered very severe winters; people were able to skate on the Thames regularly.)
(c) Do you think it is reasonable to link the main Ice Ages with a decrease in solar activity?
(d) Other theories for Ice Ages include these.

- Volcanic eruptions have filled the atmosphere with dust.
- As the Sun rotates around the galaxy, it occasionally passes through regions of dust and gas.
- Variations in the Earth's tilt and orbit have caused cooling.
- Meteorite collisions with the Earth.

Comment on each of these theories, explaining how each might account for the Ice Ages.

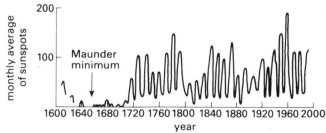

This graph shows the changes in sunspots observed over the last four centuries

Diagram (2)

SECTION D
Energy and Power

The large industries that we rely on to make our lives easier and more comfortable often produce large amounts of pollution and use a great deal of energy. Huge power stations are needed to provide electricity to power them, and these use up valuable deposits of coal and oil. Pollution is the price that we have to pay

1 What is Work?

This horse must do work in order to jump the fence

Tony works in a supermarket. His job is to fill up shelves when they are empty. When Tony lifts up tins to put on the shelves he is doing some work. The amount of work Tony does depends on how far he lifts the tins and how heavy they are.

We define work like this:

$$\text{Work done} = \text{force} \times \text{distance} = F \times d$$

Work is measured in **joules** (J). 1 joule of work is done when a force of 1 newton moves something through a distance of 1 metre, in the direction of the applied force.

$$1\,J = 1\,N \times 1\,m$$

Example. How much work does Tony do when he lifts a tin with a weight of 20 N through a height of 0.5 m?

$$W = F \times d$$
$$= 20\,N \times 0.5\,m$$
$$= 10\,J$$

Tony does the same amount of work when he lifts a tin with a weight of 10 N through 1 m.

Figure 1

Does a force always do work?

Does a force always do work? The answer is no! In Figure 1 Martin is helping Salim and Teresa to give the car a push start. Teresa and Salim are pushing from behind; Martin is pushing from the side. Teresa and Salim are doing some work because they are pushing in the right direction to get the car moving. Martin is doing nothing useful to get the car moving. Martin does no work because he is pushing at right angles to the direction of movement.

In Figure 2 Samantha is doing some weight training. She is holding two weights, but she is not lifting them. She becomes tired, because her muscles use energy, but she is not doing any work, because the weights are not moving. To do work you have to do something useful, like lifting a load or pushing a car.

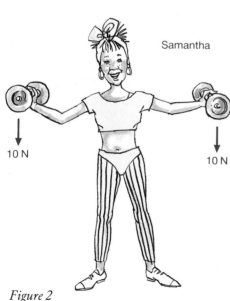

Figure 2

Finding the fuel

When you want a job done you have to pay for it. This is because you have to buy fuel. This provides **energy** for the job to be done. The supermarket manager has to pay Tony to do his work. The most important thing that Tony buys with his money is food. Food gives him the energy to do his work.

76

Figure 3 shows a crane at work on a building site. The crane runs on diesel fuel. The table shows the amount of diesel used for some jobs. You can see that the amount of diesel used is proportional to the amount of work done lifting a load.

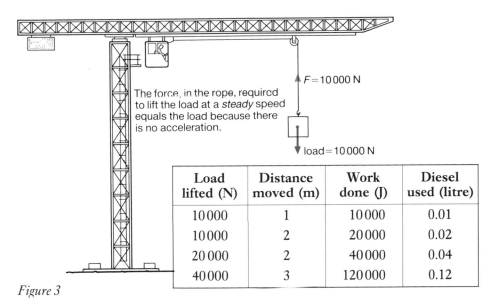

The force, in the rope, required to lift the load at a *steady* speed equals the load because there is no acceleration.

$F = 10\,000$ N

load $= 10\,000$ N

Load lifted (N)	Distance moved (m)	Work done (J)	Diesel used (litre)
10 000	1	10 000	0.01
10 000	2	20 000	0.02
20 000	2	40 000	0.04
40 000	3	120 000	0.12

Figure 3

To do work a source of energy is needed. Rechargeable batteries provide the energy for this fork lift truck to do its work

Questions

1 The table below shows some more jobs done by the crane in Figure 3. Copy the table and fill in the missing values.

Load lifted (N)	Distance moved (m)	Work done (J)	Fuel used (litre)
5000	2		0.01
10 000	4		
	10	40 000	
6000			0.06
	5		0.1
25 000		90 000	

2 Joel is on the Moon in his spacesuit. His mass (and the suit) is 80 kg. The gravitational field strength on the Moon is 1.6 N/kg.
(a) What is Joel's weight?
(b) Joel now climbs 30 m up a ladder into his space craft. How much work does he do?
3 Mr Hendrix runs a passenger ferry service in the West Indies. He has three ships, which are all the same. Sometimes he has problems with the bottoms of the ships, when barnacles stick to them. This increases the drag on the ships, and they use more fuel than usual.
(a) Explain why a larger drag makes the ships use more fuel.
(b) The table below shows the amount of fuel used by Mr Hendrix's three ships, on recent journeys. Which one has barnacles on her bottom?

Boat	Journey	Distance (km)	Fuel used (litre)
Island Queen	Vieux Fort – Bridgetown	175	1050
Windward Beauty	Plymouth – Kingstown	220	1100
Caribbean Princess	Bridgetown – St. George's	255	1275

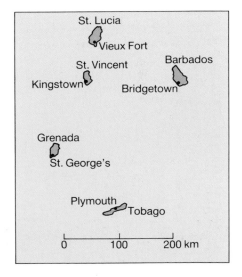

(c) Use the map and the data in the table to answer these questions: (i) how much fuel would the *Caribbean Princess* use to travel from Vieux Fort to Plymouth? (ii) how much fuel would the *Island Queen* have used for the same journey?
(d) The drag force on the *Caribbean Princess* is 50 000 N. How big is the drag on the *Island Queen*?

2 Energy

In this fun-fair pirate ship, kinetic energy is changed into potential energy and then back into kinetic energy

What is energy?

You have to put fuel into a machine to make it work. We say that the fuel has got some **energy**. For example, 1 litre of petrol has 40 million joules (MJ) of energy stored in it. This means that the greatest amount of work that we can get out of 1 litre of petrol is 40 MJ. Notice that work and energy are both measured in joules.

> Energy can be used to do work

Different types of energy

We cannot make energy and we cannot lose energy. This means that once we have got some energy we always have it. But energy often changes from one form to another. This is called the principle of conservation of energy. Several different types of energy are listed below:

- **Chemical energy**. Food has chemical energy stored in it, which is released by chemical reaction inside our bodies. Food provides energy to keep us warm and to do work. The stored energy in petrol is released when it is burnt inside a car's engine. (Burning is a chemical reaction.)
- **Gravitational potential energy**. When a rock is at the top of a hill, it has some stored energy. This sounds odd, because a rock at the top of a hill looks the same as a rock at the bottom of the hill. But when the rock is sent rolling down the hill, it can do some work. Water stored in a high-level dam helps us to produce electrical energy. The water has the potential to do work.
- **Kinetic energy** is the name given to the energy of motion. All moving objects have kinetic energy. A moving hammer has kinetic energy; it can do the job of knocking in a nail.
- **Heat energy**. When something is hot it possesses heat energy.
- **Strain energy**. If you have ever used a bow, you will know that you have to pull hard on the string before you can fire the arrow. You have done work to stretch the string. The bow is now strained, and it stores some energy. The strain energy is used to give the arrow kinetic energy.
- When a battery makes a current flow, **electrical energy** has been produced.
- You will also meet **nuclear**, **sound** and **light energy** later in the book.

Energy conversions

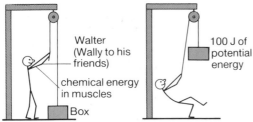

Figure 1

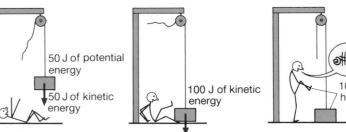

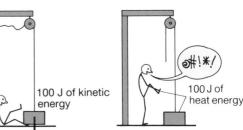

Chemical energy in Wally's muscles allows him to do work to pull up the box (Figure 1). This work gives the box potential energy. When Wally slips the potential energy turns into kinetic energy. When the box hits the ground the kinetic energy turns into heat energy.

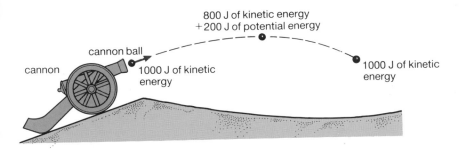

Figure 2

Small particles of rock called meteors, which are found in space, burn up in our atmosphere. They approach the Earth at about 40 000 kilometres per hour. The friction between the meteors and the atmosphere turns the meteors' kinetic energy into heat and light energy. This photo was taken with a long exposure. Most of the trails are stars. Which two are the meteors?

In Figure 2 you can see how kinetic energy can be turned into potential energy, then back to kinetic energy again.

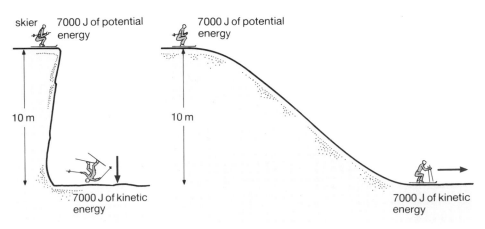

Figure 3

In Figure 3 a skier falls through a height of 10 m. He ends up with the same kinetic energy whether he falls straight down or accelerates more smoothly down the slope. In each case he loses the same potential energy.

Figure 4

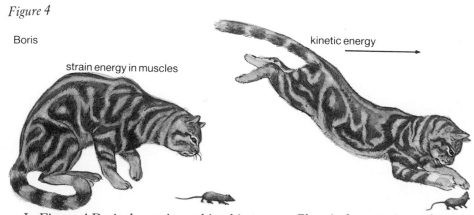

In Figure 4 Boris the cat is catching his supper. Chemical energy is used to create strain energy in Boris' muscles. This strain energy is converted to kinetic energy when Boris pounces on the mouse.

Questions

1 Judy pushes against a wall. Is she doing any work? Are her muscles using any energy?

2 A plane takes off in London and lands in New York. List and explain all the energy changes that happen during the flight.

3 Below, a ball falls to the ground.
(a) What is its kinetic energy when it hits the ground?
(b) What is its potential energy at B?

A ● potential energy = 50 J
  kinetic energy = 0

B ● kinetic energy = 30 J

C ● potential energy = 0

4 Paul kicks a stationary football. His foot is in contact with the ball over a distance of 0.1 m, and provides an average force of 500 N. How much kinetic energy does the ball have, just after it has been kicked?

3 Calculating the Energy

Noberto is lifting 110 kilograms. How much work does he do?

Potential energy

Noberto is a weightlifter. When he lifts his weights he does some work. This work increases the gravitational potential energy (PE) of the weights. The mass on the bar is now *m*. How much work does Noberto do when he lifts the bar a height *h*? (*m* is in kilograms and *h* is in metres.) The pull of gravity on the bar (its weight) is $m \times g$. We call the pull of gravity on each kilogram *g*; this is 10 N/kg.

$$\text{Work done} = F \times d$$
$$= mgh$$

> The work done is equal to the increase in potential energy.
>
> $$\text{Potential energy} = mgh$$

Kinetic energy

> The kinetic energy of a moving object is given by the formula:
>
> $$\text{kinetic energy} = \tfrac{1}{2}mv^2$$
>
> *m* is the mass of the object in kg, and *v* is its velocity in m/s.

This formula is very important for working out the stopping distance of a moving car. When cars are travelling very quickly, they need a large distance to stop in. The table below shows the stopping distances for a car travelling at different speeds. Next time you are travelling down the motorway at 40 m/s (90 mph) remember that your car needs about 120 m to stop in.

Speed (m/s)	Thinking distance (m)	Braking distance (m)	Total stopping distance (m)
10	6	6	12
20	12	24	36
30	18	54	72
40	24	96	120

Table 1. The faster the car travels, the greater the distance it needs to stop.

Can you estimate the total kinetic energy of the vehicles in this photograph?

When a driver sees a hazard, there is a small delay between taking his foot off the accelerator and putting it on the brake. In this time the car moves forward at its original speed. This is the *thinking distance*, which is proportional to the car's speed. When the brakes are applied, work is done by the braking force to take away the car's kinetic energy. Since the car's kinetic energy depends on v^2, the *braking distance* also depends on v^2. This means that when the car's speed doubles from $10\,\mathrm{m/s}$ to $20\,\mathrm{m/s}$, the braking distance increases by a factor of 4.

Strain energy

When a spring is stretched it stores strain energy. This energy can be obtained from the spring when it is released, provided that it has not been stretched past its elastic limit.

Figure 1 shows a force/extension graph for a spring. How much energy is stored when the spring is stretched $0.1\,\mathrm{m}$?

Energy stored = work done in stretching the spring

$$= \text{average force} \times \text{distance}$$

$$= 5\,\mathrm{N} \times 0.1\,\mathrm{m}$$

$$= 0.5\,\mathrm{J}$$

When the spring is stretched the force pulling it changes, so we have to average the force.

Converting kinetic energy to potential energy

Sean throws a ball into the air with an upwards speed of $20\,\mathrm{m/s}$. How high will it go? Using the principle that energy is conserved, we can say: the kinetic energy of the ball as it leaves Sean's arm, $\frac{1}{2}mv^2$, turns into potential energy, mgh, at its highest point.

So $\frac{1}{2}mv^2 = mgh$

$$h = \frac{v^2}{2g} = \frac{(20\mathrm{m/s})^2}{20\ \mathrm{m/s}^2} = 20\,\mathrm{m}$$

Work is being done to increase the potential and kinetic energy of the Harrier jet

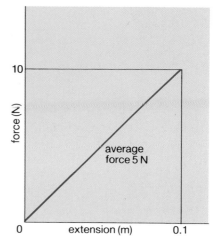

Figure 1

Questions

1 A charging rhinoceros moves at a speed of $15\,\mathrm{m/s}$, and its mass is $1000\,\mathrm{kg}$. What is its kinetic energy?

2 A car of mass $750\,\mathrm{kg}$ slows down from $30\,\mathrm{m/s}$ to $15\,\mathrm{m/s}$ over a distance of $50\,\mathrm{m}$.
(a) What is its change in kinetic energy?
(b) What is the average braking force that acts on it?

3 This question is about the stopping distances in table 1.
(a) When the driver of the car sees a hazard, how long does it take him to start braking?
(b) In a built-up area the driver is travelling at $15\,\mathrm{m/s}$. A child rushes out from behind an ice cream van $25\,\mathrm{m}$ in front of him. Use the data to decide whether he misses the child or not.

4 A catapult stores $10\,\mathrm{J}$ of strain energy when it is fully stretched. It is used to fire a marble of mass $0.02\,\mathrm{kg}$ straight up into the air.
(a) Calculate how high the marble rises.
(b) How fast is the marble moving when it is $30\,\mathrm{m}$ above the ground? (Ignore any effects due to air resistance.)

4 Power

David is carrying many more bricks than Alice and he is working about 20 times faster than her

The hind legs of this locust are extremely powerful. The insect takes off with a speed of 3 m/s. The jump is fast and occurs in a time of 25 milliseconds. The locust's mass is about 2.5 g. What power is generated? What is the power/mass ratio of a locust compared to a human?

The photographs on the left show Alice and David lifting some bricks. They each lift 20 bricks through a height of 5 m. This means that they do the same amount of work. However, David is large and powerful and he lifts all of his bricks in one go. Alice who is rather smaller, lifts her bricks one at a time. We use the word powerful to describe someone who can do the work quickly. **Power** is the rate of doing work or converting energy.

$$\text{Power} = \frac{\text{work done or energy converted}}{\text{time taken}}$$

From the equation above you can see that power is measured in J/s. But we give power its own unit called the **watt**. 1 watt is equal to a rate of working of 1 J/s. Power ratings can be very large; then we use the units of kW (kilowatt or 1000 W) and MW (megawatt or 1 000 000 W).

Body power

Alice measures her personal power output by running up a flight of steps. She takes 8.4 s to run up a flight of steps. Her mass is 14 kg. What power does she develop? She has done some work to lift her weight of 140 N through a height of 6 m.

$$\text{Work done} = \text{force} \times \text{distance}$$
$$= 140 \, \text{N} \times 6 \, \text{m} = 840 \, \text{J}$$
$$\text{power} = \frac{\text{work done}}{\text{time}}$$
$$= \frac{840 \, \text{J}}{8.4 \, \text{s}} = 100 \, \text{W}$$

Alice is converting energy at about the same rate as an electric light bulb.

Power of an express train

The photograph on the next page shows an inter-city express train moving along at a steady speed of 20 m/s. When the train moves at a constant speed, the driving force from the wheels is exactly balanced by opposing frictional forces. These opposing forces are caused by friction in the axles of the wheels and by wind resistance. So the train does work against these frictional forces.

How much power does the train have to produce when it is running at 20 m/s?

$$\text{Power} = \frac{\text{work done}}{\text{time}} = \frac{\text{force} \times \text{distance}}{\text{time}}$$

$$= F \times \frac{d}{t}$$

So the power developed is equal to the driving force × the distance travelled per second. But the distance travelled per second is the speed, v.

$$\boxed{\text{Power} = F \times v}$$

The resistive force on the train travelling at 20 m/s can be found from Figure 1.

Power = 8 kN × 20 m/s

= 8000 N × 20 m/s

= 160 000 W

or 160 kW

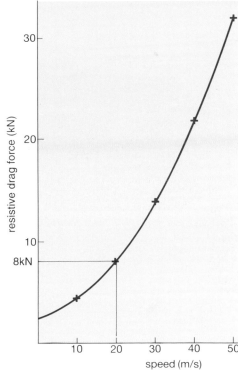

Figure 1
Graph to show the resistive force acting on a 125 train as the speed increases

This engine can provide 2 megawatts to drive the train at full speed

Questions

1 What is the unit of power?

2 David runs up the same flight of steps as Alice in 4 s. His mass is 100 kg. Roughly how many times more powerful is he than Alice? (Alice's power = 100 W)

3 On July 6th 1986, Leonid Taranenko lifted 265 kg above his head. In the last stages of his lift he raised the bar through 1 m in 2 s.
(a) How much work did he do?
(b) What power did he develop?

4 (a) Use the graph to estimate the resistive drag force on a 125 train when it is travelling at: (i) 40 m/s, (ii) 55 m/s.

(b) Show that the power that the engine produces to pull the train at 40 m/s is 880 kW.
(c) Calculate the power the engine produces when the train runs at 55 m/s.

5 (a) Use the formula $W = F \times d$ to calculate the work done against resistive forces when the train goes on a 200 km journey travelling at (i) 20 m/s (ii) 40 m/s. Give your answers in MJ.
(b) Now explain why, in 1973, when there was a temporary shortage of petrol, the government imposed a speed limit of 50 mph on all roads.

6 Here is part of an answer written by a student to explain the motion of a cricket ball. She has made some mistakes. Rewrite what she has written, explaining and correcting her errors.
'When a batsman hits a cricket ball he gives it a certain force. As the ball rolls over the grass this force is used up and eventually the balls stops. All the power that the batsman used in hitting the ball ends up as heat.'

5 Machines

Force multipliers

Machines are clever devices that multiply forces for us. They help us to lift up loads that we are not strong enough to lift up directly. But do they multiply energy for us as well; do we get more energy out of the machine than we put in?

Figure 1 shows the principle of the **lever**, a simple machine. A 300 N downwards force acts 1 m to the left of the pivot. This force is balanced by a force of 100 N, that acts downwards at a distance of 3 m to the right of the pivot:

$$300\,\text{N} \times 1\,\text{m} = 100\,\text{N} \times 3\,\text{m}$$

So we can lift a load of 300 N by applying a force (effort) of only 100 N. The work done in lifting the load is

$$300\,\text{N} \times 0.5\,\text{m} = 150\,\text{J}$$

The work done by the person applying the effort is

$$100\,\text{N} \times 1.5\,\text{m} = 150\,\text{J}.$$

Energy is conserved, so we cannot get more out than we put in.

A car jack can multiply a force about 20 times

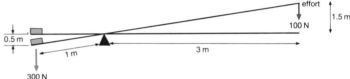

Figure 1
The principle of the lever

An **inclined plane** is another example of a common machine. Figure 2 shows a man rolling a barrel up to the top of the slope. The work he would do in lifting the barrel straight up through a height of 1 m is $1000\,\text{N} \times 1\,\text{m} = 1000\,\text{J}$. By rolling the barrel along the slope the effect of gravity is diluted. We can work out what force needs to be applied to the barrel.
Work $= F \times d$;   work done $= 1000\,\text{J}$;   distance $= 5\,\text{m}$.

$$1000\,\text{J} = F \times 5\,\text{m}$$

$$F = \frac{1000\,\text{J}}{5\,\text{m}} = 200\,\text{N}$$

This way the man will avoid serious damage to his back.

It is thought that the pyramids of Ancient Egypt were built by gangs of slaves pulling slabs of stone along a slope. Tree trunks were used as rollers.

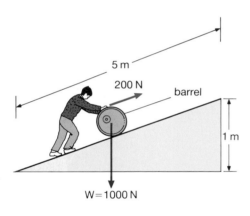

Figure 2
An inclined plane

Stonehenge consists of huge slabs of rock with masses of up to 50 tonnes. Can you suggest how the lintel stones (those across the top of the arches) were put into place?

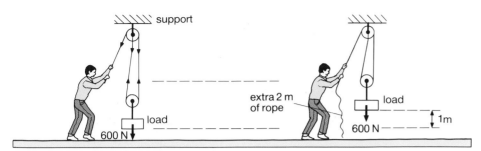

Figure 3
A pulley system

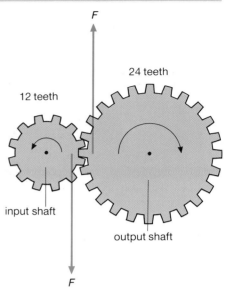

Figure 4
*Where the gear teeth touch, equal and
opposite forces act
Couple on input shaft = $F \times r$
Couple on output shaft = $2F \times r$*

A **pulley system**, another simple machine, is shown in Figure 3. The load is
600 N and the lower pulley block is supported by two ropes. So if the tension in
the rope is 300 N the load will be supported. The man can lift the load by
applying a smaller effort, but he does not win from the energy point of view. If he
lifts his load 1 m, each of the supporting ropes is shortened by 1 m, so he pulls the
rope through 2 m. He applies half the force, but pulls through twice the distance,
thereby doing the same work as if he had lifted the load straight up.

Gears

The machines that we have looked at so far increase forces; **gears** will increase
(or decrease) couples (see page 15). Figure 4 shows two gear wheels in contact;
the input shaft has to be turned twice to make the output shaft turn once.
However, the couple from the input shaft is twice as big as that applied to the
input shaft.

The bicycle is an example of a machine that is a distance multiplier. The force
needed to push a bicycle along a road is small. So a cyclist applies a large force on
the pedals, but moves his feet a small distance in one rotation of the pedals. This
work is converted into a small force acting to push the bicycle forward, but the
distance that the wheel rotates in one revolution is a lot more than the pedals'
rotation. So the bicycle has increased the distance we have moved – that is the
idea, to get there faster (Figure 5).

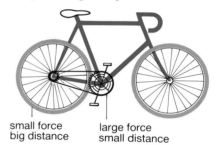

small force   large force
big distance   small distance

Figure 5
A distance multiplier

Questions

1 (a) Why is it useful to have machines
to multiply forces for us?
(b) Can a machine save us energy?
Explain your answer.

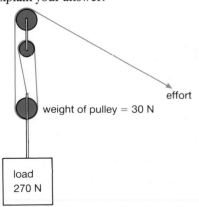

effort

weight of pulley = 30 N

load
270 N

2 The diagram on the left shows a
pulley system. Calculate the smallest
effort you need to apply to lift the load.
3 Soraya is pedalling her bicycle along
level ground.
(a) How many times does the wheel
turn for each rotation of the pedals?
(b) How far does the bicycle move for
one rotation of the pedals?
(c) How much work does Soraya do in
one rotation of her pedals?
(d) All the work Soraya does is used to
push her along the road. Calculate F.
(e) Soraya turns her pedals twice per
second. Calculate (i) the speed of the
bicycle (ii) the power her legs produce.

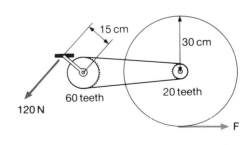

- distance moved by the pedal
  in one rotation = 0.94 m
- circumference of wheel
  = 1.88 m

6 Efficiency of Machines

Three hot machines. The heat that is lost has not done any work, and so these machines are inefficient

Figure 1

heat loss 65%

chemical energy stored in coal

useful electrical energy 35%

(a) How energy is used in a power station. The efficiency is 35%

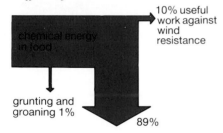

10% useful work against wind resistance

chemical energy in food

grunting and groaning 1%

89%

heat loss from body and lungs

(b) How an athlete's energy is used. The efficiency is 10%

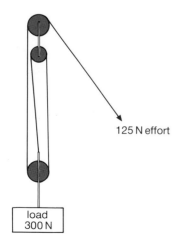

125 N effort

load 300 N

Figure 2

Unwanted heat

In the last unit we met some machines. It was assumed that we got as much work out of these machines as we put in. However, this is not usually true. Energy must be conserved, but it is often converted to a form that we do not want. The photographs in this unit are of three hot machines. The power station is producing electrical energy from chemical energy. The light bulb and athlete are producing light and kinetic energy, respectively, from their fuel. But in all cases heat energy is produced too. Figure 1 illustrates the principle for the power station and the athlete.

This leads us to the idea of **efficiency**, which is defined like this:

$$\text{Efficiency} = \frac{\text{useful energy (or work) out of machine}}{\text{energy (or work) put into a machine}}$$

$$or\ \text{Efficiency} = \frac{\text{power out}}{\text{power in}}$$

Example. Figure 2 shows a pulley system being used to lift a load. What is its efficiency?
If the load moves 1 m, the effort rope will have to be pulled 3 m.

$$\text{Efficiency} = \frac{\text{work done on load}}{\text{work done by effort}}$$

$$= \frac{300\,\text{N} \times 1\,\text{m}}{125\,\text{N} \times 3\,\text{m}}$$

$$= 0.8\ \text{or}\ 80\%$$

We can express efficiency either as a fraction or a percentage.
The main causes for unwanted energy losses in the pulley system are
● we are lifting the lower pulley block
● there are frictional forces on the pulley axles which produce heat.

Human efficiency

We get our energy from food by the process of **respiration**. During respiration, food reacts with oxygen in our bodies forming carbon dioxide and water. We obtain the oxygen when we breathe in, and when we breathe out we get rid of the carbon dioxide formed during respiration.

food + oxygen → carbon dioxide + water + energy
(containing carbon
and hydrogen)

When we take exercise we breathe faster. This means we get more oxygen to provide us with the mechanical energy. But the process is inefficient and our muscles make lots of heat too.

Figure 3 shows Philip doing pull-ups in the gym. While he does this his body uses 30 000 J of energy every minute. In one minute Philip does 20 pull-ups, before stopping to rest. Each pull-up was through a height of 0.5 m, his weight is 750 N; what is his efficiency?

Work out $= 20 \times 750\,\text{N} \times 0.5\,\text{m}$

$= 7500\,\text{J}$

$$\text{Efficiency} = \frac{\text{work out}}{\text{work in}}$$

$$= \frac{7500\,\text{J}}{30\,000\,\text{J}} = 0.25 \text{ or } 25\%$$

Figure 3
Philip doing pull-ups

Table 1 shows the efficiencies of two power stations. You can see that the efficiencies have improved over the years. But will we ever be able to do better than 37%? Surely we ought to be able to make a power station nearly 100% efficient? The answer is no. We are producing electricity (an ordered form of energy) from heat, which is associated with the disordered movement of molecules. Making order is far harder than making disorder, so the production of electricity will always be inefficient.

Producing electricity in a coal-fired power station is an example of an irreversible process. Most processes are irreversible, because the **entropy** of molecules always increases. Entropy is a measure of the disorder or randomness of molecular motion.

Station	Date	Efficiency
Battersea A	1933	16%
Drax	1975	37%

Table 1. Data from CEGB

Questions

1 Why do you get out of breath if you run too fast?

2 A 60 W electric light bulb uses 60 J of electrical energy each second. It produces 2 W of light. Calculate its efficiency. What happens to the other 58 W?

3 You have just got a job on a building site, as a bricklayer's mate. It is your job to carry bricks up a ladder 10 m high. In three hours you have to carry up 1000 bricks. Each brick weighs 40 N.
(a) How much work will you do in three hours?
(b) To do this work, your body will have to produce about four times as much energy. Explain why.
(c) Use the table in the next column to choose a breakfast that gives you enough energy to do the work.

Food	Energy value (kJ)
1 Weetabix	800
1 slice of bacon	700
1 egg	400
1 sausage	600
beans	150
1 slice of toast, butter and marmalade	600

4 The diagram (right) shows a wheel and axle. A load of 600 N is lifted a distance of 1 m by an effort of 200 N.
(a) How far does the effort force move to lift the load up 1 m? (Hint: look at the radius of each part of the machine.)
(b) How much work is done on the load?
(c) How much work is done by the effort?
(d) What is the efficiency of the machine?

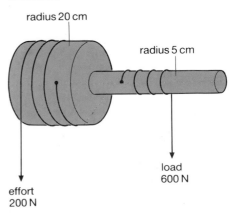

radius 20 cm

radius 5 cm

load
600 N

effort
200 N

7 Electrical Energy Production

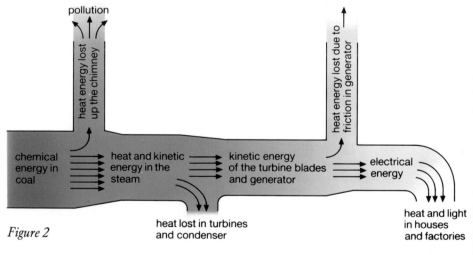

Figure 1
The principle of a power station

This is a 660 megawatt generator inside the coal-fired Drax power station. Superheated steam is used to drive the generator

Figure 1 shows the principle behind the production of electrical energy in a power station. Most of our power stations use coal as their source of energy. When coal is burnt its stored chemical energy is released as heat energy. This heat energy boils water at high pressure to make superheated steam at temperatures of about 700°C. The kinetic energy in the superheated steam is used to drive **turbines**. These are connected to the electricity generator by large coils rotating inside a strong magnetic field.

You can see in Figure 2 a chart showing how energy is used in this process. There is a lot of heat lost in the power station. The electrical energy itself is also converted to heat energy in factories and houses.

Figure 2

There are two problems that arise from the production of electricity that worry a lot of people.

- **Pollution**. Burning coal makes the gases carbon dioxide and sulphur dioxide. These pollute the atmosphere. When sulphur dioxide dissolves in water, an acidic solution is formed containing sulphurous acid.

water + sulphur dioxide → sulphurous acid

So when sulphur dioxide gets into rain, the rain becomes acidic. We call this **acid rain**. Acid rain damages stonework in buildings. It is thought that acid rain is also killing trees in Scandinavia. It is likely that sulphur dioxide produced in Britain is blown across to Scandinavia by the prevailing south-westerly winds.

Heat energy is always produced when we make electrical energy. People worry that we will warm the Earth up, as we increase our use of electrical energy. This warming is caused by (1) the extra heat from power stations, factories and homes; (2) extra carbon dioxide in the atmosphere trapping heat in, so that the Earth's temperature will rise slowly. An **increase in the Earth's temperature** could have very serious consequences. An increase of 1°C or 2°C to the Earth's average temperature would probably melt a large amount of ice from the polar ice caps. Although we get great benefits from electrical energy, we have to consider its effect on the environment. If we are not careful we will damage the world that we live in.

The turbine blades at the Drax power station

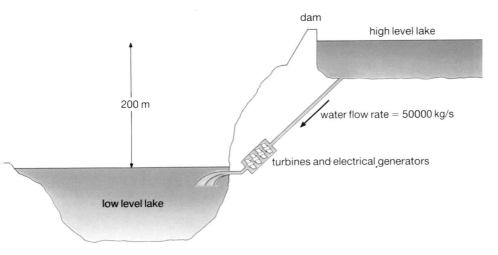

Figure 3
A pumped storage power station

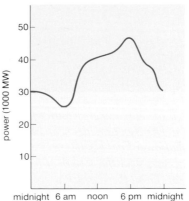

Figure 4
The typical use of electrical energy on a winter's day in Britain

Questions

1 (a) What is smog? Why is it usually found in large cities?
(b) Why do large cities have smokeless zones?
2 (a) What is acid rain?
(b) How is acid rain formed?
(c) What effects does acid rain have on the environment?
(d) What would you do to reduce the problems caused by acid rain?
3 Figure 3 (above) shows the layout of a pumped storage power station. Water from the high level lake produces electrical energy by flowing through the turbine generators. These are placed just above the low level lake. When

there is a low demand for electricity, the generators are driven in reverse to pump water back into the high level lake. This means there will be enough water to generate electricity again, when demand is high.
(a) Why is this sort of power station useful to the Central Electricity Generating Board?
(b) Would this pumped storage power station pollute the atmosphere when it generates electricity?
(c) Where does the energy come from to pump the water back up the hill again?

(d) What energy changes occur as water flows from the high level lake to the low level lake?
(e) Calculate the loss of potential energy of the water in 1 second. (Use the information in Figure 3.)
(f) The turbines are 60% efficient. Calculate the electrical power output of this station. Give your answer in MW.
(g) For the next two questions, look at Figure 4. At what times of day will the pumped storage power station be producing electricity?
(h) Explain the shape of the graph in Figure 4. Why does it reach a maximum at about 6 pm?

8 The World's Energy Resources

A large city like Los Angeles might use $10^{14}J$ of energy in one day. You need to burn 100 000 tonnes of coal to provide this amount of energy

The Sun is the major energy source for the Earth. The Sun provides heat and light to make our lives possible, as well as providing energy for the growth of foods that we eat. Fossil fuels, which we burn, originally derived their energy from the Sun too

Fossil fuels

At the moment the world faces an energy crisis. This may seem surprising; you have learnt that energy cannot be lost. The problem is that we are burning fuels like coal, oil and gas. These fuels produce electricity, warm our houses and provide energy for transport. The end product of these fuels is heat energy. We cannot recapture the heat and turn it back into coal or oil. These fuels are known as **non-renewable energy sources**. Once we have burnt them they have gone forever.

Coal, oil and gas are known as *fossil fuels*. They are the remains of plants and animals that lived some hundreds of millions of years ago. Supplies of these fuels are limited. We use coal mainly for the production of electrical energy. If we go on using coal at its present rate, it will last us for about another 300 years.

We get petrol from oil. So this energy source is vital for running cars and aeroplanes. At our present rate of use, oil will last us about 60 years. Gas will last for about the same time. Twenty years ago Britain started to drill for oil in the North Sea. This supply of oil is going to run out, in the 1990s. So this crisis is not something that is going to happen a long way ahead. It is going to happen in your lifetime. It is important that we use our fossil fuels carefully, and we must also look for other sources of energy.

Other energy sources

- **Nuclear power** is being used more and more to generate electricity. By the year 2000 France will generate 95% of its electrical energy using nuclear power. The nuclear fuel that is used is uranium. This is also a non-renewable energy source. But nuclear power could provide our energy for a few thousand years. Nuclear power also worries people because of the radioactive waste that is produced. This is discussed further in Section L.

Figure 1
How long will our fossil fuels last?

Some further sources of energy are described below; these are **renewable energy sources**. In these cases we extract energy from the environment and these energy sources will always be available to us.

- **Biomass** is the name we use when talking about plants in the sense that we can extract energy from them. The most common way of extracting energy from biomass is to cut down a tree and burn it. Trees are a renewable energy source provided forests are looked after properly. Unfortunately, in Africa, Asia and South America trees are being cut down far faster than they are being replanted. It is thought that cutting down large forests has changed our climate by reducing rainfall in some parts of the world.

  In some countries vegetable oils are already in use to drive farm machinery. So the world's biomass may be used in future to power our cars.

- **Tidal power**. Figure 2 shows a map of the Severn estuary, which is a suitable position for a tidal barrage. The idea is that water flows in through the sluice gates at A at high tide. At low tide water flows out through the turbogenerators at B. We are using the potential energy of the water to generate electrical energy. If the Severn estuary barrage is built it will produce about 7000 MW of power.

- **Hydroelectric power** is widely used in Scandinavia, where a lot of water flows down the mountains from melted snow. The principle is the same as tidal power: the potential energy of the water is used to generate electrical energy.

- **Wind power**. Energy from the wind can also be used to generate electricity. The photograph opposite shows a wind farm in Scotland.

- **Geothermal power**. Energy can be obtained from a hot spring. When the water in the spring boils, the steam formed can be used to drive electrical turbines. In China, warm water (50°C) is pumped directly into factories and houses to provide central heating.

- **Solar power** can be used directly to warm up water in panels in the roofs of houses. Energy from the sun can also be used to generate electricity using photocells. A lot of us now use solar-powered calculators.

A small wind turbine on Scoraig on the west coast of Scotland. Scoraig has no mains gas or electricity and its dozen or so houses are entirely dependent on wind turbines and storage batteries

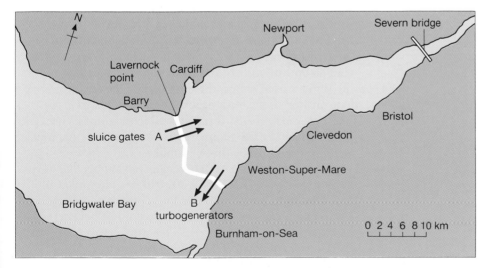

Figure 2
The Severn Barrage. A report from the Severn Barrage Committee suggested that it was technically feasible to build the barrage. At the time of the report (1981) the cost was estimated at £7.5 billion. The barrage would be able to produce 7000 MW of power. This amounts to about 6% of the national need

Questions

1 Explain the difference between a renewable and a non-renewable fuel.
2 Why are air fares likely to be very expensive in 50 years' time?
3 Why is petrol wasted if a car accelerates or brakes rapidly?
4 Discuss whether wood is a renewable energy source.
5 Which of the following countries do you think are high users of energy, and which are low users: USA, Norway, India, Britain, Thailand, Ethiopia? Is there a connection between wealth and a country's use of energy?

SECTION D: QUESTIONS

1 (a) Explain the energy transfers that occur when a car stops, without skidding, when the brakes are applied.
(b) What energy transfers occur when the car does skid?

2 An electric motor on a building site lifts a load of bricks through a height of 5 m. The weight of the bricks is 800 N.
(a) Calculate the work done on the load.
(b) It took 16 s to raise the load through 5 m. Calculate the power output of the motor.
(c) The motor is only 30% efficient. Calculate the power input to the motor.
(d) Explain why the motor is inefficient.

3 *Voyager 1* was launched on 5 September 1977. After a journey of nearly two years it reached Jupiter. On 30 January 1979 *Voyager 1* was 35.1 million km away from Jupiter. The data below show how *Voyager* approached Jupiter during the next month until passing close by Jupiter on 5 March. The times recorded are from 00.00 h GMT on 30 January 1979.
 During all of the approach to Jupiter, the motors on *Voyager* were turned off. So the only force acting on *Voyager* was the gravitational pull of Jupiter.
(a) Explain why *Voyager's* speed increased as it approached Jupiter.
(b) Work out the acceleration of *Voyager* between these two pairs of points: (i) 1 and 1A, (ii) 7 and 7A. Express your answer in m/s². (Hint: 1 hour = 3600 s.)
(c) Now work out the force acting on *Voyager* at points 1 and 7. *Voyager's* mass is about 2000 kg. Explain why the force changed.

	Distance from Jupiter's centre R (millions of km)	Velocity v (m/s)	Time from 00.00 h GMT Jan 30 1979 days hours		Date
1	35.1 0	10 900	0	00	Jan 30
1A	34.7 1	10904	0	10	
2	28.0 8	11 000	7	10	Feb 6
2A	27.6 8	11 006	7	20	
3	21.0 6	11 135	14	18	Feb 13
3A	20.66	11145	15	4	
4	1.10 4	11 402	21	23	Feb 20
4A	13.8 8	11 411	22	3	
5	7.0 2	12 165	28	21	Feb 27
5A	6.9 3	12 184	28	23	
6	3.5 1	13 564	32	1	Mar 3
6A	3.4 6	13 601	32	2	
7	2.8 1	14 212	32	15	Mar 3
7A	2.7 6	14 271	32	16	

(d) Work out *Voyager's* increase in kinetic energy as it moved from point 2 to point 6. Where did this increase in kinetic energy come from?

4 The Eiffel Tower in Paris is 300 m high and can be climbed using its 1792 steps. Jacques decides to climb the tower; he took 15 minutes to do it and he has a mass of 60 kg.
(a) How much work did he do climbing the steps?
(b) What was his average power output during the climb?
(c) A croissant provides Jacques with 400 kJ of energy when digested. How many croissants should Jacques eat for breakfast, if his body is 20% efficient at transferring this energy into useful work?
(d) Explain where most of the energy from Jacques' food goes.

5 (a) The diagram shows a windmill with a rotor 100 m across. This could be used to generate electricity.

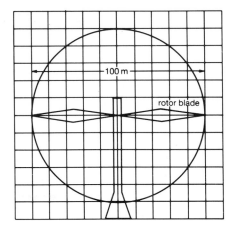

(i) Estimate the total area of the rotor blades.
(ii) The wind has a velocity of 10 m/s. The force it can cause on a surface of 1 m² is 90 N. How much work can this wind do on 1 m² in 1 s? (Work = force × distance.)
(iii) Wind is used to turn the rotor blade. The effective work done on 1 m² of the blade is 50% of that calculated in (ii). Find how much energy is transferred to the rotor in 1 s.
(iv) Calculate the maximum electrical power, in watts, which would be generated by this windmill.
(v) A conventional power station generates about 900 MW of electrical power (1 MW = 1 000 000 W). How many of these windmills would be needed to replace a power station?
(b) In unit 8 you saw where a barrier could be built across the estuary of the River Severn. This would make a lake with a surface area of about 200 km² (200 million m²).
The diagram on the next page shows that the sea level could change by 9 m between low and high tide; but the level in the lake would only change by 5 m.
(i) What kind of energy is given up when a cubic metre of water falls from position A to position B?

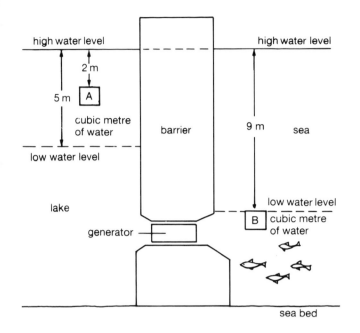

(ii) A cubic metre of water has a mass of 1000 kg. The acceleration due to gravity is 10 m/s². Calculate the force of gravity on a cubic metre of water. Calculate the work that this can do as it falls from A to B. (Force = mass × acceleration.) (Work = force × distance.)
(iii) How many cubic metres of water could flow out of the lake between high and low tide?
(iv) Use your answers to parts (ii) and (iii) to find how much energy could be obtained from the tide. Assume that position 'A' is the average position of a cubic metre of water between high and low levels in the lake.
(v) The time between high and low tide is approximately 20 000 seconds (about six hours). Use this figure to estimate the power available from the dam. Give your answers in megawatts.
(c) Explain the advantages and disadvantages of the wind-mill and the Severn Barrier as sources of power.

ULEAC

6 The diagram shows a motor that is used to lift up a pile driver. The driver is dropped from a height of 10 m on top of the pile. The driver and pile each have a mass of 500 kg.
(a) How much potential energy does the driver have when it is 10 m above the pile?

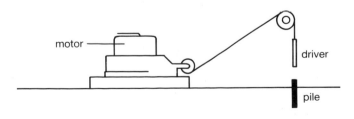

(b) Calculate the speed of the driver just before it hits the pile.
(c) The driver and pile stick together on impact. Now calculate the speed immediately after impact. (Hint: use the principle of conservation of momentum.)
(d) Calculate how much kinetic energy is lost on impact. What has happened to this energy?
(e) The driver and pile come to rest after a time of 0.1 s. Calculate the average force that the ground exerts on the pile during this time.
(f) Calculate how far the pile goes into the ground.

7 Boris has caught Mikhail the mouse and has set him to work in the tread-wheel. When Mikhail runs he can just lift Boris at a speed of 0.03 m/s.
(a) How fast is Mikhail running relative to the treadwheel?
(b) What force (F) does Mikhail exert on the treadwheel? Assume the machine is 100% efficient.

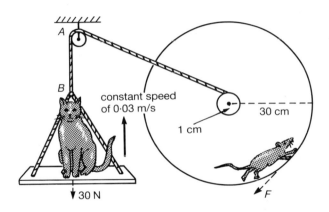

8 A steel spring and a rubber band have force/extension graphs as shown below. When the band is stretched to the maximum extension shown on the graph, the energy stored in the rubber is 20 J.

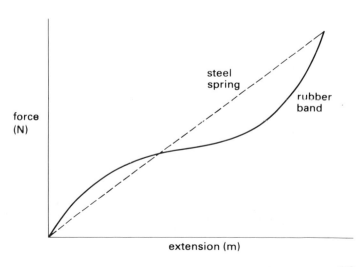

(a) When both the spring and band are stretched to the maximum extensions shown, which stores more energy? Explain your answer.
(b) The rubber band is now used in a catapult. 75% of the maximum stored energy is given to a stone as kinetic energy. The stone, which has a mass of 0.1 kg, took 0.02 s to leave the catapult. What is the maximum speed of the stone?
(c) How high could the stone be fired?

9 The Highway Code states how the braking distance, for a car with good brakes on a dry road, changes as the speed increases. The braking distance is the distance travelled by the car from the instant the brakes are applied until it comes to rest. The table shows the braking distance for different speeds.

speed (m/s)	0	9	13.5	18	22.5	27	31
braking distance (m)	0	6	14	24	38	54	75

(a) Plot a graph of braking distance (y-axis) against speed (x-axis).
(b) From the graph determine the braking distance for a car travelling at (i) 15 m/s, (ii) 30 m/s.
(c) Predict, giving your reasons, the braking distance for a car travelling at 36 m/s.

The time taken for a driver to react to a situation before actually putting on the brakes is called the thinking time, and the distance travelled during the thinking time is called the thinking distance. The thinking time is usually taken to be $\frac{2}{3}$ second.
(d) (i) What is the thinking distance for a car travelling at 30 m/s?
(ii) What is the total stopping distance for a car travelling at 30 m/s?
(iii) What is the minimum distance a driver should allow from another car, also travelling at 30 m/s?

On a foggy day on a motorway the limit of visibility may be only 50 m.
(e) (i) Select and justify a suitable speed limit for these conditions.
(ii) What difference would it make to your chosen speed limit if the road were wet? Explain your answer.
(f) In any collision the wearing of a seat belt is likely to reduce the injuries suffered by a passenger. Explain in terms of momentum how the belt achieves this.
(g) A pupil put forward the theory that the braking distance is proportional to the (speed)². Describe an experiment which you could carry out in a school laboratory to test the theory. (Hint – use a level sand tray to brake the model car.)
MEG

10 Diagram 1 shows a ski-jump on a mountain. Two ramps are covered with smooth, hard ice. The ramp WX curves near its lower end so that it is horizontal at X. An empty sledge of mass 15 kg slips from rest at W, slides down the ski-jump and eventually hits the mountainside at Y.

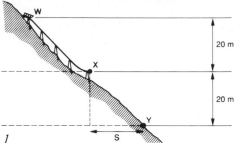

Diagram 1

In answering parts (a) and (b) you may ignore the effects of friction and air resistance.
(a) (i) What is the change of the potential energy of the sledge as it moves from W to X?
(ii) With what speed does the sledge leave the ramp at X?
(iii) How long does the sledge take to travel from X to Y?
(iv) How big is the distance marked S on Fig. 1 ?
(b) The ramp in diagram 1 is replaced by a steeper one between the same levels (diagram 2). Say what effects this will have, if any, on
(i)   the time taken for the sledge to travel from W to X
(ii) the speed with which the sledge leaves the ramp
(iii) the position of the point of impact.

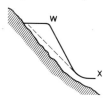

Diagram 2

(c) The sledge will probably be damaged by its impact, yet a skier can land safely. Explain how the skier manages to do this.
(d) In order to make a longer jump, a skier (i) pushes himself off when leaving W, (ii) keeps his knees bent but suddenly straightens them as he reaches X then,
(iii) positions himself with his body held forward and his arms at his sides and, (iv) with the points of his skis slightly above the horizontal (diagram 3).

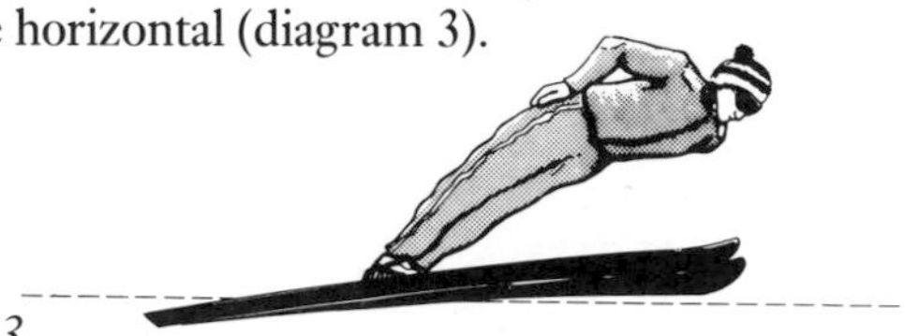

Diagram 3

For each of (i) to (iv) give a reason, or reasons, why the action might increase the length of the jump.
MEG

SECTION E
Matter

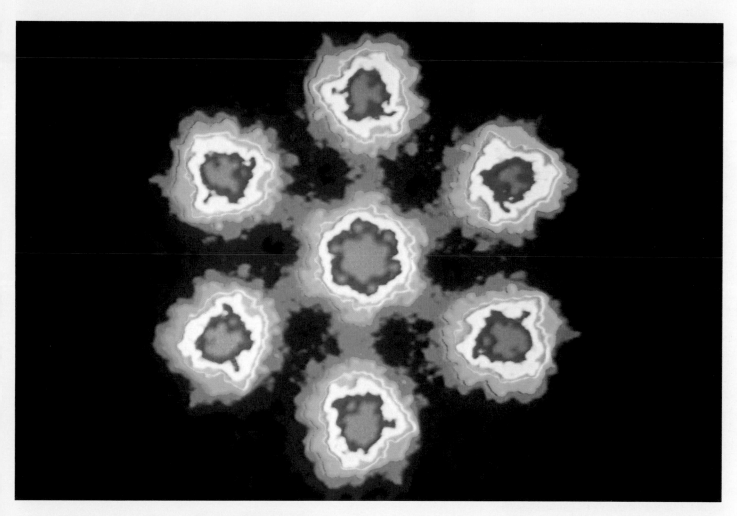

A false colour photograph of a uranyl microcrystal taken with a scanning electron microscope. The image shows the uranium atoms arranged in a perfect hexagonal shape around a central atom. Each atom is spaced about 0.000 000 000 32 metres from its neighbour, which means that the magnification in this photograph is about 100 million times!

1 Small Particles

This male moth has large antennae for detecting small particles of the female moth's scent

Everything we touch, swallow and breathe is made out of tiny particles. The smallest particles are **atoms**. There are only about 100 different types of atom. Materials that are made from only one type of atom are called **elements**. For example, aluminium contains only aluminium atoms. Some other common elements are oxygen, hydrogen, nitrogen and carbon.

The small particles in lots of materials are **molecules**. Atoms combine chemically to make molecules. For example, a water molecule is made up of two hydrogen atoms and one oxygen atom, while a carbon dioxide molecule contains one carbon and two oxygen atoms.

Atoms and molecules are far too small for us to see directly. But the photograph shows you some molecules seen through a powerful electron microscope. Electron microscopes show us that oil molecules are about 0.0000001 cm (10^{-7} cm) long. That means that $10\,000\,000$ molecules put end to end would be about 1 cm long.

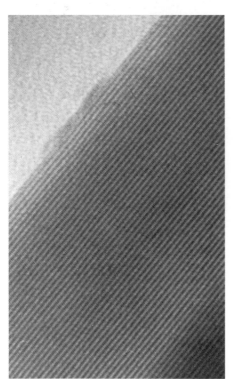

This is an electron microscope photograph of a manganese compound. The dark lines are rows of manganese atoms neatly arranged in the crystal

Discovering molecules

The idea that things were made from atoms and molecules was thought of over a hundred years ago, long before electron microscopes were invented. Below are some examples which suggest that matter is made up of small particles. **Matter** is the name that we use to describe all solids, liquids and gases.

This is an X-ray diffraction photograph of DNA. The dark spots are caused by X-rays being diffracted by DNA molecules. A photograph like this one played a vital part in the unravelling of the complex structure of the DNA molecule

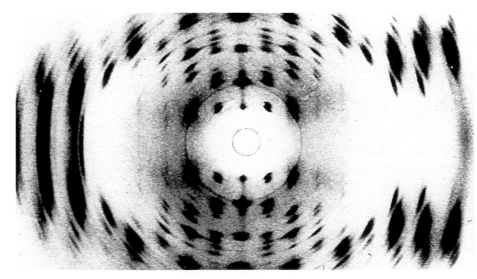

- The photograph opposite shows a male Emperor moth. He uses his very large antennae to detect the scent of a female. A male Emperor moth can be attracted by the scent of a female at a distance as far as 10 km. The female produces only a small quantity of scent. This suggests that tiny particles of her scent must spread out through the air.
- There is a simple experiment that you can do for yourself, which shows that liquids are also made up of very small particles. You can take a small drop of blue ink and put it into a glass of water (see page 99). If you stir the water it will turn a very pale blue. The small particles in the ink have now been spread further apart.
- Growing crystals. Figure 1 shows how you can grow a crystal. You dissolve some copper sulphate in water to make a strong solution. Then a small crystal of copper sulphate is placed in the solution. During a week or so, this crystal grows very slowly into a larger one. This can be explained by saying that the solution contains very small particles of copper sulphate. These particles stick to the crystal so that it grows larger.

Growing a crystal is a bit like making a neat pile of oranges. As another orange is added the pile grows. You can see that the shape of a pyramid of oranges is the same as the shape of an alum crystal.

When oranges are piled together in neat rows, they form a pyramid shape

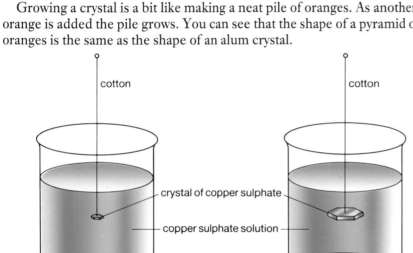

Figure 1
(a) A small crystal of copper sulphate is placed in a copper sulphate solution

(b) A week later a large crystal has grown

The shape of this alum crystal suggests that the molecules in it are also stacked neatly in rows

⚠ Copper sulphate is harmful. Wash your hands after handling chemicals

Questions

1 When you cook a curry the smell spreads right through the house. Why does this suggest that the curry powder is made up of small particles?

2 Why do we think that molecules in crystals are packed into neat rows?

3 In the Straits of Hormuz in the Persian Gulf, an oil tanker is holed by machine-gun fire and develops a small leak. The tanker spills 10 m³ of oil. The diagram opposite shows how the oil spreads out over the sea's surface.

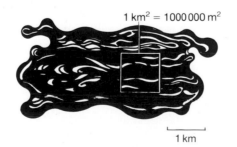

1 km² = 1 000 000 m²

1 km

(a) Estimate the area of the spill in km².
(b) What is the area of the spill in m²?
(c) Calculate the oil spill's thickness.

(d) Does the thickness of the oil spill tell you the size of an oil molecule? Explain your answer.

4 The South American golden dart-poison frog produces the world's most lethal poison, batrachotoxin. Only 0.002 g of this substance is enough to kill an average adult (mass 70 kg).
(a) Work out the lethal dose of this poison per gram of human flesh.
(b) Why do you think that this poison is made up of small particles?

2 Molecules on the Move

Diffusion

In the last unit you read how a male Emperor moth can detect a female moth's scent at a great distance. This supports the idea that the scent is made up of molecules and also the idea that the molecules in a gas are moving. Here are some simple experiments to show you more about the movement of molecules. In Figure 1 you can see a long tube with all the air pumped out of it. It is closed with a tap. On the other side of the tap is attached a small capsule of bromine inside some rubber tubing. The capsule is broken and then the tap opened. As soon as the tap is opened the bromine vapour fills the long tube. This tells us that the bromine molecules are moving quickly. They are actually moving at a speed of about 200 m/s.

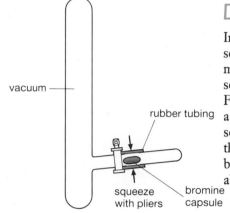

vacuum

rubber tubing

squeeze
with pliers

bromine
capsule

(a) tap closed

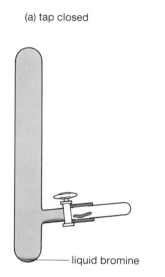

liquid bromine

(b) tap open–the tube fills immediately

Figure 1
Bromine vapour filling a vacuum

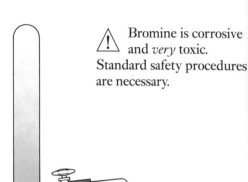

air

squeeze
with pliers

(a) tap closed

(b) twenty minutes later bromine is just
reaching the top of the tube

⚠ Bromine is corrosive and *very* toxic. Standard safety procedures are necessary.

Figure 2
The bromine molecules bump into the air molecules, and are slowed down

Figure 2 shows what happens when the experiment is repeated with air inside. This time when the tap is opened the bromine does not fill the tube quickly. After about 20 minutes the bottom half of the tube is coloured dark brown, but the top is only light brown. So although the bromine molecules travel very quickly it takes a long time for them to reach the top.

The reason for this is that the air molecules are also moving quickly. The air molecules get in the way of the bromine molecules. When two molecules bump into each other they will change direction. The bromine molecules keep changing direction and so take a long time to reach the top.

> This process of one substance spreading through another is called **diffusion**.

The essential nutrients for a plant diffuse through the soil to its roots. The roots are split up into many branches that spread out so that the plant can absorb as much of the nutrients as possible

Diffusion also occurs in liquids, but only very slowly in solids. Diffusion is very important for all living things. When animals have eaten a meal, food is digested and diffuses into the blood. Blood then carries the food all round the body. Plants need nitrogen, potassium, phosphorus and other elements. These diffuse through the soil to the plants' roots.

microscope

glass lens to focus
the light onto the cell

light bulb

smoke cell

Figure 3
Place some ink in a glass of water. After an hour or so the water is a light blue colour.
Diffusion works more slowly in liquids, which suggests that molecules move more slowly in
liquids than in gases.

Figure 4

Brownian motion

Robert Brown, a botanist, looked through his microscope at some grains of pollen
which were in water. He noticed that the grains of pollen were moving around
randomly. They jiggled from side to side.

You can see the same sort of **Brownian motion** if you look through a
microscope at smoke particles. Figure 4 shows a typical experimental
arrangement. A small puff of smoke is put into a smoke cell under the
microscope. Light from a small bulb shines on to the cell. The smoke particles
show up as tiny specks of light through the microscope. Figure 5 shows the sort of
path a smoke particle follows.

The smoke particles move randomly because they are always being knocked by
air molecules. These air molecules are too small to see under the microscope, but
they are moving so quickly that they can deflect a much larger smoke particle.

Figure 5
The path of a smoke particle as seen
through the microscope

Questions

1 The diagram shows two bromine
molecules moving in a jar full of air.
Both molecules reach the top after a
while.
(a) Copy the diagram and complete the
paths of *A* and *B*, to show how they
reached the top.
(b) Explain why bromine molecules
move in this way.
(c) Further up the jar is molecule *C*.
Will it reach the top before *A* and *B*?

2 When molecules are warmed up they
travel faster. Explain what effect a high
temperature would have on
(a) the motion of smoke particles in a
smoke cell.
(b) the rate at which gases diffuse.
3 Below, you can see some data
collected in a series of experiments on
diffusion. An approximate time for
different gases to diffuse through a
distance of 10 cm is recorded. The
temperature of the gases in all
experiments was 20°C. The purpose of
the experiments was to study how the

rate of diffusion of a molecule depends
on its mass.
(a) Plot a graph of the time of diffusion
(*y*-axis) against the relative mass of the
molecule (*x*-axis).
(b) Ammonia has a relative molecular
mass of 17. Use your graph to predict
how long, on average, ammonia
molecules will take to diffuse 10 cm
through air.
(c) What conclusion can you draw
about the relationship between speed
and mass of a gas molecule?

Substance	Relative molecular mass	Time to diffuse 10 cm through air (s)
Hydrogen	2	30
Carbon dioxide	44	160
Chlorine	71	200
Bromine	160	300
Iodine vapour	254	380

3 The Kinetic Theory of Matter

Water appears in different forms . . . Ice

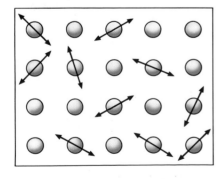

. . . Water

When something is moving it has **kinetic energy**. In the last unit you met the idea that molecules are always on the move.

This idea is called the **kinetic theory**. The most important points in the theory are:

- Every kind of material is made of small particles (molecules or atoms).
- The sizes of particles are different for different materials.
- The particles themselves are very hard. They cannot be squashed or stretched, but the distance between particles can change.
- The particles are always moving. The higher the temperature of a substance, the faster its particles move.
- At the same temperature all particles have the same energy. So heavy particles will move slowly and lighter particles will move quickly.

The last point helps to explain why hydrogen can diffuse through air much more quickly than bromine.

Solids, liquids and gases

Ice, water and steam are three different states of the same material. We call these three states solid, liquid and gas. You will now see how kinetic theory helps us to understand them.

- **Solid**. In a solid the particles are packed into rows just like apples or oranges stacked in shops (Figure 1(a)). The particles can never move out of their rows, but they vibrate around their fixed positions. As the temperature of a solid is raised the particles vibrate more and the material expands. This means that the distance between particles has increased a little.

  In a solid the particles are very close to each other and held in position by very strong forces. This makes it very difficult to change the shape of a solid by squashing it or pulling it.

- **Liquid**. In a liquid the particles are still very close to each other, which means that liquids are also very difficult to compress (Figure 1(b)). The particles can now move around from place to place. A liquid can change shape to fit into any container, but its volume will remain constant at a given temperature. As a liquid warms it also expands a little.

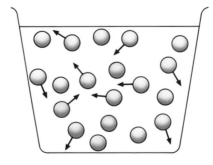

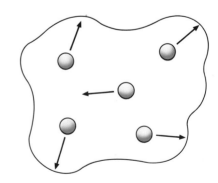

Figure 1
(a) Particles in a solid are arranged in neat rows. The particles can vibrate but do not leave their positions in the rows

(b) Particles in a liquid are close together, but free to move around

(c) Gas molecules are free to move into any available space. They move quickly and there are large distances between them

• **Gas**. In a gas the particles are separated by big distances (Figure 1(c)). The forces between gas particles are very small. It is easy to compress a gas because there is so much space between the particles. The gas particles are in a constant state of rapid random motion. This makes a gas expand to fill any available space.

Gas pressure

A gas in a container exerts a pressure on the container walls. This is because the gas particles are always hitting the walls. The pressure that the gas causes depends on:
• The number of collisions that the particles have with the container walls per second.
• How hard the particles hit the walls.
The pressure inside a container of gas can be increased in three ways:
• Putting more molecules in the container. The number of collisions that the molecules make with the walls each second is now larger.
• Making the volume of the container smaller. The same number of molecules make more collisions with the walls, because they travel less distance between collisions.
• Heating the container. The molecules travel faster and they now hit the container walls harder and more often.

... Vapour and steam. Geysers are formed when water deep in the Earth is heated under pressure. The water finds a route to the surface and boils to form a jet sometimes as much as 70 metres high

Questions

1 Use the kinetic theory to help you explain the following:
(a) Some solids, like metals, are very stiff and difficult to bend.
(b) A liquid can be poured.
(c) If a gas pipe has a small hole in it, gas will escape from the pipe and you will be able to smell it a few metres away from the leak.
2 The data in the table below shows how the pressure in a gas cylinder varies with the mass of the cylinder.

Mass of cylinder (kg)	Pressure in the cylinder (Pa × 10⁵)
8.7	1.5
9.0	2.7
9.6	5.0
10.2	7.3

(a) Plot a graph of pressure (*y*-axis) against mass (*x*-axis).
(b) Use the graph to determine the mass of the empty gas cylinder.

(c) The greatest pressure the cylinder can take safely is 20×10^5 Pa. What will the mass of the cylinder be then?
3 Diagram *(i)* shows a box with two molecules in it. One molecule, *X*, moves parallel to the line *BC*. The molecule bounces backwards and forwards and hits the pink wall 100 times a second. The other molecule bounces backwards and forwards and hits the blue wall 100 times a second. The volume of the box is now halved by pushing in the pink wall as shown in diagram *(ii)*.
(a) How many times a second does *X* hit the pink wall now?
(b) How many times a second does *Y* hit the blue wall now?
(c) (i) Suppose the box is filled with a lot of molecules. When the volume is halved as shown in diagram *(ii)* the pressure acting on the pink wall of the box doubles. Explain why. (ii) What happens to the pressure acting on the blue wall?

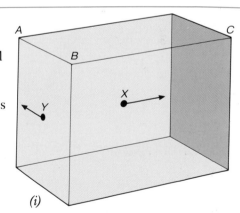

(i)

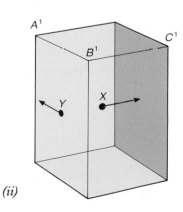

(ii)

4 Melting and Evaporation

When water freezes it expands, sometimes causing pipes to burst

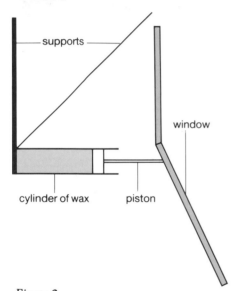

Figure 2
Automatic window opener

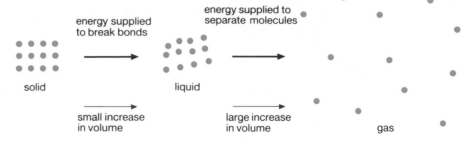

Figure 1

Changing solid to liquid

When a solid is heated the molecules inside it vibrate more and more quickly. If enough heat is supplied the molecules will break away from their fixed positions and start to move around. The solid has melted. The temperature at which this change happens is called the **melting point** of the material (Figure 1).

Most materials expand when they melt, because molecules are a little further apart in the liquid state. Figure 2 shows how we can use this in a greenhouse. A cylinder is filled with wax that melts at about 25°C. When air in the greenhouse reaches this temperature the wax melts and expands. This pushes the window open. This automatic window opener prevents damage to crops due to overheating.

Water is an unusual substance because it expands when it freezes (see Figure 3). Water is at its most dense at 4°C, so when a pond freezes over there is a layer of dense water at 4°C at the bottom. Ice is less dense than water which is why we see ice on the surface of ponds.

Changing the melting point

The expansion of water on freezing can be a nuisance. It causes pipes to burst in houses or in cars. To prevent freezing in car cooling systems, drivers add antifreeze. This lowers the melting point of water. When salt is added to water the melting point is lowered. This is why salt is put onto our roads in winter.

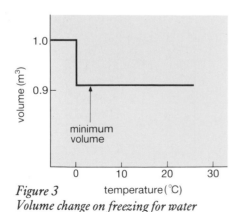

Figure 3
Volume change on freezing for water

This road gritter is used in icy conditions to make roads less slippery

The melting point of ice can also be lowered a little by applying very large pressures to it. An extra pressure equal to that of the atmosphere will lower the melting point of ice by about 0.01°C.

Evaporation

As a liquid warms up, the average speed of the molecules in it gets larger. However, not all of the molecules in the liquid will be travelling at the same speed. Figure 4 shows that some will be travelling slowly and some more quickly. Some molecules near to the surface of the liquid have enough energy to escape. They evaporate from the liquid to form a vapour. Evaporation from the surface of a liquid can happen at any temperature. But as the temperature gets higher more molecules have enough energy to escape. So evaporation happens at a faster rate.

Eventually the temperature rises to the **boiling point** of the liquid. At this point, evaporation also happens inside the liquid. Bubbles of vapour form inside the liquid and rise to the surface. Heat applied to a boiling liquid gives the molecules enough energy to evaporate.

Evaporation causes cooling. This is because it is the faster (hotter) molecules that escape, leaving behind the slower (colder) molecules. The evaporation of sweat from your skin keeps you cool on a hot day. But the evaporation of water from us can also be very dangerous. You can lose heat from your body very rapidly in wet and windy conditions.

This wine cooler has been soaked in water before the wine bottle was placed in it. Evaporation of the water through the porous pot takes heat away from the wine and keeps it cool

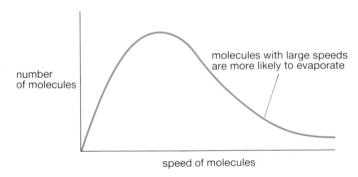

Figure 4
The molecules in a liquid do not all travel at the same speed

With the lid in place this coffee would stay hot longer. Why is this?

Questions

1 (a) Why does washing dry faster on warm, windy days?
(b) Why do you hang out washing, rather than leaving it in a pile to dry?
(c) Why does the amount of water in the atmosphere (humidity) affect the drying rate?
2 Use Figure 3 to help you sketch a graph of the way in which the density of water changes as it freezes.
3 When you go ice skating a layer of water forms between the bottom of your skate and the ice. This layer of water is important for reducing friction, so that you can skate quickly. Two theories have been suggested to explain how this water forms:

I When pressure is applied to ice its melting point is lowered. So the pressure from an ice skate will melt some ice.

II When the skate moves over the ice, heat is generated due to frictional forces.

Use the following data to help you decide which theory is more likely.

- Weight of skater = 700 N
- Area under 1 ice skate = 10^{-3} m^2
- Temperature of the ice = -5°C
- Melting point of ice = 0°C
- The melting point of ice will be lowered by 0.01°C by a pressure of 100 000 N/m^2

4 Explain why a saucepan will come to the boil more quickly if it is covered by a lid.

5 Physics in the Kitchen

As the air pressure is lower up a mountain, water boils at a lower temperature and food takes longer to cook

Changing boiling points

When water boils, bubbles of water vapour form inside the liquid. The pressure inside these bubbles is equal to the pressure of the air above the water. So when the air pressure changes the water boils at a different temperature. If the air pressure is greater than 1 atmosphere, water boils above 100°C; if the pressure is lower than 1 atmosphere, water boils below 100°C.

Figure 1 shows a graph of how the boiling point of water changes with pressure. You would notice this effect if you lived somewhere like Mexico City which is 3000 m above sea level. The air pressure there is only about 0.7 atmospheres, and water boils at 90°C. At the top of Mount Everest, 9000 m above sea level, a kettle would boil at only 70°C.

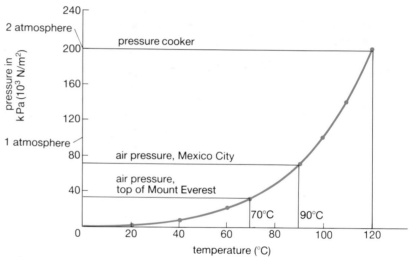

Figure 1
Graph showing how the boiling point of water changes with pressure

In a pressure cooker the air pressure in increased to about 2 atmospheres. The cooker has an airtight lid except for a small hole at the top. Weights are put on top of this hole so that the air pressure inside must be greater than atmospheric pressure before steam will escape. The advantage of cooking at high pressure is that the boiling point of water is raised and cooking times are considerably reduced (Figure 2, Figure 3).

Figure 2
Graph to show how the cooking time for potatoes decreases as the temperature is raised

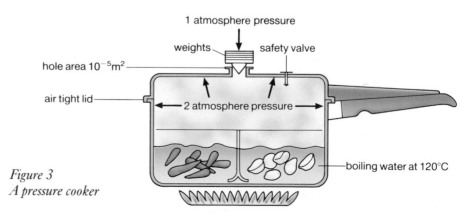

Figure 3
A pressure cooker

When you go to the doctor for an injection, your arm gets cleaned with some alcohol. This evaporates quickly making your arm cold. A refrigerator in your kitchen cools your food in a similar way.

Figure 4 shows a refrigerator. The cooling in a refrigerator is done by chemicals called freons, which boil at low temperatures (about −30°C). The cooling occurs in the ice box where boiling freons evaporate. The freon gas then reaches the compressor which squashes the gas to a high pressure. The compression also warms the gas. The gas turns back into liquid freons as it cools.
On the back of your refrigerator you will see cooling fins which cool the liquid back to room temperature. Then the freon is allowed to expand. When the pressure is low, the freon starts to boil and its temperature falls. The freon now passes through the ice box to start the cooling cycle again.

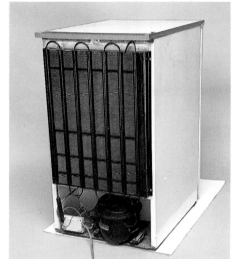

The cooling fins on the back of a fridge. Warm freon is pumped through the tubes which pass through the fins. As it circulates, the freon loses heat and cools down

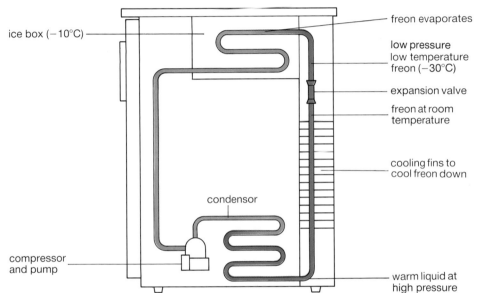

ice box (−10°C)

freon evaporates

low pressure
low temperature
freon (−30°C)

expansion valve

freon at room
temperature

cooling fins to
cool freon down

condensor

compressor
and pump

warm liquid at
high pressure

*Figure 4
A refrigerator*

Questions

1 When salt is added to water, its boiling point is increased. You often add salt to water when you cook vegetables. Does this make the vegetables cook more slowly or more quickly?

2 (a) The area of the hole in the top of the pressure cooker in Figure 4 is 10^{-5} m. What weight must be put on top of the hole to keep the pressure inside the cooker at 2 atmospheres (2×10^5 N/m^2)? Remember the pressure outside is 1 atmosphere.
(b) What is the pressure inside the cooker when a weight of 0.5 N is used?

3 This question is about the time taken to cook your potatoes as you climb Mount Everest. The higher you get the longer it takes.

(a) The table below shows you how the air pressure drops as you climb up the mountain. Copy the table and then use Figure 2 to fill in the column of boiling points.
(b) Use Figure 3 to fill in the column for cooking times.
(c) Now plot a graph to show how your cooking times (y-axis) change with your

height above sea level (x-axis).
(d) Use your graph to predict cooking times at (i) 5000 m (ii) 8500 m.
(e) Would your cooking times have changed so much if you used a pressure cooker to cook your potatoes? Explain your answer carefully.

Height above sea level (m)	Air pressure kPa (10^3 N/m^2)	Boiling point of water (°C)	Cooking time for potatoes (minutes)
0	100	100	14
2000	79		
4000	62		
6000	47		
8000	36		

6 Internal Combustion Engines

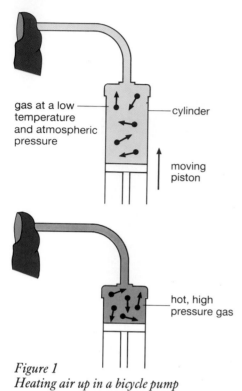

gas at a low temperature and atmospheric pressure — cylinder

moving piston

hot, high pressure gas

Figure 1
Heating air up in a bicycle pump

Figure 2
The four-stroke petrol engine

Petrol engines

If you put your finger over the end of a bicycle pump, and pump the cylinder up and down a few times, you find that the air inside the cylinder gets hot. As the piston moves it collides with moving molecules. The molecules rebound off the moving piston at a greater speed. This means the gas gets hotter. The pressure inside the bicycle pump cylinder increases for two reasons. First the air has been compressed, and then this compression warms the air up. If you compress the air slowly the temperature will stay the same because heat will escape through the sides of the cylinder (Figure 1).

Figure 2 shows how the idea is used in a **four-stroke petrol engine**.

(1) On the **intake** stroke a petrol vapour and air mixture is fed into the cylinder.

(2) On the **compression** stroke both the inlet and exhaust valves are closed. The piston moves up rapidly to compress the air/petrol mixture to about $\frac{1}{10}$ of its original volume; the *compression ratio* is 10:1. The pressure inside the cylinder will now be about 20 atmospheres (2 MPa) and the temperature about 350°C.

(3) The **working** stroke. Once the gas has been compressed the sparking plug produces a spark which starts the fuel burning. The temperature in the cylinder now rises to about 1000°C and the pressure rises to about 50 atmospheres (5 MPa). The higher pressure now forces the piston back. Although energy is used to squash the gas during the compression stroke, the working stroke gives out far more energy.

(4) In the **exhaust** stroke, the exhaust valve opens to allow the high pressure gases to escape.

Most cars have four **cylinders** using a four-stroke cycle. The cylinders are arranged so that they take turns in performing one of the four strokes (Figure 3). At any instant, one of the cylinders will be producing power. A car with only one cylinder would produce a very jerky ride.

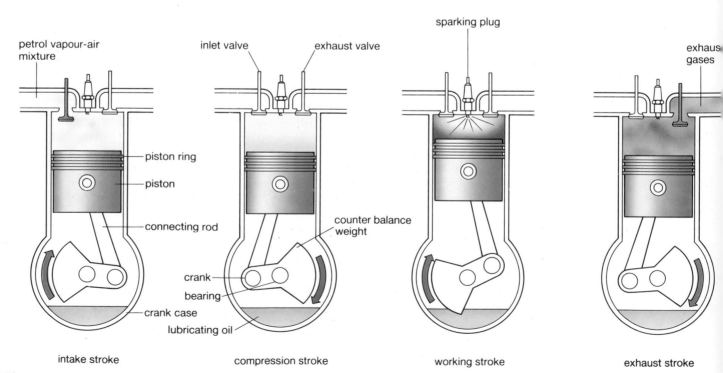

petrol vapour-air mixture

inlet valve   exhaust valve

sparking plug

exhaust gases

piston ring

piston

connecting rod

counter balance weight

crank

bearing

crank case

lubricating oil

intake stroke   compression stroke   working stroke   exhaust stroke

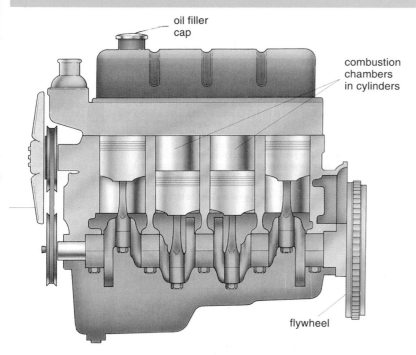

oil filler cap

combustion chambers in cylinders

flywheel

Figure 3
A car engine cut away to show its four cylinders

Diesel engines

A four-stroke diesel engine works in a similar way to the petrol engine you have just read about. The differences are these:

- During the intake stroke only air is taken into the cylinder.
- The compression ratio of diesel engines is higher than that of petrol engines. At the end of the compression stroke the air has been squashed to about 1/20 of its original volume. At this point the temperature of the air is about 700°C, and its pressure is about 35 atmospheres (3.5 MPa).
- The diesel engine has no sparking plug. At the end of the compression stroke fuel is forced into the cylinder under high pressure. The temperature of the air is so hot that the fuel burns as soon as it has mixed into the air. When the fuel burns, the temperature inside the cylinder reaches about 2500°C and the pressure is about 100 atmospheres (10 MPa)
- Such an engine using diesel oil has a higher **efficiency** than a petrol engine.

Questions

1 What is meant by the term 'compression ratio'?

2 At the beginning of the working stroke in a petrol engine the pressure in a cylinder is 5 MPa (5×10^6 N/m²). The area of the piston is 0.008 m².
(a) Calculate the force acting on the piston.
(b) Explain why cylinders and pistons in a petrol engine have to be well built. Why do they have to be even stronger in a diesel engine?

3 Why does a diesel engine not have a sparking plug?

4 Why do most cars have four cylinders?

5 (a) You were told in the text that diesel engines have higher efficiencies than petrol engines. What does that mean?
(b) Use the data opposite to work out the cost of fuel to drive: (i) a petrol-driven car 10 000 km, (ii) a diesel-driven car 10 000 km.
(c) How far do you have to drive a diesel-fuelled car before you have recovered the extra cost of its engine?
(d) If you were buying a car which sort of engine would you choose?

- cost of diesel engine: £1000
- cost of petrol engine: £700
- 1 litre of petrol will take the car 10 km
- 1 litre of diesel oil will take the car 14 km
- petrol costs 60p per litre
- diesel costs 55p per litre
- cars with diesel engines are noisier and more sluggish than petrol driven cars

7 The Gas Laws

Pressure (atmospheres)	Volume of air (l)	$\frac{1}{V}$ (l^{-1})
0.5	2.0	0.5
1.0	1.0	1.0
2.0	0.5	2.0
4.0	0.25	4.0
10.0	0.1	10.0

Table 1

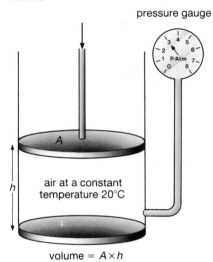

Figure 1
Testing a piston and cylinder

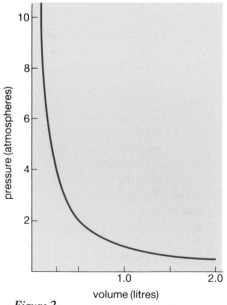

Figure 2
As the volume of the gas is halved, the pressure doubles

Boyle's Law

Figure 1 shows a cylinder and piston. The pressure of the air in the cylinder can be changed by moving the piston up and down. Provided the piston is moved slowly the temperature of the gas stays the same. This apparatus can be used to see how the pressure, P, changes as the volume, V, of the air is increased or decreased.

Table 1 shows the sort of results you can get if you do the experiment. When the gas is compressed to half its volume the pressure is doubled. We say that the pressure of the gas is *inversely proportional* to its volume:

$$P \propto \frac{1}{V}$$

The table of results also shows that when you multiply together the pressure and volume, you always get the same number. (What is it?)

When the mass of a gas stays the same, and its temperature does not change:

$$\boxed{P \times V = \text{constant} \qquad \text{This is } \textbf{Boyle's law}}$$

Law of pressures

Figure 3 shows you the same apparatus, placed in a large tank of water. This time the volume of the gas is kept constant. The pressure is measured at a lot of different temperatures.

The results of such an experiment are shown in Figure 4. As the temperature, T, is raised from 0°C to 100°C the pressure rises steadily from 1 to about 1.4 atmospheres. We can draw a straight line through these results. If this line is extended below 0°C you can predict what will happen to the pressure at lower temperatures. You can see that the pressure will be about 0.5 atmospheres at −136°C. At a temperature of −273°C the pressure due to the gas is nothing.

A gas exerts a pressure on the walls of its container because of its moving molecules. If a gas no longer exerts a pressure it is because the molecules have stopped moving. This means that −273°C is the lowest possible temperature. At that temperature molecules have stopped moving so we cannot cool them any further. We call −273°C **absolute zero**.

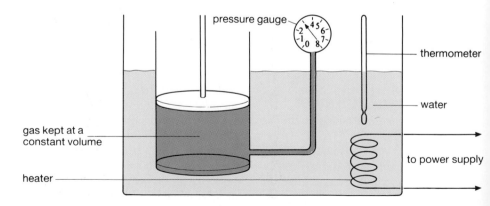

Figure 3

Absolute temperatures are measured in *degrees Kelvin*, or K. Absolute zero is 0 K and 0°C is 273 K. The size of 1 degree Kelvin is the same size as 1°C. Notice that no degree sign is used with degrees Kelvin. We write 77 K, *not* 77°K.

Figure 4 shows that the pressure of a gas is proportional to its absolute temperature, provided its volume stays the same:

$$P \propto T \qquad \text{This is the \textbf{law of pressures}}$$

Charles' Law

The apparatus shown in Figure 3 can be used for a third experiment. The pressure of the gas is now kept constant. The gas is allowed to expand when it is heated.

Experimental results show that the volume of the gas is proportional to its absolute temperature, provided its pressure remains the same:

$$V \propto T \qquad \text{This is \textbf{Charles' law}.}$$

Gas equation

The three gas laws may be combined into one equation, which holds true for a fixed mass of gas:

$$\frac{PV}{T} = \text{constant}$$

Example. A cylinder of gas of volume 1 m³ is initially at a temperature of 300 K and a pressure of 2 atmospheres. It is compressed to a volume of 0.5 m³ and warmed to a temperature of 450 K. What is the pressure of the gas now?

$$\frac{P_1 V_1}{T_1} = \frac{P_2 V_2}{T_2}$$

$$\text{So} \quad \frac{2 \times 1}{300} = \frac{P \times 0.5}{450}$$

$$P = 6 \text{ atmospheres}$$

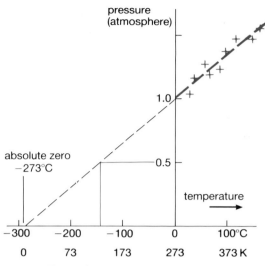

Figure 4
Results of an experiment using the apparatus of Figure 3

Questions

1 This question refers to the experiment shown in Figure 1.
(a) Use Table 1 or Figure 2 to work out the volume of the air when the pressure was: (i) 3 atmospheres, (ii) 8 atmospheres
(b) Use the data to plot a graph of the pressure, P, against the reciprocal of the volume, $1/V$.
(c) A student took a measurement of the volume and pressure and recorded 0.16 litres and 7.0 atmospheres. Plot this point on your graph. The student made an error in measuring the pressure. Suggest what the value should have been.
2 This question refers to the experiment described in Figure 3.
(a) Use the graph in Figure 4 to predict the pressure in the cylinder at:
(i) a temperature of 150 K,
(ii) a temperature of 150°C.
(b) What will the pressure in the cylinder be at 600 K?
3 The diagram (right) shows a gas storage vessel of volume 160 000 m³. The pressure in the main gas grid pipelines is about 800 kPa. In the storage vessel the pressure is just greater than 100 kPa (atmospheric pressure). Calculate the volume of gas from the grid pipelines required to fill the gas storage vessel.

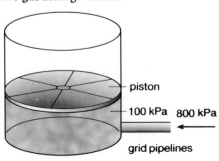

SECTION E: QUESTIONS

1 Smoke particles in air are seen through a microscope. The smoke particles move and make frequent changes in direction because they
A repel each other.
B attract each other.
C are able to move themselves.
D are colliding with each other.
E are colliding with invisible air particles.

MEG

2 (a) In this diagram the bottom circle represents the molecules in the liquid. Copy and complete the top circle to represent the molecules in the gas.

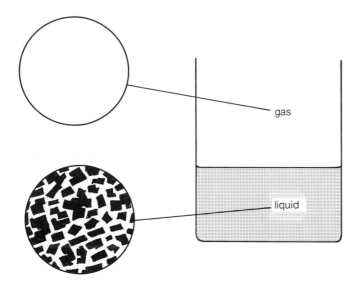

(b) With reference to the diagram explain why gases can be compressed easily but liquids cannot.
(c) Brownian motion is usually shown in a school laboratory by looking at smoke specks under a microscope.
(i) Describe the motion of the smoke specks.
(ii) What causes this motion?

ULEAC

3 A drop of oil, of volume 0.4 mm^3, was placed on a water surface and spread out to a maximum area of 1200 mm^2. Assuming the oil patch was one molecule thick, what is the best estimate of the diameter of an oil molecule, in mm?
A 0.00004   B 0.00008   C 0.000012
D 0.00033   E 0.00048

NEAB

4 Which one of the following will cause the boiling point of water to rise?
A Reducing the pressure over the surface of the water.
B Dissolving some salt in the water.
C Boiling the water faster by increasing the supply of heat.
D Using an immersion heater instead of an external supply of heat.
E Boiling the water at a greater height above sea level.

NEAB

5 The Kinetic Theory assumes that gases are made up of rapidly moving molecules. Which of the following facts does **not** support the idea that gas molecules are moving?
A A tiny smoke particle undergoes 'Brownian Motion' when in a gas.
B Gases exert a pressure on the walls of their containers.
C Gases have lower densities than solids and liquids.
D When placed together, two gases will diffuse into each other.
E When released into an empty container, a gas will rapidly fill it.

MEG

6 The two strokes of a four-stroke engine cycle during which only **one** valve is open are
A induction and compression.
B compression and exhaust.
C induction and exhaust.
D compression and power.

SEG

7 When a drop of ether is put onto the skin, the skin feels cold. This is because
A liquids like ether are poor conductors of heat.
B ether is a good conductor of heat.
C the ether is an anaesthetic.
D the ether evaporates by taking heat from the skin.
E heat cannot reach the skin through the ether.

NEAB

8 A sealed can contains air. If the can is heated the pressure of the air inside the can increases. Which of the following statements best describes why the pressure increases?
A When heated, air molecules expand and press harder on the can.
B More air molecules are created and so more molecules press on the can.
C When heated, air molecules slow down and collide less frequently with the can.
D Air molecules are attracted to the heated sides of the can.
E When heated, air molecules move faster and collide with the can more often.

ULEAC

9 At the beginning of the compression stroke in the cylinder of a diesel engine, the air is at a temperature of 127°C, at a pressure of 0.1 MPa. During the compression stroke, the volume of the air is squashed to 1/20th of its original volume, and its temperature rises to 727°C. What is the pressure in the cylinder at the end of the stroke?

NEAB

10 The diagram shows Boris in a diving bell. He is just about to explore the depths of the black lagoon. As the bell is lowered into the sea, water rises to fill the bell.
(a) Explain why water starts to fill the bell as it is lowered.
(b) the pressure below the surface of the sea is given by the formula: $P = (100\,000 + 10\,000\,d)$ N/m^2; d is the depth below the surface in metres. What is the pressure of air in the diving bell at a depth of 10 m?
(c) How deep can the bell go, before Boris gets his feet wet?

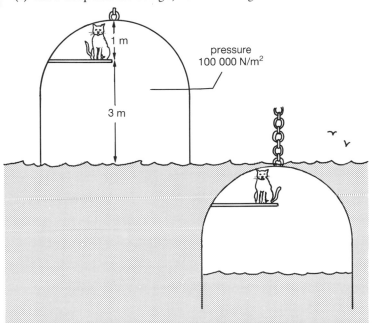

11 A petrol engine repeats the same sequence of four strokes continuously. Each stroke involves the following parts of the engine: *inlet valve*, *outlet valve*, and the *piston*. The *spark plug* produces the spark needed to ignite (set fire to) the fuel. The diagrams on page 106 show the sequence of the four strokes. Use them to help you answer the questions which follow.

(a) Copy and complete the following table to show the sequence of events during the four strokes.

Stroke number	1	2	3	4
Direction piston is moving (*up* or *down*)	down			up
Inlet valve position (*open* or *closed*)		closed	closed	
Outlet valve position (*open* or *closed*)		closed	closed	

(b) (i) In which stroke does air and petrol *enter* the cylinder?
(ii) In which stroke do the exhaust gases *leave* the cylinder?
(c) (i) Copy this flow chart. Complete the boxes to show the main types of energy transferred in the engine.

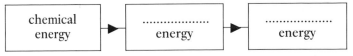

(ii) Suggest what type of energy provides the spark.
(d) (i) Explain why air is needed as part of the fuel mixture.
(ii) Explain why it is important that the fuel and air are well mixed.
(e) (i) Petrol is a mixture of compounds. Most of these compounds contain the elements carbon and hydrogen and are called hydrocarbons. Suggest two substances which are likely to be produced when a hydrocarbon burns in oxygen.
(ii) What process in living organisms is similar to this reaction?

(f) A diesel engine has the same moving parts as a petrol engine but differs in the way it works. Some of these differences are given in this table.

	Type of engine	
	Petrol	Diesel
fuel	petrol	diesel
greatest pressure just before fuel ignites	about 900 KPa	about 2200 KPa
spark needed	yes	no
efficiency	24%	40%

(i) In both types of engine the temperature of the gas mixture in the cylinder increases as the piston moves upwards with both valves closed. Use your understanding of particles to explain why this increase in temperature happens.
(ii) In which type of engine will this increase in temperature be greater?
(iii) Suggest and explain one important difference between the general structure of the two engines.
(iv) Explain what is meant by saying that the petrol engine is *24% efficient*.
(v) Explain what happens to the remaining 76% of the energy.

SEG

12 Which substance is a liquid at room temperature?

substance	melting point (°C)	boiling point (°C)
A	−218	−183
B	− 39	357
C	44	280
D	119	444
E	1038	2336

MEG

13 The diagram shows some bubbles rising in a glass of a fizzy drink. Can you explain why the bubbles get larger as they rise to the surface?

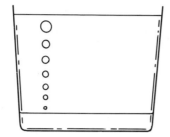

14 The picture below shows part of a chemical works in which a tanker is being filled with a chemical. The chemical is normally a gas. It is invisible, has no smell and is heavier than air. It is a non-ionic compound which is non-flammable and non-toxic.

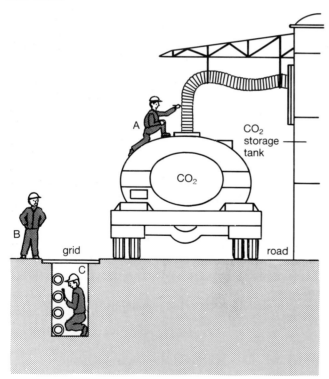

(a) What is the *name* of the chemical?
(b) If there is a leak of the chemical, which worker will be *most* affected? Give **two** reasons for your answer.
(c) The same chemical can be used in fire extinguishers designed to put out fires involving live electrical equipment.
(i) Suggest and explain **one** reason for this.
(ii) Explain why water is unsuitable for this type of fire.
(d) Although the chemical in the tank is normally a gas, it is stored and transported as a liquid. This is because each kilogram of the chemical takes up far less space as a liquid.
(i) Explain why the liquid chemical takes up less space.
(ii) Explain why it may be an advantage to store and transport the chemical as a liquid.
(iii) Suggest and explain one method by which the gas might be turned into a liquid.
(e) Suggest **two** safety features that you would make part of the tanker's design. Give reasons for your choice.

SEG

15 A sealed test tube contains air at atmospheric pressure and room temperature. When it is surrounded by ice, which of the following statements about the air in the test tube is **not** correct?
A The pressure becomes lower.
B The average kinetic energy of the molecules becomes lower.
C The average momentum of the molecules becomes less.
D The molecules collide less often with the test tube.
E The average number of molecules in each cm^3 becomes less.

MEG

16 (a) Explain in terms of molecules, why a gas exerts a pressure on the walls of its container.
(b) The air inside an old syrup tin is heated with a bunsen burner. The lid is pressed firmly on. Explain why the pressure in the tin increases.
(c) If the tin is heated enough the lid flies off. Explain why.
(d) The experiment is now repeated, but with a small amount of water placed in the tin. Explain why the lid flies off at a lower temperature.

17 In the diagram below, you can see the planet Mars in an elliptical orbit around the Sun. At its closest (position A) Mars is about 200 million km away from the Sun. At its most distant (position B) Mars is about 250 million km away from the Sun.

Mars has a thin atmosphere which is mostly carbon dioxide. The pressure of the atmosphere changes from winter to summer. The table below shows some details of the Martian climate. The data was taken by the Viking Lander, near to the equator in 1977. Can you explain why the pressure of the Martian atmosphere changes so much?

	Average daily temperature (°C)	Average pressure (N/m²)
Position B (winter)	−73	50
Position A (summer)	−13	65

SECTION F
Heat

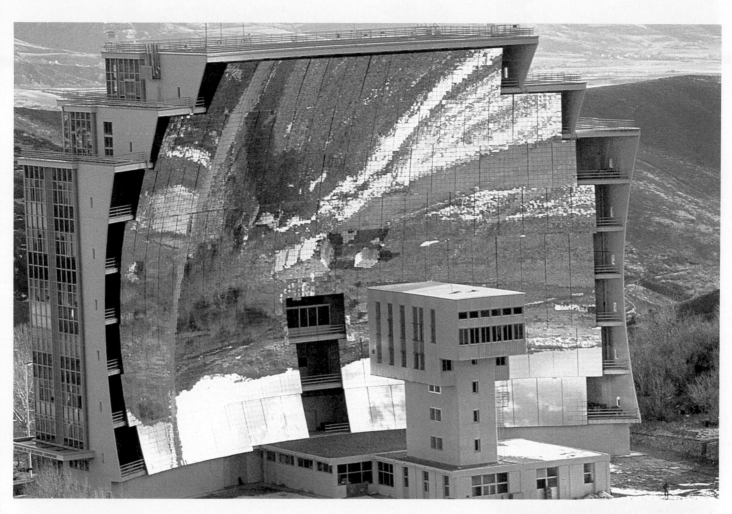

Solar furnaces like this one at Odeillo in France, are power stations that can produce up to 10 megawatts of electricity using energy from the Sun. To find out how they work, turn to page 126

1 Hot and Cold

The Namib desert: daytime temperatures of 58°C have been recorded here in midsummer. However, at night, temperatures can reach freezing point. This range of temperature makes the desert a very hostile environment for living things

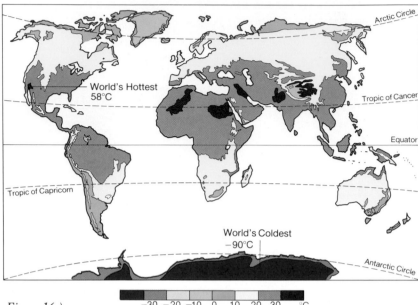

Figure 1(a)
Mean temperature in July

The Antarctic: during the continuous darkness of mid-winter, the temperature here can be as low as −87°C

Everybody knows the meanings of the words hot and cold. If you pick up a hot plate you burn your fingers. On a cold winter's day when you go outside without gloves, your fingers freeze. When you get too hot or too cold life becomes very unpleasant. Your body works normally at a temperature of about 37°C. You will become very ill if your body temperature rises as high as 40°C or falls as low as 34°C.

We are only comfortable wearing light clothing when the air temperature is around 10°C to 20°C. The maps in Figures 1a) and b) show the world's average temperatures in January and July. As you can see there are many places where it is too hot or too cold. In winter the average temperature in Verkhoyansk (Siberia) is below −40°C; the lowest temperature recorded there is about −70°C. People only survive the winter by staying indoors for months at a time. If you lived in Ethiopia you would face a different problem. The average yearly temperature in some parts of Ethiopia is about 35°C. In July the midday temperature can reach 50°C. Under such scorching conditions there is little water and food is hard to grow. Millions of Ethiopians have died over the past few years through shortage of food and water.

Life can also be uncomfortable in other places. In Bombay, you might prefer to live in an air-conditioned flat. In Madrid, where midday temperatures reach 45°C in the summer, banks open from 9.00 am−12 noon and then again from 4.30 pm−6.00 pm. Most Spaniards have a siesta in the middle of the day, simply because it is too hot to be working outside.

Heat and temperature

It is a very common mistake to confuse the two words heat and temperature. When we talk about **heat** we mean the heat energy that we have had to put into a material to warm it up. Heat energy is measured in **joules**. The word **temperature** is used to describe how hot something is. Temperatures are usually measured in **degrees celsius** (°C).

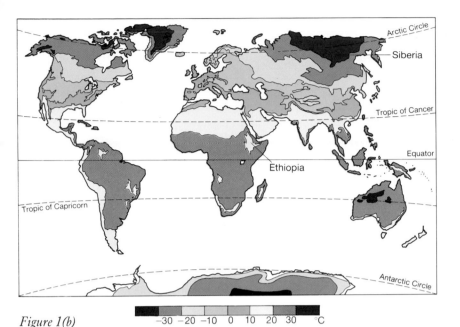

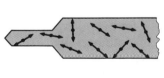

Figure 1(b)
Mean temperature in January

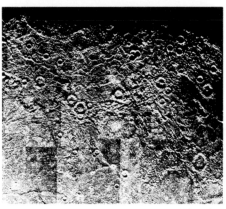

The solar system also has extremes of temperature. The daytime temperature on Mercury is about 330°C — hot enough to melt lead!

On Uranus, the daytime temperature is a cool −216°C

Peter is a plumber and he uses a soldering iron which has a mass of 0.6 kg. Elsie is an electrician and she uses a soldering iron which has a mass of 0.1 kg. They each plug their irons in and wait for them to warm up. After one minute Peter's iron has been given 20 000 J of heat and it has reached a temperature of 50°C. In that time Elsie's iron has been given 10 000 J of heat and it has reached a temperature of 100°C. Peter's iron has more heat energy than Elsie's but Elsie's iron is at a higher temperature.

Figure 2 shows the molecules in the two irons after they have been warmed up. The length of the arrows in the diagrams shows how fast the molecules are moving. The temperature is a measure of the energy of *each* molecule. The heat energy that you have to supply to warm an iron up is a measure of the energies of *all* of the molecules.

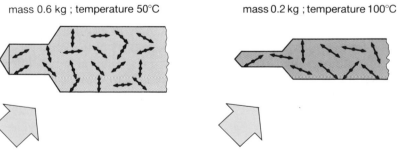

mass 0.6 kg ; temperature 50°C

mass 0.2 kg ; temperature 100°C

Figure 2

(a) 20 000 J of heat energy warmed up Peter's soldering iron to 50°C

(b) 10 000 J of heat energy warmed up Elsie's soldering iron to 100°C

Questions

1 Use the maps of the world to help you answer the following questions.
(a) Why do very few people live in:
(i) Greenland, (ii) the centre of Australia?
(b) Which part of the world has the greatest difference in average temperatures, between January and July?
(c) What other factors, besides temperature, affect how many people live in a country?
2 Which will cause the worse burn: (i) a pan of hot water at 70°C falling on your foot, (ii) a spark from a bonfire at about 500°C landing on your hand? Explain your answer.
3 Criticise this statement: 'The heat of the inside of the Sun is about 15 million degrees celsius'.

2 Expansion of Solids

In the photo on the right you can see how the engineers who built the Humber Bridge overcame the problem of the bridge expanding in hot weather. When the bridge expands, the plate on the left slides under the plate on the right, so that the surface of the bridge stays flat

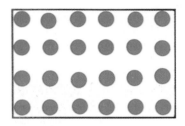

Figure 1
(a) Cold solid: atoms vibrate slowly

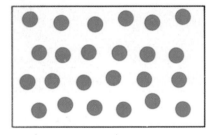

(b) Hot solid: atoms vibrate quickly and push each other apart. Expansion results

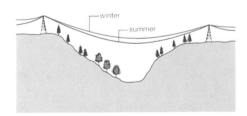

Figure 2
Engineers must allow for the expansion of overhead power lines. In summer, the lines will be longer and they will sag more

The molecules in solids and liquids are always vibrating. If the temperature falls, molecules vibrate less and they move closer together (Figure 1 (a)). Then a substance will **contract**. When the temperature rises, molecules vibrate more and push each other apart. This causes substances to **expand** (Figure 1 (b)). If you try to stop something expanding then very large forces may result. Sometimes these forces cause us serious problems but at other times they may be useful.

Expansion causing problems

In Saudi Arabia the temperature at midday in July can rise to as high as 40°C. But on a winter's night in January, the temperature can drop below 0°C. These large temperature changes cause problems with oil pipelines. A long straight length of pipe would soon buckle under the forces caused by expansion. The problem is solved by putting a series of expansion loops into the line. These loops make the line more flexible, so that it is free to expand or contract as the temperature changes. Figure 2 shows how the expansion of overhead power lines is tackled.

Putting expansion to use

Not all materials expand by the same amount when heated. Table 1 opposite shows you the increase in length of various materials when they are heated through 1°C. You can see that brass expands nearly twice as much as iron for the same temperature rise. Some materials such as invar and Pyrex glass hardly expand at all. This difference in expansion between materials allows us to make a very useful device called a **bimetallic strip** (Figure 3). A strip is made out of two different metals such as iron and brass. The two metals are fixed together so that they cannot move separately. When the strip is cooled the brass contracts more than the iron so the strip bends upwards. When the strip is heated it bends the other way because the brass is now longer than the iron. This principle is used in some fire alarms and thermostats.

● **Fire alarm.** Figure 4 shows the principle behind a simple fire alarm. Inside the fire detector itself there is a bimetallic switch. When the bimetallic strip gets hot it bends upwards and completes an electrical circuit. A current now flows and makes a fire alarm sound.

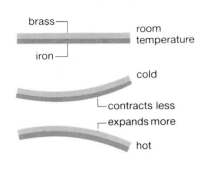

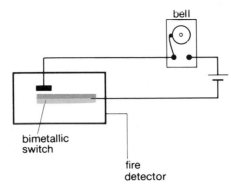

Material	Increase in length (mm) for 1 m heated through 1°C
aluminium	0.025
brass	0.019
iron	0.012
steel	0.011
glass (common)	0.009
glass (Pyrex)	0.003
invar (an alloy of nickel and steel)	0.001

Figure 3
A bimetallic strip

Figure 4
A switch for a fire alarm

Table 1

- **Thermostat.** A bimetallic switch can also be used to keep the temperature of your clothes iron constant (Figure 5). When the iron is cold the switch is up and current flows to the heating element. However, when the iron is hot the bimetallic strip bends downwards and breaks the circuit. A control knob allows you to choose your iron's temperature; hot for cotton, cooler for nylon. The hotter you want the iron to be, the more you turn the control knob downwards. This means that the bimetallic strip has to bend more before the circuit is broken, so it must be hotter.

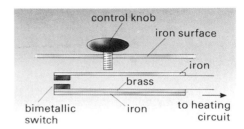

Figure 5
A thermostat in an iron

Questions

1 Below is a diagram of a gas oven thermostat. When the oven is hot enough the valve controlling the gas supply closes. When the oven cools down the valve opens again.
(a) Explain how it works.
(b) How would you add a control knob to adjust the oven temperature?

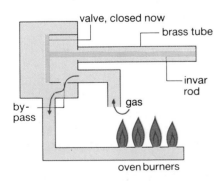

2 The graph shows how the current flowing in the heating element of a clothes iron changes with time.

(a) Explain why it behaves as it does.
(b) Sketch a graph to show how the temperature of the iron changes with time.

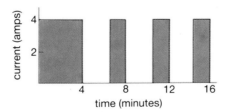

3 Look at the diagram of a bridge, and read the data below it. The bridge was built in the middle of a cold Russian winter when the temperature was −20°C. The ends of the main steel girders were fixed in cement and no room was left for expansion.
(a) By how much does a steel rod of length 10 m expand when it warms up by (i) 1°C, (ii) 40°C.
(b) Explain why the girders in the bridge are under compression in a hot summer when the temperature is 20°C.

(c) Calculate the force require to compress the bridge girder by 1 mm.
(d) Now calculate the force compressing the girder in the summer when the temperature is 20°C.
(e) Explain what might happen to the girder or the concrete in the summer.

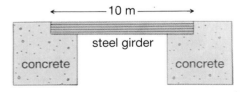

- A rod of steel 1 m long expands by 0.01 mm when it warms up by 1°C.
- A steel rod 10 m long with a cross-sectional area of 1 cm² needs a force of 2000 N to compress it by 1 mm.
- The cross-sectional area of the girder in the bridge is 100 cm².

3 Supplying Heat

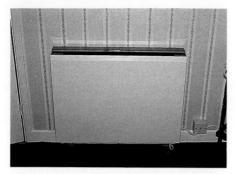

Night storage heaters use concrete to store heat energy. The concrete is heated up at night when electricity is cheap and it then releases heat during the day.

When you want a cup of coffee you put some water in the kettle and switch it on. The water gets hotter because heat energy is being supplied which makes the molecules vibrate more rapidly. The hotter the water is, the faster the molecules move.

The amount of heat energy that must be supplied to warm up the kettle depends on two things:
- More energy is needed for a large temperature rise.
- More energy is needed if there is a lot of water in the kettle.

Warming 200 g of water from 20°C to 80°C needs 50 000 J of heat. Warming 200 g of wine from 20°C to 80°C (for making fondue) needs 40 000 J of heat. The amount of energy needed depends on the substance.

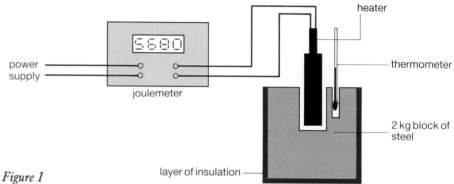

Figure 1
An experiment to measure the heat energy required to warm up a block of steel

Specific heat capacities

You can see in Figure 1 an experiment to measure the heat energy required to warm up a block of steel. A thermometer records the temperature rise, while a Joulemeter measures the number of joules that the heater has supplied. The block of steel is wrapped in thick insulating material to make sure that no heat can escape to the surrounding air. Figure 2 shows how the temperature of the block rises as heat energy is supplied.

The amount of heat required to warm 1 kg of a substance by 1°C is called its **specific heat capacity**. It is measured in units of **J/kg°C**. We can work out the specific heat capacity of steel like this. Figure 2 shows us that 18 kJ warmed the block from 20°C to 40°C.

$$18000\,J \text{ warmed } 2\,kg \text{ by } 20°C$$

$$so \quad 9000\,J \text{ would warm } 1\,kg \text{ by } 20°C$$

$$and \quad \frac{9000\,J}{20} = 450\,J \text{ would warm } 1\,kg \text{ by } 1°C$$

The specific heat capacity of steel is 450 J/kg°C. As you can see from table 1, different substances have very different specific heat capacities. In general you can work out the heat required to warm up a substance using the equation below.

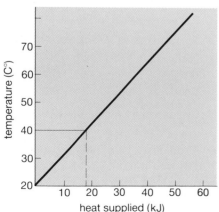

Figure 2
A graph to show how the temperature of the steel block rises as heat energy is supplied

> Heat supplied = mass × specific heat capacity × temperature change
>
> $$joules = kg \times \frac{joules}{kg°C} \times °C$$

Example. How much heat is required to heat up the tip of Elsie's soldering iron by 400°C? The tip of the soldering iron is made of copper and has a mass of 30 g.

$$\text{Heat supplied} = m \times s \times (T_2 - T_1)$$
$$= 0.03\,\text{kg} \times 380\,\frac{\text{J}}{\text{kg}°\text{C}} \times 400°\text{C}$$
$$= 4560\,\text{J}$$

Water has a very high specific heat capacity. This means that it absorbs a lot of heat energy when it warms up. Conversely, water gives out a lot of heat when it cools down. Several ways in which the high specific heat capacity of water is useful to us are described below:

- We are made mostly of water. This means that if we suddenly exercise, and our muscles produce a lot of heat energy, then we do not warm up too quickly.

- Water is very important for keeping our houses warm. A house central-heating system pumps hot water around the house. The hot water can release a lot of energy as it flows through radiators. If water had a low specific heat capacity, then radiators a long way from the boiler would never be warm.

- Water is capable of absorbing a lot of heat energy, so it is useful for cooling car engines (Figure 3).

- In Britain we live on an island. The fact that we are surrounded by water has a great effect on our climate. Water is so difficult to warm up and cool down that our weather is neither extremely hot nor extremely cold. Land can be warmed up or cooled down much more quickly than sea, so the temperature in the middle of a continent can change rapidly. You can see from the map on page 115 that Cananda and Russia have worse winters than we do, although they are no further north.

Substance	Specific heat capacity (J/kg°C)
water	4200
alcohol	2400
ice	2100
concrete	800
glass	630
steel	450
copper	380
lead	130

Table 1

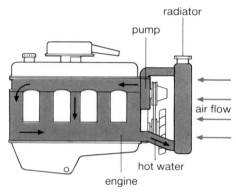

Figure 3
A car cooling system

Questions

1 This question refers to Figure 2 which shows how the temperature of a 2 kg steel block changes as heat is supplied to it.
(a) Make a copy of the graph.
(b) Mark on the graph how the temperature would have changed if:
(i) the mass was 4 kg, (ii) the mass was 1 kg, (iii) the insulating material was removed from the metal. Label these graphs (i), (ii), (iii).
2 A night storage heater contains 60 kg of concrete. How much heat is required to warm the concrete from 10°C to 40°C?
3 Chen and Sally are sitting beside a lake. Chen is throwing stones into the lake. Here is an extract from their conversation.

Chen: Just think of all that kinetic energy the stone has. It is all converted into heat energy when it makes a splash in the lake. If I keep this up the lake will get warmer.
Sally: Don't be stupid, it would take all day!
Evaluate their conversation.
4 Julie has an outdoor swimming pool which is 20 m long, 10 m wide and 2 m deep.
(a) When the pool is full, what is the volume of water in it?
(b) The density of water is 1000kg/m³: what mass of water is there in the pool?
(c) It is necessary to warm the water from 17°C to 22°C. How much heat energy does that require?

(d) During a hot summer spell the sun shines on the pool for an average of 8 hours a day. While the sun shines, the pool absorbs energy at a rate of 60 kW. How much energy is absorbed each day?
(e) Roughly how many days does it take the pool to warm up to 22°C? (Ignore any heat losses at night.)
(f) Julie decided to warm the pool with electrical heaters. How much would it cost her to warm up her pool by 5°C? The electricity board charges 2p for 1 MJ of energy.

4 Latent Heat

During its manufacture, steel is toughened by rapid cooling. Water, evaporating from the steel's surface, takes heat away.

In the last unit you learnt that when heat energy is supplied to a substance it warms up. However, this is not always the case. Heat energy must be supplied to melt ice or to boil water. You have seen pans of water boiling on a cooker. The cooker gives heat to the water but the boiling water never gets any hotter than 100°C. The energy that is supplied is used to evaporate molecules. The energy required to turn 1 kg of a liquid at its boiling point into 1 kg of vapour is known as **the specific latent heat of vaporisation**. It is measured in joules per kilogram (J/kg).

Figure 1 shows a simple experiment that allows you to measure how much heat energy is used to boil water. A beaker of water is boiling on top of an electric balance. The heater is supplying heat energy at a rate of 1000 W. In a time of 250 s the reading on the balance dropped from 1000 g to 900 g. From this we can see that 100 g (or 0.1 kg) boiled away in that time.

How much energy is needed to turn 1 kg of boiling water at 100°C into steam at 100°C (Figure 2)?

$$\text{Heat supplied} = 1000\frac{\text{J}}{\text{s}} \times 250\,\text{s}$$

$$= 250\,000\,\text{J}$$

This boiled away 0.1 kg. So ten times as much energy, 2 500 000 J, will boil away 1 kg of water (Figure 2).

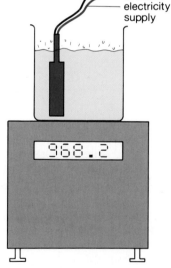

to mains electricity supply

968.2

Figure 1
An experiment to measure how much heat energy is used to boil water

⚠ Care is to be taken with this method. A safer way would be to use a kettle.

1 kg of water at 100°C

2.5 MJ of heat supplied

1 kg of steam at 100°C

Figure 2
The specific heat of vapourisation of water is 2.5 MJ/kg

Melting and casting

Casting is a very important process which is used in many different industries. For example, pipes, valves, guns and pistons are all manufactured by casting metal. Figure 3 shows you how this process works. Molten metal is poured into a mould which has been made exactly to the shape of a pipe. When the metal has solidified and cooled again the mould can be removed. Usually the moulds are made from sand which has been made rigid with a resin.

If you were in business making metal castings you would find you had a large electricity bill. You would be using a very large amount of energy to heat up and melt your metal. Figure 4 shows a graph of how the temperature of a piece of aluminium changes as heat is supplied to it in a furnace. Over the region AB the heat energy is used to warm up the aluminium. At B the aluminium begins to melt. By C the aluminium has melted completely and the temperature begins to rise again. You can see that it takes almost as long to melt the aluminium as it does to warm it up. So the energy required to melt the aluminium is large. This energy is called the **latent** (or hidden) **heat of fusion**. The *specific latent heat of fusion* of aluminium is the heat energy required to melt 1 kg of aluminium.

Ice also needs a lot of energy to melt it, which is why if you want a cold drink you will put a couple of ice cubes into your glass. Also an ice pack is a very effective way of removing heat from body tissues. If you have a bad bruise, try an ice pack to help reduce the swelling.

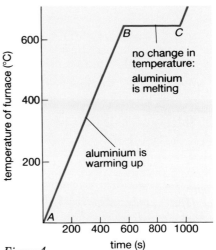

Figure 4

Figure 3
Metal casting
(a) The finished pipe

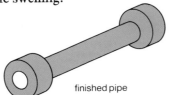

finished pipe

(b) How it is cast

Questions

1 When is it possible to supply heat to a substance without a temperature change?

2 This question refers to the aluminium that was heated in the furnace. How its temperature changed is shown in Figure 4. Some additional data is shown in the table below.

- Mass of aluminium heated = 100 kg
- Time of heating = 1000 s
- Power of heaters in the furnace = 100 kW

(a) At what temperature did the aluminium melt?
(b) How long did it take to warm the aluminium up to its melting point?
(c) How much heat was supplied to the aluminium in that time?

(d) Use the equation:
Heat supplied = $m \times s \times (T_2 - T_1)$
to calculate the specific heat capacity of aluminium.
(e) How much heat was supplied to melt the aluminium?
(f) How much heat energy is required to melt 1 kg of aluminium?
(g) Figure 4 shows how the temperature changes in an imaginary ideal furnace, which loses no heat to its surroundings. Real furnaces do lose heat. Make a copy of Figure 4 and add to it a second graph to show how the temperature would change in a real furnace.

3 Icebergs come from the Arctic or Antarctic ice caps; they are made from fresh water. It has been suggested that icebergs could be towed from the Arctic to North Africa to supply water to countries that suffer from drought.

Comment on this proposal.

4 A sample of molten wax is put in a boiling tube and allowed to cool. A graph of the temperature of the wax against time is shown below.
(a) What is the melting point of the wax?
(b) Explain why the temperature remains constant for about 15 minutes.

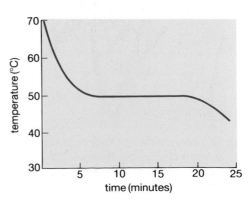

5 Conduction

When you walk around your house in bare feet you will notice that your feet feel warm as long as you stay on a carpet. But if you go into a kitchen which has tiles on a concrete floor your feet will soon feel cold. You notice this effect because the tiles are good **conductors** of heat. A good conductor of heat will carry heat away from your body quickly; this makes you feel cold.

Table 1 shows you materials that conduct heat well and those that are poor conductors or **insulators**. All metals are good thermal conductors and things like plastic, wood and air are insulators. When heat is transferred by conduction, hot atoms pass some of their kinetic energy to colder neighbouring atoms. Metals contain electrons that are free to move. When one end of a metal rod is heated, energy can be carried away from the hot end of the rod by fast moving electrons (Figure 1(a)). In a thermal insulator there are no electrons which are free to move, and heat is transferred more slowly by hot atoms (Figure 1(b)) bumping into colder atoms.

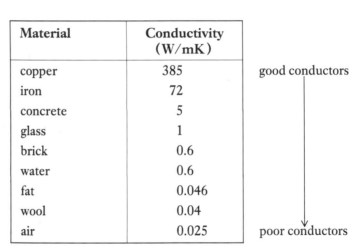

Figure 1
(a) Heat is conducted quickly by fast-moving electrons

(b) Heat is conducted slowly by atoms bumping into each other

Material	Conductivity (W/mK)	
copper	385	good conductors
iron	72	
concrete	5	
glass	1	
brick	0.6	
water	0.6	
fat	0.046	
wool	0.04	
air	0.025	poor conductors

*Table 1. This table showing **conductivities** allows you to compare how well two different materials will conduct heat. A concrete floor will take heat away from your feet far faster than a woollen carpet*

Keeping warm

If you look at Table 1 you can see that fat, wool and air are all poor conductors of heat. It is no surprise then, that these three substances are important for keeping warm-blooded animals (including us) at the right temperature. If you have ever been swimming in the North Sea you will know how rapidly you get cold. Your skin is in contact with cold water and you are losing heat by conduction. In really cold Arctic waters you could not survive more than a few minutes. Seals, however, spend all of their lives in cold water. They are protected from losing heat by conduction by a very thick layer of fat (blubber) which surrounds all of their body.

Birds have feathers and other animals have fur or hair. Fur and feathers are poor conductors of heat, but the way they reduce heat loss is by trapping a thick insulating layer of air.

We have few hairs to keep us warm. If we stand naked, moving air currents soon take heat away from us. However, wearing clothes traps air, so keeping us warm.

This song thrush has fluffed out her feathers to trap a layer of air. Air is a poor conductor of heat and so she manages to keep warm even in cold weather

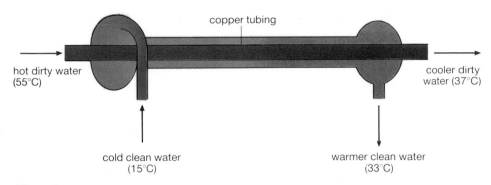

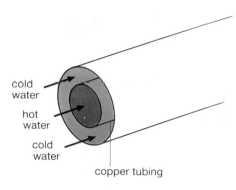

Figure 2
A heat exchanger for a laundry

Figure 3
Heat exchanger wall made of metal

Heat exchange

Figure 2 shows a double pipe heat exchanger that is being used in a laundry. Hot dirty water from the washing goes through the central pipe. Clean cold water flows round the outside of the pipe. The idea is to warm up the clean water using heat from the dirty water. This helps to reduce the laundry's heating bill.

Figure 3 shows part of the wall of another heat exchanger. The following points make sure that as much heat is removed from the hot liquid as possible:
- The wall of the container should be made from a good conductor (usually metal).
- The wall should be thin, so that heat is conducted rapidly.
- The surface area should be large to conduct more heat.
- The temperature difference between the cooling water and the hot liquid should be large.
- More heat is extracted from the hot liquid if it flows slowly through the heat exchanger.

Questions

1 Here is part of a conversation between Peter, Ramesh and Lorraine, who are talking about keeping warm in winter.

Peter: I think string vests are ridiculous, they are just full of holes. How can they possibly keep you warm?

Ramesh: I heard on the radio that it was better to wear two vests than one. How can they do any good?

Lorraine: It's not the vest that keeps you warm, it's the air underneath them.

Evaluate this conversation.

2 Opposite you can see two hot water tanks full of water at 40°C. The walls of each tank are made from the same material.

(a) One face of the small tank loses heat at a rate of 1000 J/s. In total how much heat does the small tank lose per second?

(b) Tank A has a larger surface area so it loses more heat per second. How much heat per second does this larger tank lose?

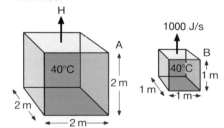

(c) For each tank work out the ratio:

$$\frac{\text{heat lost per second}}{\text{volume of tank}}$$

Which tank will cool down more quickly?

(d) In winter small animals are more likely to freeze to death than larger animals. Explain why.

3 This question refers to the laundry's heat exchanger shown in Figure 2. What effect will the following changes have on the final temperature of the clean water that is being warmed up?

(a) Iron tubing is used instead of copper.

(b) Thinner copper tubing is used for the wall of the heat exchanger.

(c) The length of the heat exchanger is increased.

(d) The hot dirty water is made to run faster.

4 Why do power stations need heat exchangers?

6 Convection

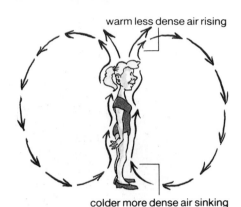

Figure 1
Convection currents near a person in a room

In July 1987 Richard Branson made an adventurous crossing of the Atlantic travelling in a hot air balloon. After a 3000 mile journey from America he landed in the Irish Sea. To make a hot air balloon fly you need a burner which heats up the air inside the balloon. When air is heated it expands and its density becomes less. When hot air is surrounded by colder, more dense air it rises. This is the same principle as a cork floating to the surface of water.

Figure 1 shows how air circulates if you stand, with most of your clothes off, in the middle of a room. Your body heats up air next to it, which then rises. Colder air flows down near the walls to replace the warmer air. The currents of air that flow are called **convection currents**.

Heat can be transferred by convection in *liquids* and *gases*, but not in solids. Although most liquids and gases do not transfer heat very well by conduction, they do transfer heat quickly by convection. Convection currents allow large quantities of a hot liquid or gas to move and give heat to a colder part. If we wore no clothes our bodies would lose lots of heat by convection. By wearing clothes we trap a layer of air which acts as an insulator.

Exactly the same idea is used to reduce heat losses from your house. Your loft is full of air and convection currents there can cause a large amount of heat loss. Lofts can be insulated with felt or glass fibre. To stop convection in the cavity between inner and outer walls, it is possible to pump in polystyrene foam. The foam has lots of air trapped in it and so is a poor conductor of heat. Double glazing also traps air, in a layer between two pieces of glass (Figure 2).

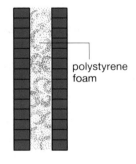

Figure 2
(a) Cavity wall insulation

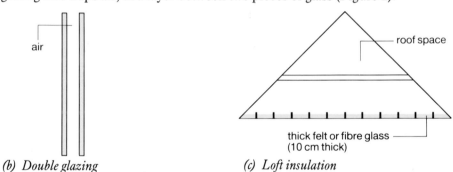

(b) Double glazing

(c) Loft insulation

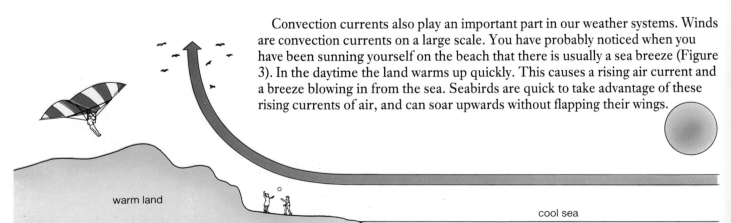

Convection currents also play an important part in our weather systems. Winds are convection currents on a large scale. You have probably noticed when you have been sunning yourself on the beach that there is usually a sea breeze (Figure 3). In the daytime the land warms up quickly. This causes a rising air current and a breeze blowing in from the sea. Seabirds are quick to take advantage of these rising currents of air, and can soar upwards without flapping their wings.

Figure 3
The large specific heat capacity of the sea means that it stays cool. The land warms quickly, and warm air rises above the land

Convection in water

When water is heated it too will transfer heat energy by convection. Figure 4 shows a simple way that you can show this in a laboratory. Place some potassium permanganate crystals at the bottom of a flask of water. When heat is supplied from underneath the dissolved potassium permanganate rises showing the path of the currents.

It is possible to circulate water around the heating system of a house using convection currents. Figure 5 shows such a heating system. The boiler must be put at the lowest point in the house. The hot water rises upwards to the roof. The water then feeds the radiators and finally the cool water is returned to the boiler.

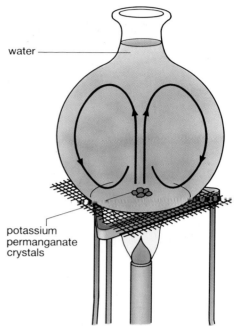

Figure 4
A laboratory experiment to show convection currents in water

⚠️ Potassium permanganate is dangerous; wash your hands after use.

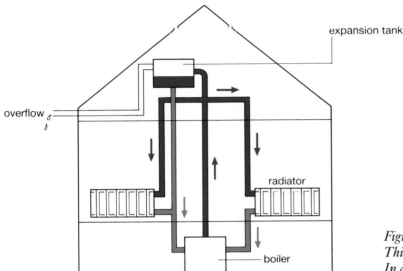

Figure 5
This system of circulating water is suitable for a small house only. In a large house a pump is used to help push the water round

Questions

1 Heat cannot be transferred by convection in a solid. Explain why.
2 Figure 3 shows the direction of a sea breeze in daytime. At night, land cools down more quickly than the sea. Draw a diagram to show the direction of the sea breeze at night.
3 About a hundred years ago in Cornish tin mines, fresh air used to be provided for the miners through two ventilation shafts shown below.
(a) To improve the flow of air, a fire was lit at the bottom of one of the shafts. Explain why.
(b) One day, Mr Trevethan, the owner of the mine, had an idea. To make the ventilation even better he lit another fire at the bottom of the second shaft. Comment on this idea.

4 This question is about the cost of insulating a house. At the moment the house has no loft insulation, cavity wall insulation or double glazing.
(a) Use the data provided to decide which method of insulation provides the best value for money.
(b) How many years do you have to wait before the double glazing has paid for itself?

- Yearly heating bill for house: £700
- Cost of double glazing: £3000
  This would reduce the fuel bill by 20% a year
- Cost of loft insulation: £450
  This would reduce the fuel bill by 30% a year
- Cost of cavity wall insulation: £1200
  This reduces the fuel bill by 25%

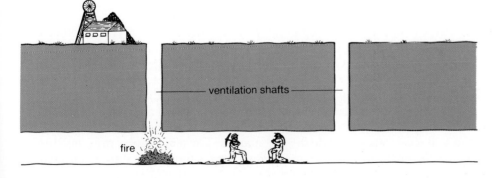

7 Radiation

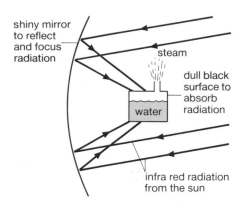

Figure 1
A solar furnace

Radiation is the third way by which heat energy can be transferred from one place to another. Heat energy from the Sun reaches us by radiation. The Sun emits electromagnetic waves which travel through space at high speed. Light is such a wave, but the Sun also emits a lot of **infra-red** waves (or infra-red radiation). Infra-red waves have longer wavelengths than light waves. It is the infra-red radiation that makes you feel warm when you lie down in the sun. Infra-red radiation travels at the same speed as light; as soon as you see the Sun go behind a cloud you feel cooler.

Good and bad absorbers

Infra-red radiation behaves in the same way as light. It can be reflected and focussed using a mirror. Figure1 shows the idea behind a solar furnace. The shiny surface of the mirror is a poor absorber of radiation, but a good reflector. Radiation is absorbed well by dull black surfaces. So the boiler at the focus of the solar furnace is of a dull black colour.

Good and bad emitters

Figure 2 shows an experiment to investigate which type of teapot will keep your tea warm for a longer time. One pot has a dull black surface, the other is made out of shiny stainless steel. Radiation that is emitted from the hot teapots can be detected using a thermopile and a sensitive ammeter. When you do the experiment you will find that the black teapot emits more radiation than the shiny surface. So a shiny teapot will keep your tea warmer than a black teapot.
- Black surfaces are good absorbers and good emitters of radiation
- Shiny surfaces are bad absorbers and bad emitters of radiation

Figure 2
A sensitive instrument called a thermopile can detect radiation

This butterfly loses heat very rapidly because it has a very large surface area and a relatively small volume. To warm itself up it opens its wings to absorb sunlight. To cool off, it folds them shut

The greenhouse effect

In Italy, where the average temperature in summer is about 5°C higher than in Britain, tomatoes grow very well. It is a great help to a tomato grower in Britain if he uses a greenhouse. On a warm day the temperature inside a greenhouse can be 10°C or 15°C higher than outside (Figure 3). Infra-red radiation from the sun passes through the glass of the greenhouse, and is absorbed by the plants and soil inside. The plants radiate energy, but the wavelength of the emitted radiation is much longer. The longer wavelengths of radiation do not pass through the glass and so heat is trapped inside the greenhouse. The temperature rises until the loss of heat through the glass by conduction balances the energy absorbed from the Sun.

Some people worry that a similar greenhouse effect could be happening in our atmosphere. As we continue to burn fossil fuels we are filling our atmosphere with carbon dioxide and other chemicals. As these chemicals absorb long wavelength radiation emitted from the Earth's surface, the average temperature of the Earth increases. The planet Venus shows the greenhouse effect in a big way. Its atmosphere is mostly carbon dioxide, and its average surface temperature is about 460°C, hot enough to melt some metals.

Figure 3

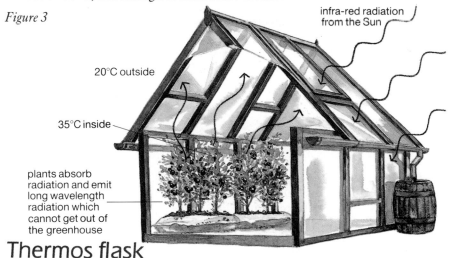

infra-red radiation from the Sun

20°C outside

35°C inside

plants absorb radiation and emit long wavelength radiation which cannot get out of the greenhouse

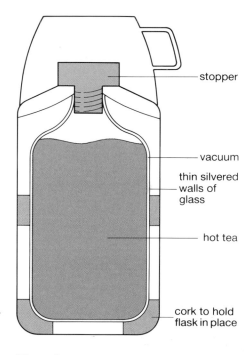

stopper

vacuum

thin silvered walls of glass

hot tea

cork to hold flask in place

Figure 4
A thermos flask

Thermos flask

A thermos flask keeps things warm by reducing heat losses in all possible ways. The flask is made with a double wall of glass and there is a vacuum between the two walls. Conduction and convection cannot take place through a vacuum. The glass walls are thin, so that little heat is conducted through the glass to the top. Heat can be radiated through a vacuum, but the glass walls are silvered like a mirror so that they are poor emitters of radiation. The stopper at the top prevents heat loss by evaporation or convection currents (Figure 4).

Questions

1 Some casserole dishes that are used in ovens are black, but the outside of an electric kettle is shiny. Can you explain why?

2 Tracy and Abdul are talking about greenhouses. Neither of them has studied much physics, so they are not too sure how greenhouses work. Read their conversation and correct any errors in their thinking.

Tracey: I saw that Mr Brown put some aluminium foil up inside his greenhouse. I know that aluminium is a good conductor, so he has probably put it there to conduct more heat into his greenhouse.

Abdul: No. I don't agree. I think he has put the foil there to keep the greenhouse cooler, but I don't understand why it works.

3 This question is about controlling the temperature of a greenhouse. The greenhouse can lose heat by radiation and conduction when its windows and doors are closed. Heat can also be lost by convection when a window is opened. Information about heat losses on a particular day is shown in the graph. On this day the greenhouse is absorbing 10 kW of power from the Sun.
(a) Explain why the graph (opposite) shows that the air temperature outside the greenhouse is 20°C.
(b) At what rate must the greenhouse be losing heat if the temperature inside it is constant?

(c) At 38°C how much heat is lost by:
(i) radiation, (ii) conduction? Explain why 38°C is the steady temperature with the door and windows closed.
(d) The window is now opened. Use the graph to work out the new steady temperature of the greenhouse.

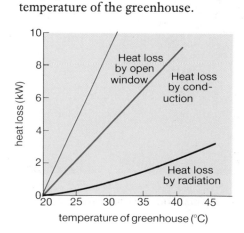

Heat loss by open window

Heat loss by conduction

Heat loss by radiation

heat loss (kW)

temperature of greenhouse (°C)

8 Warming-Up Exercises

- Ground floor area = 100 m²
- Ceiling area = 100 m²
- Area of brick wall = 200 m²
- Area of windows and doors = 40 m²

U-values (W/m²°C)

Uninsulated house	Uninsulated roof	2
	Cavity wall	1
	Floor without carpets	1
	Windows, single glazed doors	5
Insulated House	Insulated roof	0.3
	Cavity filled wall	0.5
	Floor with carpets	0.3
	Windows, double glazed doors	2.5

Keeping warm in the house

Suppose you are a heating engineer who wants to know what size of central heating system to install in a house. You can calculate the heat losses from a house using **U-values**. If a material transfers heat well it has a *high* U-value; if the material is a good insulator it has a *low* U-value. You can see some values for U-values above. An engineer can calculate the rate of loss of heat using this equation:

$$\text{Rate of heat loss} = U\text{-value} \times \text{area} \times \text{temperature difference}$$
$$(\text{Watts})$$
$$= \frac{W}{m^2 \, °C} \times m^2 \times °C$$

We can use the data provided to calculate the rate at which heat is lost from a poorly insulated house. The outside temperature is 0°C and the house is to be kept at 15°C.

Heat loss per second, $H = U \times A \times T$

- Ceiling area: $H = 2 \times 100 \times 15 = 3000\,\text{W}$
- Walls: $H = 1 \times 200 \times 15 = 3000\,\text{W}$
- Floor: $H = 1 \times 100 \times 15 = 1500\,\text{W}$
- Windows: $H = 5 \times \ \ 40 \times 15 = 3000\,\text{W}$

Total 10 500 W

So you would need to install a heating system capable of producing about 10 kW. If the house owner used a gas central heating system continuously at this rate it would cost about £25 per week.

Keeping cool

If you have ever done any long-distance running, you will know that you get very hot indeed and that your efforts cause you to sweat a lot. You will see from the data provided on the next page that marathon runners can suffer severe problems due to temperature changes of the body or loss of water (dehydration).

Double glazing is needed to reduce heat loss in buildings that have large areas of glass

- First we will use the data (bottom right) to calculate how much water one runner, Sarah, loses in a three hour marathon when the temperature is 18°C. At this temperature she loses heat due to evaporation at a rate of 600 W.

  Over 3 hours the heat lost is:

  $$\text{Heat lost} = 600\frac{J}{s} \times (3 \times 3600)s$$

  $$= 6\,500\,000\,J$$

  $$= 6.5\,MJ$$

  But 2.5 MJ of heat evaporates 1 kg of water, so she loses $\frac{6.5}{2.5} = 2.6$ kg of water in the race.

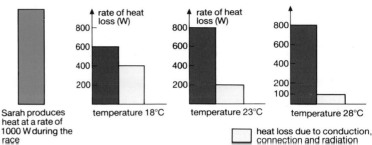

Figure 1

- When marathon runners finish a race they are wrapped up in bags to stop them cooling too quickly. When Sarah finishes her race and rests, her body will only be *producing* heat at a rate of 100 W, but she is still *losing* it at a rate of 1000 W (600 W due to evaporation, 400 W due to other processes). So she loses heat from her body at a rate of 900 W. If this is allowed to happen even for 10 minutes, she could become very cold. In 10 minutes:

  $$\text{heat lost} = 900\,J \times (10 \times 60)s = 540\,000\,J$$

  We can calculate her loss in temperature using:

  $$\text{Heat lost} = \text{mass} \times \text{specific heat capacity} \times \text{temperature drop}$$

  $$so\ \text{temperature drop} = \frac{H}{ms} = \frac{540\,000\,J}{60\,kg \times 4200\,J/kg°C}$$

  $$= 2.1°C$$

Questions

House heating

1 Use the data provided to work out the rate of heat loss from the same house, if it is insulated well.

2 Some people think that it is more important to insulate the walls than to put in double glazing. This seems sensible because the windows are small. Comment on this idea.

3 How much would it cost to heat a well-insulated house per week, if the heating runs for only six hours per day? Assume that the temperature difference between inside and outside is still 15°C.

Marathon running

4 Explain why as the air temperature rises, a marathon runner loses more heat by evaporation of sweat, and less by conduction, convection and radiation.

5 If Sarah loses as much as 3 kg of water, she will suffer severe effects from dehydration.
(a) Explain why she could be in danger if the temperature is more than 23°C.
(b) What action can a runner take to avoid dehydration?

6 (a) What will happen to Sarah's body temperature if she runs when the air temperature is 28°C? Assume that she has no means of cooling other than those suggested in the graphs.
(b) How much extra heat energy is she producing: (i) per second, (ii) per hour?
(c) Calculate by how much her body temperature will have increased after one hour.
(d) Explain why Sarah is unlikely to finish a marathon if the temperature is as high as 28°C.

Keeping warm at the end of a marathon. The shiny blanket is a poor radiator of heat

Marathon running data

- Mass of runner = 60 kg
- Specific heat capacity of water = 4200 J/kg°C
- Specific latent heat of vaporisation of water = 2.5 MJ/kg
- At rest an athlete produces heat energy at a rate of 100 W
- Average body temperature = 37°C
- The body can only work normally between 33°C and 41°C

SECTION F: QUESTIONS

1 An electric heater, which has a power rating of 6 kW, is used to produce hot water for a bathroom shower. Cold water from the main water supply flows directly to the shower heater at a temperature of 10°C. Water flows through the shower at a rate of 2.4 litres per minute.

(a) Calculate the temperature of the hot shower water.
(1 litre of water requires 4200 J to warm it through 1°C.)
(b) Calculate the cost of the shower, if electricity costs 7p per kWh. Assume a shower lasts 5 minutes.
(c) Another shower has a heater rated at 5 kW, but the water flowing has a temperature of 50°C. Why can this happen?
(d) What happens to the water temperature if the water pressure drops?
(e) Explain any safety features that should be built into the shower.

2 This question is about the comparative costs of heating a bath by using gas and electricity. Below you can see an old gas bill charging the householder £128.47 for 290 therms of gas. The 'therm' is a unit of energy. The calorific value of the gas gives the amount of energy released when 1 m³ of gas is burnt; this is 38.5 MJ/m³ or 1032 Btus per cubic foot. A 'Btu' is a 'British thermal unit'.

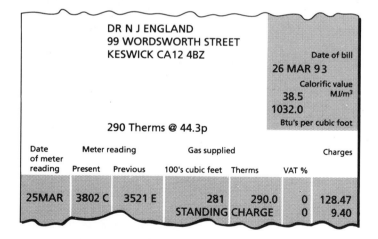

(a) How many different units of energy appear on the gas bill?
(b) Use the information on the bill to show that 10^5 Btu equal 1 therm.
(c) For this part of the question you need to know that 1 m³ = 35.3 ft³.
Work out the calorific value of the gas in:
(i) MJ/ft³, (ii) Therms/ft³.
(iii) Now show that 1 therm equals 106 MJ.
(iv) Calculate the cost of providing 1 MJ of energy by burning gas.
(d) (i) A bath requires about 100 kg of water. We will assume that the water needs to be warmed from 10°C to 50°C. Calculate the energy required to warm the water.

(1 kg of water needs 4.2 kJ to warm it through 1°C. Give your answer in MJ.)
(ii) Calculate the cost of heating using gas. Assume the gas boiler is 60% efficient.
(iii) Calculate the cost of heating the water using electricity. Assume the immersion heater warms the water with 100% efficiency. The cost of electricity is 6.6p per kWh.
(iv) Why is electrical heating more efficient than gas heating?
(e) The gas board now charges 1.477p per kWh. How does this compare with the old price?

3 The diagrams show a simple solar panel. Water is pumped slowly through thin copper tubing which has been painted a dull black colour. The tubing has been embedded in an insulating material.

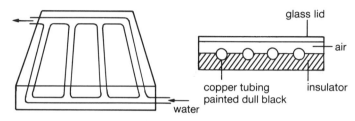

(a) Why is thin copper tubing used and why has it been painted black?
(b) Why does a glass lid improve the efficiency of the solar panel?
(c) Which material would you choose to insulate the bottom of the panel?
(d) The graph shows how the power of sunlight falling on 1 m² of solar panel varies throughout a bright summer's day.
(i) Explain the variation through the day. What do you think happened round about noon?
(ii) Use the graph to estimate the total energy absorbed by the panel during the day. Assume the panel absorbs 20% of the incident radiation; express your answer in kWh.
(iii) A typical house needs about 10 kWh of energy per day. Estimate the size of panel needed to supply energy to a house through the summer months.
(c) Comment on the limitations of solar power in Britain.

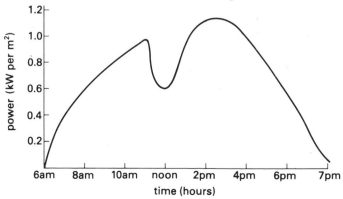

4 Lucy finishes watching television at midnight and goes to bed. She switches off the central heating as she goes. The graph shows how the temperature of the house fell during the night; the average temperature outside was 4°C.

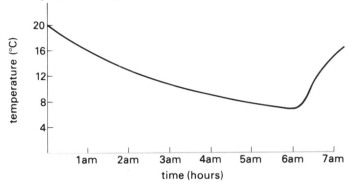

(a) Explain the shape of the graph in as much detail as possible.

(b) The thermal conductance of the house is defined using this equation: Power loss through walls (or ceiling etc.) is P.

$$P = \text{thermal conductance } (C) \times \text{temperature difference across the walls } (T_2 - T_1)$$

$$P = C \times (T_2 - T_1)$$

The thermal conductance of the house is 0.3 kW/°C. Explain what this means.

(c) What is the rate of loss of heat from the house when the temperature is (i) 18°C, (ii) 11°C?

(d) Use the graph to calculate the rate of temperature drop in °C/h at temperatures of 18°C and 11°C. Comment on your answers.

5 Mr Adams is a neurosurgeon. Recently he had to operate on Polly. She had a giant aneurysm. An aneurysm is like a balloon that has blown out from the side of an artery in the brain. Aneurysms occur when the wall of the artery is weak. They are very dangerous since they could burst and bleed into the brain.

Polly's aneurysm was so large that it was necessary to stop the blood flowing into the brain. At normal body temperatures the brain dies in a few minutes without a blood supply. But at 15°C the brain can last for nearly an hour without a blood supply. To cool her down Polly's blood was passed through a heart/lung machine. The graph shows how her body temperature changed during the day. The operation was successful and Polly has now recovered completely.

(a) Use the graph to work out how long it took Mr Adams to remove the aneurysm.

(b) Between 1420 h and 1520 h Polly was warmed up again. By how much did her temperature rise in that time?

(c) Polly's mass is 60 kg. How much heat was needed to warm her up again? Assume that 4000 J of heat energy warms up 1 kg of Polly by 1°C.

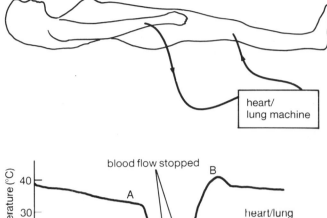

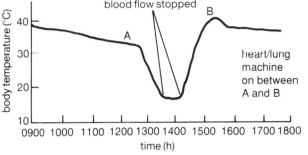

(d) What was the power of the heater in the heart/lung machine?

6 Meteors are small pieces of matter made mostly of iron. Like the Earth, meteors are in orbit around the Sun. Meteors travel very quickly and can cause a lot of damage when they hit something. The craters on the Moon were made by meteors. Fortunately few meteors hit the surface of the Earth. This is because our atmosphere slows them down in a very short time. Some data about a meteor are shown below.

- Mass of meteor = 0.01 kg
- Speed of meteor entering atmosphere = 30 000 m/s
- Specific heat capacity of iron = 500 J/kg°C

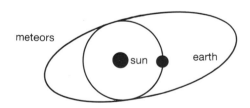

(a) Calculate the kinetic energy of the meteor as it enters the atmosphere. Use the formula:

Kinetic energy $= \frac{1}{2} \times \text{mass} \times (\text{speed})^2$

(b) Explain why the meteor slows down in the atmosphere. Andrew and Kate are having a discussion about meteors. Andrew says: 'the kinetic energy of the meteor is turned into heat energy. I worked out that the final temperature of the meteor is about 900 000°C, using the equation:

energy transformed = mass × specific heat capacity
× temperature change.'

(c) Use the data to show how Andrew got this answer. What assumption did he make?

Kate says: 'I read in a book that a meteor only reaches a temperature of about 20 000°C in the atmosphere. Your answer must be wrong. This is because you have forgotten that the meteor is losing heat.'

(d) Discuss what Kate says. Can you think of ways that the meteor will lose heat? What else will happen to the surface of the meteor at that temperature?

(e) Explain why very few meteors reach the surface of the Earth.

7 The time taken to cook an egg is:
- proportional to the mass of the yolk,
- proportional to the distance between the yolk and the shell,
- inversely proportional to the surface area of the shell.

Elaine decides to cook an ostrich egg for breakfast. Ostrich eggs are 3 times as long as hens' eggs. Elaine likes her hens' eggs boiled for 5 minutes. How long should she boil the ostrich egg for?

8 Diagram (i) illustrates an instrument used to measure the time that the Sun shines during the day. The blackened glass bulb contains mercury and is supported inside an evacuated glass case. Diagram (ii) shows how the connecting wires are arranged inside tube A.

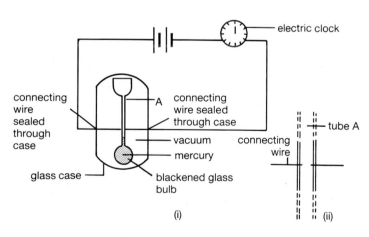

(i)

(ii)

(a) How does energy from the Sun reach the mercury? Give a reason for your answer.

(b) Explain why the clock starts when the Sun shines.

(c) Why is tube A of small cross-sectional area?

(d) Explain why blackening the bulb ensures that the mercury level falls rapidly when the Sun ceases to shine.

MEG

9 The diagram shows a circuit in which an immersion heater is used to heat a metal block. The heater supplies energy at the rate of 50 W.

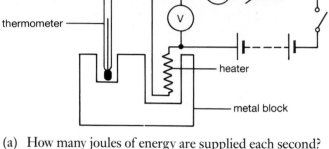

(a) How many joules of energy are supplied each second?

(b) The heater is switched on for 300 seconds. During this time the temperature rises from 20°C to 50°C.

(i) Calculate the energy supplied.

(ii) If the mass of the block is 2 kg, calculate the energy required to raise the temperature of 1 kg of the material by 1°C.

(iii) What name is given to the quantity you have calculated in (b)(ii)?

(c) When the experiment is performed as shown above, the measured rise in temperature is smaller than expected.

(i) Why is the measured rise smaller?

(ii) How would you change the apparatus to make the measured rise in temperature closer to the expected value?

NEAB

10 In an experiment in which equal masses of lead and a lead-tin alloy called solder were allowed to cool, the following cooling curves were obtained.

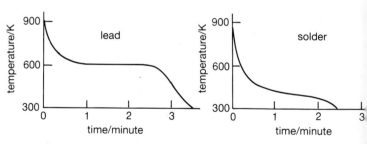

(a) For the lead:

(i) what was the physical state of the lead after 30 seconds,

(ii) what was happening to the lead at 600 K,

(iii) what is 600 K in Celsius units,

(iv) why did the temperature of the lead remain constant at 600 K for some time?

(b) State **two** important differences between the lead and the solder which can be deduced from these cooling curves.

(c) Solder is often used by plumbers to repair burst pipes by filling the fracture with solder. Why is solder better than lead for this purpose?

NEAB

SECTION G

Waves and Sound

Everyone knows about water waves, but did you know that your vision and hearing also depend on waves?

1 Introducing Waves

Waves do two important things; they carry energy and information. You have seen ocean waves crashing into a sea wall at high tide. Those waves certainly carry energy.

When you watch television you are taking advantage of radiowaves. These waves carry energy and information from the transmitting station to your house. Light and sound waves carry energy and information from the television set to your eyes and ears.

Waves on slinkies

One of the best ways for you to learn about waves is to see them moving along on a stretched 'slinky' or spring. Figure 1 shows a slinky lying on the floor. When you move your hand from side to side some humps move away from you along the slinky. Although the wave moves along the slinky, the movement of the slinky itself is from side to side. If you tie a piece of string to the slinky, you will see that it moves in exactly the same way as your hand did to produce the waves. This sort of wave is called a **transverse wave**. The particles carrying the wave in the slinky move at right angles to the direction of wave motion. Water ripples on the surface of a pond and light waves are examples of transverse waves.

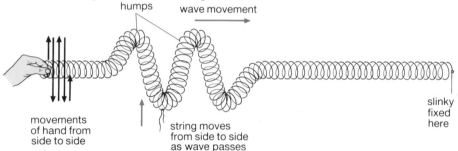

Figure 1
A transverse wave moving down a slinky

You produce a different kind of wave when you move your hand backwards and forwards along the slinky (Figure 2). Your hand compresses and then expands the slinky. The wave is made up of compressions and expansions which move along the slinky. This time a piece of string tied to the slinky moves backwards and forwards along it. Again, this is how your hand moved to produce the waves. This sort of wave is called a **longitudinal wave**. The particles carrying the wave in the slinky move backwards and forwards along the direction of wave motion. Sound waves are longitudinal.

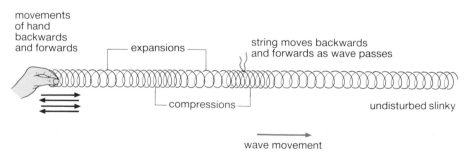

Figure 2
A longitudinal wave moving down a slinky

Aerials transmit radiowaves. These convey energy and information ...

... which your television receives

Sea waves can carry a lot of kinetic energy, particularly during a storm. This is released when the waves break against a sea wall, and the wall can be badly damaged

Describing waves

Figure 3 is a graph showing the displacement of a slinky along its length at one moment. The arrows on the graph show the direction of the motion of the slinky; a larger arrow means a larger speed.

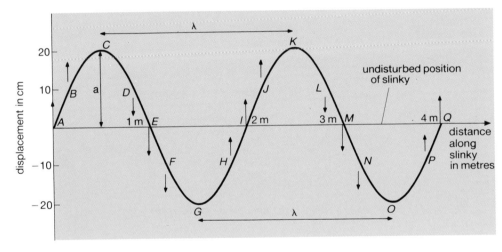

Figure 3
The displacement of a slinky along its length. The arrows on the graph show the direction of motion of the slinky

- **Phase.** Points B and J are moving in phase. They are moving in the same direction, with the same speed. They also have the same displacement away from the undisturbed position of the slinky. F has been displaced in the opposite direction to B and J, and is moving in the opposite direction. F is out of phase with B and J. However, F moves in phase with N.
- The **wavelength** of a wave motion is the shortest distance between two points that are moving in phase. You can think of a wavelength as the distance between two humps. We use the greek letter λ (*lambda*) for the wavelength.
- The **amplitude** of a wave is the greatest displacement of the wave away from its undisturbed position. You can think of the amplitude as the height of a hump.
- The **frequency**, f, of the wave is the number of complete waves produced per second. There are two complete waves in Figure 3. The unit of frequency is waves per second or *hertz* (Hz). 1 kHz means 1000 Hz.
- The **time period** of a wave, T, is the time taken to produce one complete wave.

Explosions produce shock waves which can be used to knock down buildings, as you can see in the photograph above. Sometimes the shock wave travels to other buildings, where it can smash windows

Wave velocities

The **velocity** of a wave, v, is the distance travelled by a wave in one second. The velocity of waves down a particular slinky is the same for all wavelengths. Figure 4(a) shows waves moving on a slinky with frequency 3 Hz and wavelength 0.4 m. In one second three waves have been produced, so the distance travelled by the first wave is $3 \times 0.4 = 1.2$ m. The wave velocity is 1.2 m/s. For any wave (see Figure 4(b)) we can calculate the wave velocity using the formula:

$$\text{Velocity} = \text{frequency} \times \text{wavelength}$$
$$V = f \times \lambda$$

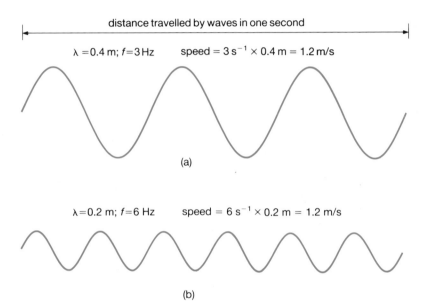

distance travelled by waves in one second

$\lambda = 0.4$ m; $f = 3$ Hz   speed $= 3\,\text{s}^{-1} \times 0.4\,\text{m} = 1.2\,\text{m/s}$

(a)

$\lambda = 0.2$ m; $f = 6$ Hz   speed $= 6\,\text{s}^{-1} \times 0.2\,\text{m} = 1.2\,\text{m/s}$

(b)

Figure 4
The speed of waves along a slinky is the same for all wavelengths

Questions

1 Explain the difference between longitudinal and transverse waves.

2 This question refers to the graph in Figure 3, on the previous page.
(a) What is the wavelength of the wave?
(b) What is the amplitude of the wave motion?
(c) The frequency of the wave motion is 2 Hz. What is the time period of the wave?
(d) Calculate the speed of the wave.
(e) Give a point moving in phase with:
(i) I, (ii) B, (iii) M.

(f) Make a sketch of the wave motion in Figure 3. Use the arrows, showing the direction of movement of the particles in the slinky, to draw in the position of the slinky a short time later.
(g) In which direction is the wave moving?

3 Make a sketch of a longitudinal wave of a slinky, and mark in a distance to show one wavelength.

4 A radio station produces waves of frequency 200 kHz and wavelength 1500 m.

(a) What is the speed of radio waves?
(b) Another station produces waves with a frequency 600 kHz. What is their wavelength?

5 Why do the wave pulses shown below contain some information?

2 Water Waves

You can learn more about the nature of waves by studying ripples that move over the surface of water. It is easy to show that these ripples are transverse waves. If you look at a floating cork you will see it bob up and down as ripples travel along the surface of the water (Figure 1).

You can produce water waves in a ripple tank, by lowering a dipper into the water (Figure 2). A motor vibrates the dipper up and down to produce waves continuously. A beam of wood produces straight waves, and a small sphere produces circular waves. If you shine a light from above the tank you will see bright and dark patches on the screen below. These patches show the positions of the crests and troughs of the waves.

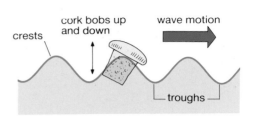

Figure 1
Ripples on water are transverse waves

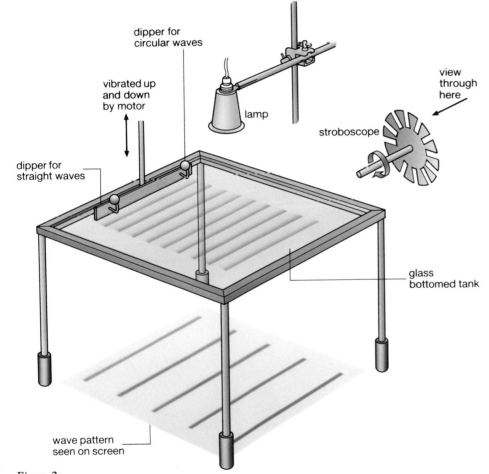

Figure 2
A ripple tank

Rain drops cause ripples to spread out in circles

Using a stroboscope

Water waves move quite quickly and it can be difficult to see them. However, if you look through a rotating stroboscope you can make the waves appear stationary. The stroboscope is a disc with 12 slits in it. If you rotate the disc twice a second you will see the ripple tank 24 times a second. The waves will appear stationary when the dipper produces waves with a frequency of 24 Hz. Each time there is a slit in front of your eye one wave has moved forwards to the position of the next wave.

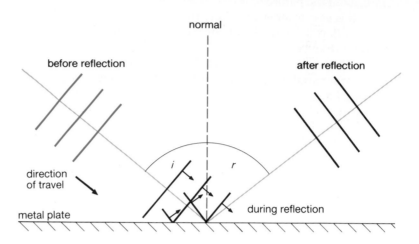

Figure 3
Reflection of waves off a plane surface; angle of incidence, i = angle of reflection, r

Reflection

In Figure 3 you can see some waves approaching a straight metal barrier in a ripple tank. On the diagram, a line is drawn at right angles to the surface of the barrier. This line is called the **normal**. The angle between the normal and the direction of travel before reflection is called the **angle of incidence**, i. The angle between the normal and the direction of travel after reflection is called the **angle of reflection**, r. When waves are reflected, i always equals r. Figures 4 and 5 show some further examples of water waves being reflected.

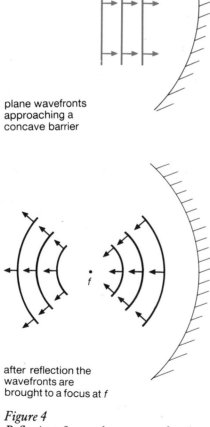

plane wavefronts
approaching a
concave barrier

after reflection the
wavefronts are
brought to a focus at f

Figure 4
Reflection of waves by a concave barrier

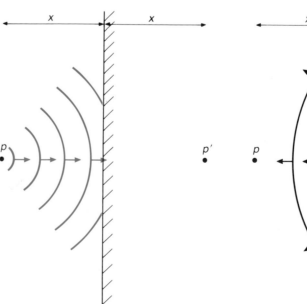

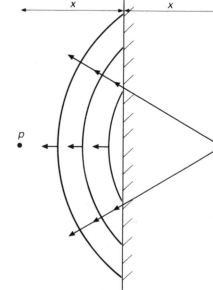

Figure 5
(a) Circular wavefronts spread out from a point

(b) After reflection, the waves appear to have come from a point behind the barrier

Refraction

In Figure 6 you can see some waves going from a region of deep water to shallow water. A region of shallow water in a tank can be made with a glass plate. In shallow water, waves travel more slowly than they do in deep water. There must be the same number of waves passing through both the deep and shallow regions. This means the frequency of the waves is the same. The speed of the waves is given by the equation $V = f \times \lambda$. So when the wave slows down in shallow water the wavelength must be less. In Figure 7 you can see some waves approaching a region of shallow water at an angle. They slow down and change direction. This is called **refraction**.

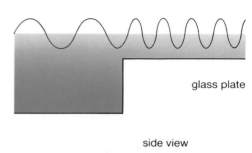

glass plate

side view

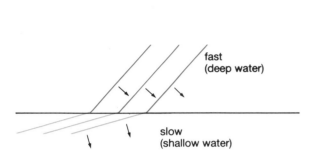

Figure 7
Water waves are refracted when they enter shallow water

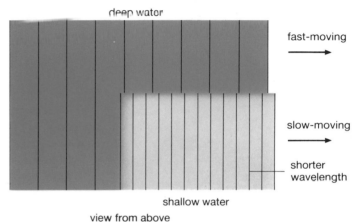

shallow water
view from above

Figure 6
Waves going from deep water to shallow water

Questions

1 Draw careful diagrams to show how the waves are reflected in the following cases.

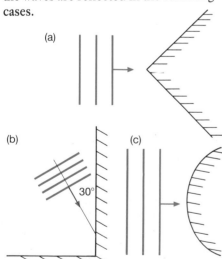

2 At the top of the next column is a bird's eye view of a ripple tank. The tank is tilted so that there is deep water one end, and shallow water at the other.

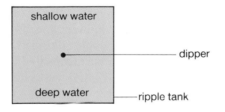

Draw a diagram to show the shape of the ripples produced by the dipper in the middle.

3 Draw careful diagrams to show how the waves are refracted in the following cases.

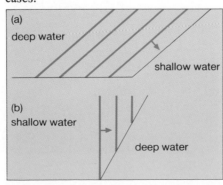

4 A propellor with two blades rotates twenty times a second. You look at it through a hand stroboscope with 10 slits.
(a) Explain why the propellor looks stationary if you rotate your stroboscope twice a second.
(b) What will you see if you rotate your stroboscope at these frequencies:
(i) 1 Hz (ii) 4 Hz (iii) 2.1 Hz?

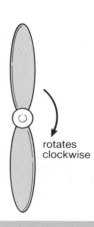

rotates clockwise

3 Diffraction and Interference

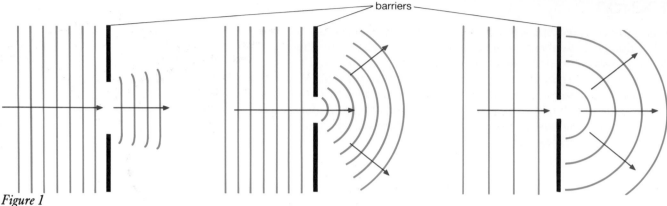

barriers

Figure 1
(a) Small wavelength, large gap   *(b) Small wavelength, smaller gap*   *(c) Large wavelength, small gap*

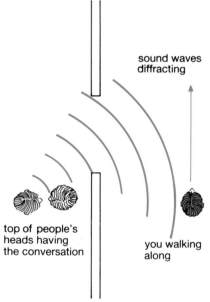

sound waves diffracting

top of people's heads having the conversation

you walking along

Figure 2
A 'bird's eye view' of a conversation overheard through a doorway

Diffraction

When waves pass through a small hole, they spread out. This is called **diffraction**. You can see the diffraction of water waves in a ripple tank. Figure 1 shows you what happens when waves go through a series of gaps in barriers.

In Figure 1(a) the gap is large in comparison with the wavelength of the waves. The waves only spread out a little. In Figure 1(b) the gap is smaller. The waves spread out more. In Figure 1(c) the wavelength is larger than the gap and the waves now spread out completely.

Sound also diffracts. This tells us that sound is carried by waves. Figure 2 shows the sort of position you might find yourself in. You are walking down a corridor and you can hear two people talking though an open door, but you cannot see them. The sound waves do not travel in a straight line. They spread out and change direction as they pass through the doorway. That is why you can hear them talking even though you are a long way down the corridor.

We also think that light is carried by waves. So why can't we see round corners? You cannot see the people talking in Figure 2, because light waves have very small wavelengths. This means that when light waves go through a doorway they hardly diffract at all. Sound waves have much longer wavelengths, so they diffract through the doorway.

In this photograph you can see water waves diffracting as they pass through a harbour entrance. Water waves have a large wavelength, and so they spread out a lot when they diffract

Interference

Some interesting patterns can be produced in a ripple tank when two small dippers make waves together. The two dippers produce a series of crests and troughs. These spread out and overlap as shown in Figure 3.

There are some places where the waves from the two dippers arrive *in phase*. This means that two crests or two troughs arrive at the same time. The two crests move the water upwards to make a larger wave. So in some places, the water moves up and down *more* than it does with only one dipper working. We call this **constructive interference**.

At other places in the ripple tank, waves from the two dippers are arriving *out of phase*. This means that when a crest arrives from one dipper a trough arrives from the other. This time the effect of the two waves is to cancel each other out. The water does not move at all; it is as if the dippers have been switched off. We call this **destructive interference** (Figure 4).

A ripple tank set up to show the interference of waves

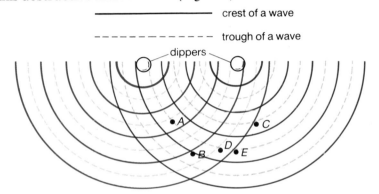

Figure 3

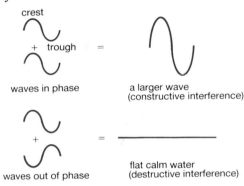

Figure 4

Questions

1 Astronomers use radio telescopes to detect radio waves from distant galaxies. A lot of radio waves have wavelengths a few metres long. Radio telescope dishes are about 100 m across. This means that the radio waves are only diffracted by a small amount. Explain why.

2 In an experiment in a ripple tank, straight water waves are produced with a frequency of 20 Hz. The waves travel at a speed of 40 cm/s through a gap in a barrier of width 1 cm.
(a) Calculate the wavelength of the waves.
(b) Draw a diagram to show how the waves would spread through the gap. Explain how you decided what to draw.
(c) Draw another diagram to show what happens when the gap is made 4 cm wide.

3 In Figure 3 which of the points *A, B, C, D, E* will show (i) constructive interference (ii) destructive interference?
4 The two diagrams below show slinkies with moving humps on them. Draw diagrams to show the shape of the slinkies (i) when the humps meet (ii) after the humps have passed through each other.

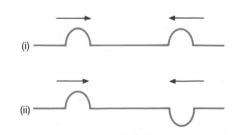

5 Use the information below to sketch a graph of the displacement of the point *A* against time. Your time axis should cover the next two seconds after the instant shown in the diagram.

- The wavelength of both sets of waves is 0.5 m.
- The amplitude of the waves is 0.2 m.

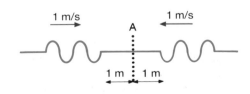

4 Sound Waves

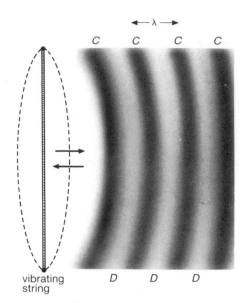

C = compression
D = decompression

Figure 1
A vibrating guitar string

Speeds of sound (m/s)	Material
330	air
1500	water
5000	steel

Table 1

The bell in the jar will make a noise as long as the jar is full of air. If we pump the air out, we cannot hear the noise any more, but we can still see the bell working

Making and hearing sounds

When you pluck a guitar string the instrument makes a noise. If you put your finger on the string you can feel the string vibrating. Sounds are made when something vibrates. The vibrations of a guitar string pass on kinetic energy to the air. This makes the air vibrate.

Sound is a longitudinal wave. Molecules in air move backwards and forwards along the direction the sound travels in. In Figure 1 you can see that when the guitar string moves to the right it compresses the air on the right hand side of it. When the string moves to the left the air on the right expands. The string produces a series of **compressions** and **decompressions**. In a compression the air pressure is greater than normal atmospheric pressure. In a decompression the air pressure is less than normal atmospheric pressure.

Compressions and decompressions travel through air in the same way that compressions will move along a slinky. Your ear detects the changes in pressure caused by sound waves. When a compression reaches the ear it pushes the ear drum inwards. When a decompression arrives, the ear drum moves out again. The movements of the ear drum are transmitted through the ear by bones. Then nerves transmit electrical pulses to the brain.

Speed of sound

The speed of sound depends on which material it travels through. Sound waves are transmitted by molecules knocking into each other. In air sound travels at about 330 m/s. In solids and liquids, where molecules are packed more tightly together, sound travels faster (Table 1). In a vacuum there are no molecules at all. Sound cannot travel though a vacuum, although light can.

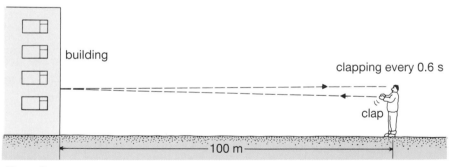

Figure 2
Measuring the speed of sound through air

Figure 2 shows a simple way for you to measure the speed of sound through air. Stand about 100 m away from a building and clap your hands. Sound waves will be reflected back to you from the building. When you hear the echo, clap again so that you clap in time with the echo. A friend, watching you clapping your hands, times 10 claps in 6 seconds. Now, you know that the sound took 0.6 s to travel 200 m (to the wall and back again).

So the speed of sound is given by:

$$V = \frac{d}{t}$$

$$= \frac{200\,\text{m}}{0.6\,\text{s}} = 330\,\text{m/s}$$

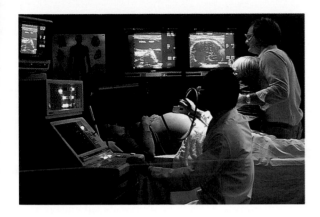

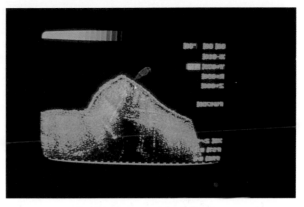

Ultrasonic depth finding

We use the name **ultrasound** to describe very high frequency sound waves. These waves have such a high frequency that we cannot hear them. Because ultrasound has a high frequency, the waves have a short wavelength. This means that it is possible to produce a narrow beam of ultrasound without it spreading out due to diffraction effects.

Ships use beams of ultrasound for a variety of purposes. Fishing boats look for fish, destroyers hunt for submarines or explorers chart the depth of the oceans. Figure 3 demonstrates the idea. A beam of ultrasonic waves is sent out from the bottom of a ship. The waves are reflected from the sea bed back to the ship. The longer the delay between the transmitted and reflected pulses, the deeper the sea is (Figure 4).

In the picture on the left, ultrasound emitted by the probe placed on the mother's stomach is reflected by the fetus. A computer builds up a picture from those reflected waves, which, unlike X-rays, are perfectly safe. In this photograph the whole family views its new member, a healthy baby boy. The photograph on the right shows the result of an ultrasound scan to check for breast tumours

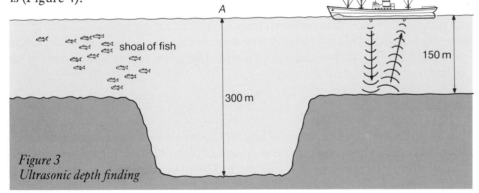

Figure 3
Ultrasonic depth finding

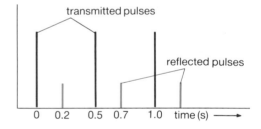

Figure 4
Ultrasound pulses for a depth of 150 m

Questions

In these questions you will need to refer to the table of speeds given on page 142.

1 The frequency of sound that comes out of your mouth is on average about 250 Hz.
(a) What is the wavelength of that sound?
(b) Explain why sound is diffracted when it comes out of your mouth.

2 The ship in Figure 3 sends out short pulses of ultrasound of frequency 50 kHz (50 000 Hz) every 0.5 s.
(a) The length of each pulse is 0.01 s. How many waves are emitted in that time?
(b) Use the information in Figure 4 to show that ultrasound travels through water at 1500 m/s.
(c) Sketch a graph to show the time interval between transmitted and reflected pulses when the ship reaches point A.
(d) What difficulty is there in

measuring the depth of the sea, when it is 500 m deep?
(e) What is the wavelength of the ultrasound waves?
(f) Why is it important to have a *narrow* beam of ultrasound waves? Explain why it is possible to produce a narrow beam with ultrasound but not ordinary sound waves.

5 Loudness, Quality and Pitch

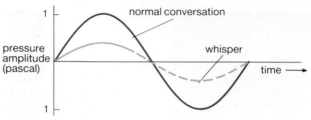

Figure 1
Sound waves caused by the human voice

Loudness

Your ears are very good detectors of energy. A disco produces sound energy at a rate of about 40 W. You find this very noisy. If you were to light a disco with one 40 W light bulb, everyone would complain that it was too dark.

Your ears detect sounds over a wide range of frequencies. You can hear frequencies as low as 20 Hz, and as high as 20 000 Hz. Yours ears are most sensitive to frequencies of about 2000 Hz. So a note at a frequency of 2000 Hz sounds louder than a note of frequency 10 000 Hz that carries the same energy. A loud noise makes your ear drums move a long way, while a very quiet noise has only a small effect on your eardrums. The loudness of a noise depends on the pressure caused by the sound wave. During normal conversation your voice will cause the air pressure to change by about 1 pascal ($1 \, N/m^2$). Your voice produces a pressure wave of amplitude 1 pascal (Figure 1). This is a small change in comparison with atmospheric pressure (100 000 pascals).

Loud noises can be very unpleasant and can damage your hearing. There are laws in force which limit the noise that industrial machinery, cars and aeroplanes are allowed to make. A sonic boom from Concorde can be heard over very great distances. Figure 2 shows roughly how the pressure caused by Concorde changes with distance. Exactly how far sound travels depends on many factors such as the strength and direction of the wind.

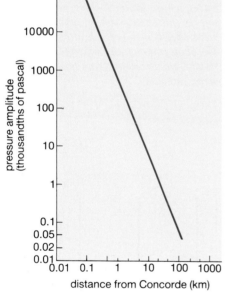

Figure 2

Pitch

We use the term **pitch** to describe how a noise or a musical note sounds to us. Bass notes are of low pitch, treble notes are of high pitch. Men have low-pitched voices, women have voices of higher pitch. The pitch of a note is related directly to its frequency. The higher-pitched notes are the notes with higher frequency.

Figure 3 shows you how you can use a *microphone* and an *oscilloscope* to show that higher-pitched notes have higher frequencies.

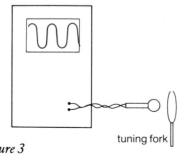

Figure 3
(a) Experiment to investigate frequencies

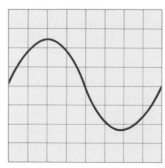

(b) A note from a tuning fork of low pitch

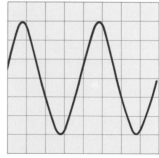

(c) A note from a tuning fork of higher pitch

If you played the same note on these instruments, they would have the same frequency (or pitch). You could still tell which instrument had played the note because each has its own special waveform

Quality

On a piano the note that is called middle C has a frequency of 256 Hz. This note could be played on a piano or a violin, or you could sing the note. Somebody listening to the three different sounds would recognise straight away whether you had sung the note or played it on the piano or violin. The **quality** of the three notes is different. The quality of a note depends on the shape of its waveform. Although two notes may have the same frequency and amplitude, if their waveforms are of a different shape you will detect a different sound (Figure 4).

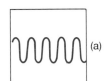 (a)   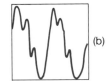 (b)   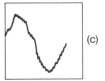 (c)   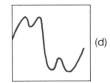 (d)

Figure 4
Different waveforms produced by four different musical instruments. The notes have different qualities

Questions

1 Use the data below, and Figure 2, to answer these questions.
(a) At approximately what distance from Concorde will the sound level be enough to interrupt a conversation?
(b) Work out roughly how far you need to be from Concorde so that you cannot hear it.

Noise level	Pressure amplitude in thousanths of a pascal
Painfully loud	100 000
Normal conversation	1000
Quiet countryside	1
Too quiet to be heard	0.02

(c) Explain why it is necessary to have laws controlling the noise levels near airports.
2 In Figure 3 the dot on the oscilloscope screen took 0.001 s to cross the screen. What is the frequency of each of the two notes?
3 Look at Figure 4. In each case the same oscilloscope settings were used.
(a) Which note has the highest frequency?
(b) Which note is the loudest?
(c) Which note is the softest?
(d) Which two notes have the same frequency?
4 An electronic synthesizer produces the two pure notes A and B as shown (right). It produces a third note C by adding the two waveforms together.
(a) Copy the two waveforms carefully onto some graph paper.

(b) Add the two waveforms together to produce the waveform of C.
(c) Does C sound louder than A or B?
(d) How does the frequency of C compare with: (i) A, (ii) B?
(e) Does C have the same quality as A or B?

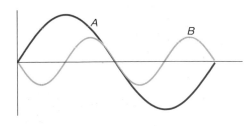

6 Electromagnetic Waves

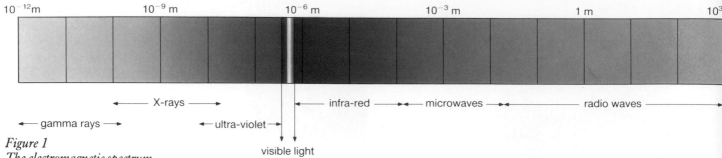

10⁻¹²m   10⁻⁹ m   10⁻⁶ m   10⁻³ m   1 m   10³

← X-rays →

← gamma rays →   ← ultra-violet →

← infra-red → ← microwaves → ← radio waves →

visible light

Figure 1
The electromagnetic spectrum

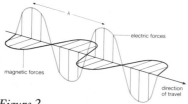

electric forces

magnetic forces

direction
of travel

Figure 2
*In an electromagnetic wave, energy is
carried by oscillating electric and magnetic
forces. These forces are at right angles to the
direction in which the wave travels*

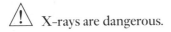 X-rays are dangerous.

*Sunbathers use ultraviolet rays from the
sun to get a tan. Staying in the sunshine (or
under a sun lamp) a long time can be
dangerous, however, as the ultraviolet rays
cause skin cancer*

You will already have heard of radio waves and light waves. These are two
examples of **electromagnetic waves**. There are many sorts of electromagnetic
wave, which produce very different kinds of effect. Figure 1 shows you the full
electromagnetic spectrum. The range of wavelengths in the spectrum stretches from
10^{-12} m for gamma rays to about 2 km for radio waves.

All the waves you have met so far travel through some material. Sound waves
travel through air, seismic waves travel through the Earth, water ripples travel
along the surface of water. Electromagnetic waves can travel through a vacuum;
this is how energy reaches us from the Sun. The energy is carried by changing
electric and magnetic forces. These changing forces are at right angles to the
direction in which the wave is travelling. So electromagnetic waves are transverse
waves (Figure 2).

Electromagnetic waves show the usual wave properties. They can be reflected
and refracted. They show diffraction and interference effects. In a vacuum all
electromagnetic waves travel at the same speed of 3×10^8 m/s. However,
electromagnetic waves do not all travel at the same speed when they travel in a
material. For example, different colours of light travel at different speeds in glass.

*This radio dish transmits international telephone calls using 0.1 metre wavelength
radiowaves*

- **Radio waves**. Radio 4 broadcasts on longwave use a wavelength of 1500 m; Capital Radio uses a wavelength of 194 m. These radio waves are produced by high frequency oscillations of electrons in aerials. Radio waves with wavelengths of a few hundred metres are used in local and national radio (see Figure 3).

  Radio waves with wavelengths of a few centimetres are used to transmit television signals and international phone calls. If you make a phone call to America your radio signals are sent out into space by large aerial dishes, like the one you can see in the photograph opposite. These signals are received by a satellite in orbit around the Earth. Then the signals are relayed to another aerial dish in America. It is important to use short wavelengths for international communications, so that a narrow beam of waves can be directed towards the satellite. Long wavelengths would be diffracted, so not much energy would reach the satellite.

- **Infra-red waves** have wavelengths between about 10^{-4} m and 10^{-6} m. Anything that is warm will lose energy by giving out infra-red radiation. You lose some heat energy by radiation. You can certainly feel the infra-red radiation given out, or *emitted*, by an electric fire. As things get hotter they also emit electromagnetic waves of even shorter wavelength. The Sun sends out light and ultra-violet waves.

  Infra-red photography can be used to measure the temperature of objects. The hotter something is, the more infra-red radiation it gives out.

- **X-rays** have wavelengths of about 10^{-10} m. These rays can cause damage to body tissues, so your exposure to them should be limited. X-rays are now widely used in medicine. X-rays of short wavelengths will pass through body tissues but will be absorbed by bone. Such rays can be used to take a photograph to see if a patient has broken a bone.

  Slightly longer wavelength X-rays are used in body scanning. These X-rays are absorbed by body tissues and doctors can build up a picture of the inside of a patient's body. This allows doctors to investigate whether a patient has cancer.

⚠ X-rays are dangerous.

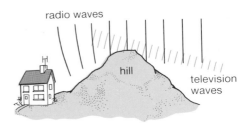

Figure 3
Long and medium wavelength radio waves will diffract around hills and houses. However, waves used for TV signals are of short wavelengths. These will not bend around hills so well; this house will have poor TV reception

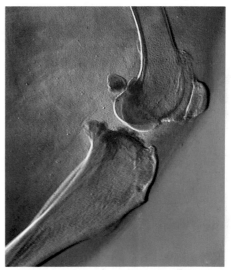

This false colour X-ray photograph shows a severely arthritic knee joint. Techniques like this are valuable for diagnosing injury or illness

Questions

1 In the table below you can see some data showing typical values of wavelengths and frequencies for different types of radio wave.

Type of radio wave	Wavelength (m)	Frequency (MHz)
Long	1500	
Medium	300	
Short	10	
VHF		100
UHF		3000

(VHF = very high frequencies, UHF = ultra high frequencies).

(a) Copy the table. Then use the equation $V = f \times \lambda$ to fill in the missing values.
(b) Which wave would you use for (i) a local radio station (ii) television broadcasts to the USA. Explain your choice.

2 The diagram (right) shows a side view of a radio dish. Explain why the receiver is placed some distance away from the dish. (A diagram may make your answer clearer.)

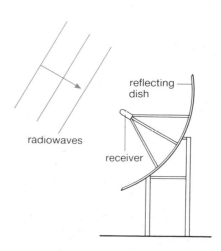

7 Communications Systems

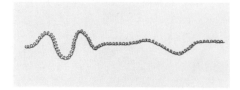

Figure 1
Information can be carried in these pulses on the slinky. The amplitude and shape of the pulses are important

Waves carry energy and information, which allow us to communicate with each other. You could communicate in code by sending pulses along a stretched spring or slinky (Figure 1). The length and shape of the pulses carry the information or message. Most of us are lucky enough to be able to communicate by talking and listening to each other.

A communications system can only work if it contains a number of **fundamental building blocks**. Figure 2 illustrates this idea, by examining what is required for two people to talk to each other.

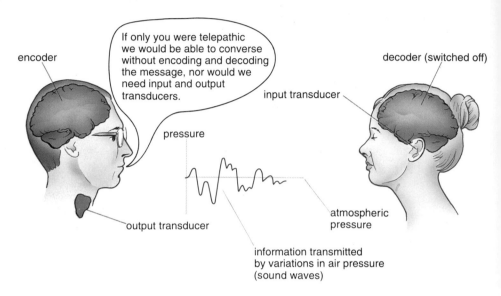

Figure 2
Speech communication system

A ship signalling with an Aldis lamp

- **Encoding.** The message is sent in code (or language). The speaker's brain encodes the message and transmits it by electrical pulses to the voice box.
- **Output transducer.** A transducer is a device than changes the way our information signal is carried. Here the voice box changes electrical pulses into sound waves.
- **Transmission.** Once the sound waves leave the voice box the signal has been sent or transmitted. The information is now carried in the pressure variations of the sound wave.
- **Input transducer.** The sound waves reach the ear of the listener. These waves make the ear drum vibrate, which then transmits electrical messages to the brain.
- **Decoding.** The brain now can decode the message; however, if you do not understand the language or you are not concentrating, the message does not get through.
- **Storage and retrieval.** Once the message has been understood the brain *stores* it in the memory. Later the message can be *retrieved* or remembered. To help the brain, you take notes in class, which are stores in a file and retrieved when it is time to revise.
- **Amplification.** Sometimes people's hearing is not so good. Then a hearing aid amplifies the sound. A hearing aid contains a tiny microphone to receive the signal, which is amplified and retransmitted by a speaker to the ear.

When you ring up a friend and speak to her on the phone, the communication system is a little more complicated. There are two more transducers: your mouthpiece turns sound waves into electrical pulses which travel along the phone line, and her earpiece turns the electrical pulses back into sound waves. When she is out, you might leave a message on the answerphone. The message is stored on magnetic tape, ready for retrieval when she comes home.

These phones transmit information using radio waves

Radio systems

This section concentrates on the broad outline of a radio system. To understand radiocommunication more fully you need to read Unit G8 too; Unit K5 tells you more about amplifiers.

Transmitting the signal

A microphone detects sound waves and turns them into audio frequency (AF) electrical pulses; this is the input transducer. The **modulator** encodes the message, by mixing in the audio frequency information, with the radio frequency carrier signal. This is amplified and transmitted from the aerial.

Receiving the signal

The process is now put into reverse. A receiving aerial picks up the radio transmission. The signal is amplified before being passed to the **demodulator** which separates out the radio and audio frequencies. The demodulator decodes the signal (see Figure 4). Finally, the audio signal is amplified, before the electrical pulses drive the loudspeaker which produces the sound. The loudspeaker is the output transducer.

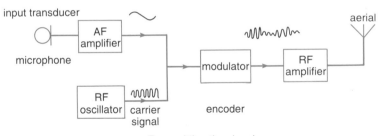

Transmitting the signal

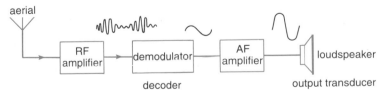

Receiving the signal

Figure 3
Radio communication system

Bar codes are used to identify almost everything that we buy in the shops

Decoding the signal

Figure 4 shows the circuit to decode or demodulate the signal. The amplitude modulated RF signal, V_{aa}, is passed into a diode. Without a capacitor or resistor in the circuit, the signal would come out like the voltage V_{bb}: the diode only conducts one way. These rectified pulses, V_{bb}, charge up the capacitor (see Unit K2), while the diode is conducting. When the diode is not conducting the capacitor partly discharges through the resistor. The value of the resistance must be carefully chosen. If R is too small, the capacitor discharges too quickly and too much of the RF signal gets through to the output. If R is too big the capacitor does not discharge fast enough to follow the AF signal properly.

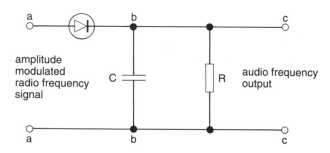

V_{aa}	V_{bb}	V_{cc}
RF modulated input		AF output

Figure 4
Decoding a signal

Questions

1 Below several means of communicating information are listed. Choose three of them and draw block diagrams to explain how the information is transmitted and received in each case. Your answer should include an account of the transducers used and the encoding and decoding processes. Explain how each message could be stored and retrieved.
(a) One ship signalling to another in morse code using flashes on a lamp.
(b) Watching television.
(c) Making an international phone call.
(d) Using a bar code reader at the checkout in a supermarket.
(e) Sending a message by electronic mail.
(f) Writing a letter.

2 Make an estimate of how long these messages took to be transmitted:
(a) Christopher Columbus reaching America, 1492.
(b) Wellington's victory at Waterloo, 1815.
(c) Sixty thousand casualties in the first day of the Battle of the Somme, 1916.
(d) Neil Armstrong landing on the Moon, 1969.

8 Radio Communications

Aerials

Aerials transmit or receive radio waves. Any conductor can act as an aerial, but for efficient transmission and reception it is necessary to design aerials carefully. In Figure 1 radio waves are being transmitted by a **dipole** aerial. This aerial consists of two vertical wires, each of which is one quarter of a wavelength long. The waves are emitted equally in all horizontal directions.

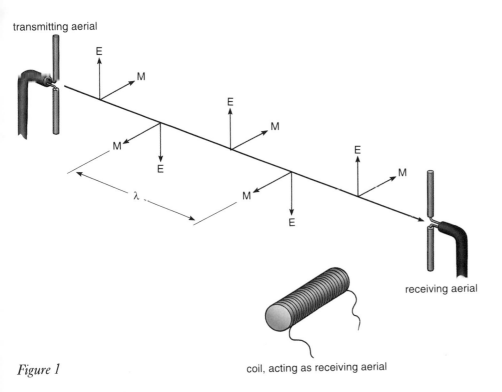

Figure 1

coil, acting as receiving aerial

The waves in Figure 1 are shown travelling towards a receiving aerial, which is identical to the transmitter. The receiving aerial only detects radio waves when it is placed vertically, when it is turned so that its wires lie in a horizontal plane it detects nothing. The waves transmitted from a dipole aerial are **polarised**. In this case they are vertically polarised, so that the electric field of the wave lies in a vertical plane; the magnetic field is at right angles to the electric field and lies in a horizontal plane. When the receiving aerial lies in a vertical plane, electrons in it are made to oscillate up and down by the changing magnetic and electric fields. Now there is a current in the aerial which can be detected by a meter. Radio waves can also be detected by a coil. Note how the coil in Figure 1 is placed so that its axis lies parallel to the magnetic field of the waves. This changing field induces a current in the coil, as you would do by pushing a bar magnet in and out.

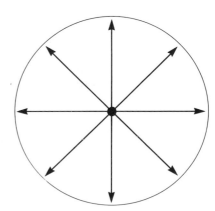

Figure 2
A bird's eye view of the aerial: waves are emitted equally in all horizontal directions

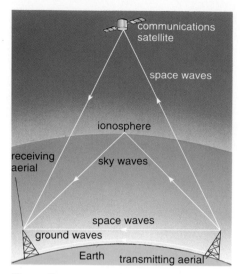

Figure 3

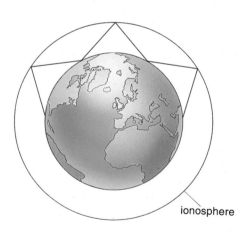

Figure 4
How sky waves can travel around the Earth

Ground, sky and space waves

Radio waves can travel to us by three different routes.

- Some waves are conducted along the ground or across an ocean. These are called **ground** or **surface waves**. Low frequency waves (long waves) can travel up to 1500 km along the ground. Higher frequency waves lose energy rapidly over the ground, so their range is much less.
- Some radio waves travel towards the sky, but are reflected back towards the Earth's surface by the **ionosphere**. These are called **sky waves**. Waves with frequencies below about 30 MHz are reflected by the ionosphere, but higher frequency waves can pass through it. The ionosphere is a region of the atmosphere stretching from about 100 km to 500 km above the Earth's surface. In this region molecules have been ionised by radiation from the sun. Sky waves of long, medium and short wavelengths can travel several thousand kilometres round the Earth due to multiple reflections.
- VHF and UHF waves and microwaves are not reflected by the ionosphere. These waves can only be transmitted round the world with the help of communications satellites. These are called **space waves**. It is also possible to transmit spaces waves directly over a distance of about 100 km provided that the transmitter is on a mountain and there is nothing to get in the way.

Frequency band	Wavelength	Use of waves
Low frequency 30 kHz – 300 kHz	Long wave 10 km – 1 km	National radio (Radio 4)
Medium frequency 300 kHz – 3 MHz	Medium wave 1 km – 100 m	National and local radio (Radio 1)
High frequency 3 MHz – 30 MHz	Short wave 100 m – 10 m	International radio Amateur and CB radio
Very high frequency (VHF) 30 MHz – 300 MHz	10 m – 1 m	Shipping, military and police communication
Ultra high frequency (UHF) 300 MHz – 3 GHz	1 m – 10 cm	TV, radiophones, aircraft guidance systems
Microwaves frequencies > 3 GHz	Less than 10 cm	Telephone calls, communications satellites, radar

This table shows the variety of radio waves and some of their uses

Modulation

A radio station might broadcast on a frequency of 3 MHz and yet the sounds we hear, voices or music, are in the audio range of frequencies 20 Hz to 20 kHz. The information about the sound is carried by modulating the radio waves. This can be done in two ways.

- **Amplitude modulation (AM).** Sound waves are first turned into electrical oscillations. These signals are then mixed in with the radio **carrier** waves. The information from the sound is then carried by varying the amplitude of the radio carrier wave (Figure 5).

● **Frequency modulation (FM).** Here the information about the sound waves is carried by varying the frequency of the carrier waves (Figure 6). The frequency of the carrier waves increases for a positive information signal, and decreases for a negative information signal.

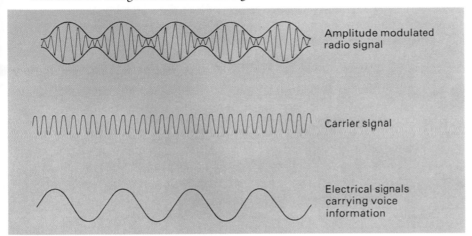

Amplitude modulated radio signal

Carrier signal

Electrical signals carrying voice information

Figure 5
Producing an amplitude modulated radio signal

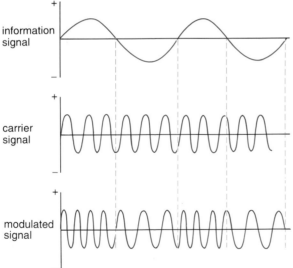

information signal

carrier signal

modulated signal

Figure 6
Frequency modulation

Questions

1 (a) Explain what is meant by polarisation.
(b) Why does the direction of your TV aerial matter?
(c) How do polaroid sunglasses help skiers?
(d) Find out how polaroid sunglasses work and write a brief explanation.
2 In 1901 Marconi succeeded in transmitting the letter 's' in Morse code across the Atlantic. However, the first television signals were not transmitted across the Atlantic until 1960. Explain how the radio waves travelled round the world in each case.
3 (a) Explain the difference between AM and FM radio.
(b) Find one station that transmits on AM and one on FM. Give the carrier frequency for each station.

Above, the driver of the car suffers a lot of glare when he looks at the road ahead. Below, this is what the driver sees through his Polaroid sunglasses – notice how the glare is reduced. All transverse waves, including light, can be polarised. Light reflected from the road is polarised. By wearing polarising sunglasses, you can cut down the glare because the sunglasses only let through light with oscillations in one direction.

9 Telephone Communications

At the beginning of the century, your great great grandparents were happy to walk several miles to see friends or relations. We still visit friends of course, but often we pick up the phone for a chat instead. Now we can call anywhere in the world. The signals our voices make are carried by electrical, light or radio waves.

Look at Figure 1. This shows you the principle of the telephone. When you speak your voice produces sound waves, which cause pressure differences in the air. These pressure changes act on the mouth piece of the telephone, making a cone move in and out. The movements of the cone squeeze some carbon powder. When the carbon is squashed, its electrical resistance becomes slightly less. This allows a larger current to flow from the battery. In this way, the information is carried along the wire by electrical waves. At the other end of the line, someone can listen to your voice. The electrical pulses are turned back into sound waves by the ear piece, which is like a small loudspeaker. The changes in current change the magnetising effect of the magnet. This moves a disc in and out to make sound waves.

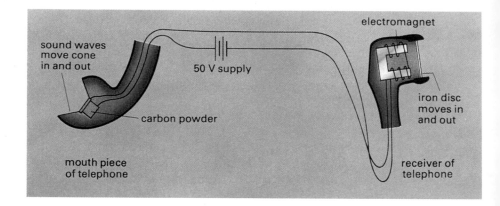

Figure 1
A conventional telephone system

Modern telephone systems transmit telephone messages by sending pulses of light down glass fibres. Optical cables have several advantages over the old copper wires.

- Optical cables are less bulky than copper cables. The resistance of a wire is proportional to its length, and inversely proportional to its cross-sectional area. So a thick wire has less resistance than a thin wire of the same length. This is why copper cables are so bulky; less energy is lost in a thick, low resistance cable.
- More messages can be carried at the same time on an optical fibre.
- There is less 'crackle' on the line. A signal on one cables does not affect its neighbour.
- Less energy is lost in optical cables. Signals travel 20 km without the need for amplification; signals in copper wires need amplification every 2 km.

Overseas telephone calls are carried by microwaves (Figure 2). The microwaves are transmitted via a satellite. These waves are electromagnetic waves with a frequency of a few gigahertz. When you speak to someone via a satellite link, you will notice a delay before they reply.

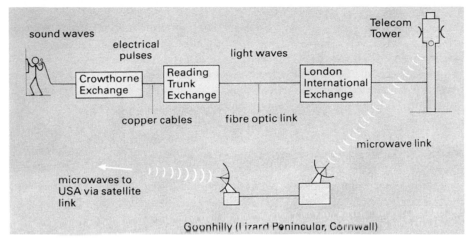

Figure 2
The path of an international telephone call as it leaves the country

Digital communications

Figure 3 shows an electrical signal; it is changing continuously. We call this an analogue signal. An analogue signal can be transmitted along an electrical cable. But before the signal can be sent along an optical cable it must be turned into a digital signal. A digital signal consists of a series of pulses – like Morse Code.

The main advantage of using digital signals is that they are less affected by unwanted voltages that are picked up ('noise'). Stray voltages distort analogue signals but have no effect on digital signals. This is because with a digital signal it is only necessary for the receiver to detect a pulse or the absence of a pulse. Analogue signals are converted to digital signals by using a process called pulse code modulation (PCM). The analogue signal is sampled 8000 times a second. This rate of sampling is adequate to reproduce speech for a telephone call. However, a greater rate of sampling is needed for music, or to transmit television signals. In Figure 3, for example, the analogue signal might be split up into 35 separate samples. The voltage levels at time t_5, t_{10}, t_{15} and t_{20} are respectively, 8, 17, 22 and 30. Each of these voltage levels can now be turned into a binary code or number (see Table 1). Next to Figure 3 you can see the pulses that make up the digital signal.

Number	Binary code or number				
(base 10)	(16)	(8)	(4)	(2)	(1)
8	0	1	0	0	0
17	1	0	0	0	1
22	1	0	1	1	0
30	1	1	1	1	0

Table 1   Some examples of binary codes or numbers

analogue signal   digital signal

|1|_|1|_|1|_|1|_|0

|1|_|0|_|1|_|1|_|0

|1|_|0|_|0|_|0|_|1|

0|_|1|_|0|_|0|_|0

voltage level

sampling time

t_5 t_{10} t_{15} t_{20} t_{25} t_{30} t_{35}

Figure 3

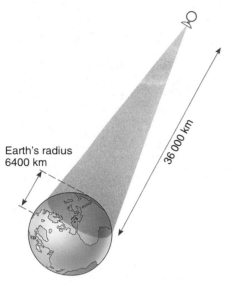

Figure 4
A geostationary orbit lies 36 000 km above the Earth's surface. Such an orbit takes exactly 24 hours

Earth's radius 6400 km

36 000 km

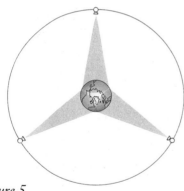

Figure 5
View of Earth from above the North Pole; three geostationary satellites can send signals to most of the Earth

Geostationary satellites

When you make a phone call to America, your call is directed via a satellite. For you to make a connection at any time, the satellite must always be in the same place relative to the Earth. Such a satellite is called a geostationary satellite. It must orbit the Earth above the equator at a height of 36 000 km. At this height the satellite takes exactly 24 hours to orbit the Earth, so it is always above the same part of the Earth. (Question 6 on page 74 helps you to understand how this height is calculated). A small number of geostationary satellites allow telephone communications across the world.

Telecommunications satellites receive signals, amplify them and then retransmit them. If the transmitting aerial is placed at the focal point of a parabolic reflecting dish, then a parallel beam may be transmitted from the dish. However, the beam is only parallel if the dish is much larger than the wavelength of the waves. If the wavelength is about the same size as the dish, the waves diffract, or spread out, rather like water waves going through a gap in a ripple tank. Therefore it is important to chose the right size of reflecting dish; the waves need to spread out to cover a large part of the Earth's surface, but not so far away that energy is wasted through the signal missing the Earth altogether.

Satellite receiver–transmitter dish at the Land Earth Station, Goonhilly, Cornwall. This satellite earth station serves the INMARSAT (International Maritime Satellite) organisation. It provides telephone, telex, data and facsimile operations to the shipping, aviation, offshore and land mobile industries. Goonhilly is one of the older satellite earth stations in England

Questions

1 Draw a diagram similar to Figure 2 to show how a telephone call is transferred from your home to a friend in New York. How many different sorts of waves are used in this phone call?
2 Explain how it is possible to use optical fibres to carry phone messages. What advantages do optical fibres have over copper wires?
3 A copper wire of thickness 1 mm is used to carry phone messages. These

messages need to be amplified by a 'repeater' every 2 km. An engineer decided to replace a faulty wire using copper of thickness 0.5 mm. Explain carefully how many repeaters he needs for each kilometre of wire now.
4 Look at the analogue signal in Figure 3.
(a) Measure the voltage levels at sample times (i) t_{15} (ii) t_{30}.
(b) Convert these voltage levels to the

appropriate binary code.
(c) Explain why music needs a higher frequency of sampling than speech.
5 (a) Explain why a geostationary orbit must lie above the Earth's equator.
(b) Use the information in Figure 4 to calculate the orbiting speed of a geostationary satellite.

10 Vibrations

A **pendulum** is an example of something that oscillates (Figure 1). For years we have used pendulums to keep the time, although now of course a lot of clocks and watches keep time electronically. A pendulum is used for timing because, provided the swings are small, the time for the oscillation does not depend on the amplitude of the swing. The time of the oscillation does depend on the length; the time increases as the pendulum gets longer.

Good vibrations

Pendulums are not the only things that oscillate or vibrate. If you pluck a tight string it vibrates. As it moves backwards and forwards it causes compressions and decompressions in the air. This makes a sound. If the string is made shorter or the tension in it is increased, the pitch or frequency of the sound increases. But the pitch of the sound does not depend on the amplitude of the vibrations. A lot of musical instruments are made by using strings of different tensions and masses; a heavier string produces a lower pitched note. You make different notes by changing the length of the strings by altering the position of your fingers. As well as string instruments, there are wind instruments. A wind instrument works by setting up vibrations in a column of air. Different notes can be made by changing the length of the air column, as in a trombone; lower pitched notes are produced by longer columns of air. Or you can make different notes by placing your fingers over holes in the instrument, in a recorder for example.

Bad vibrations

Vibrations can also occur in bridges, ships or buildings. But these are unwanted, because vibrations in large structures can cause serious damage. When wind blows over a large structure, **vortices** are produced (Figure 2). This is called **vortex shedding**. These wind vortices rotate at a frequency that depends on the wind speed and the shape of the structure. The structure can be made to vibrate at certain natural frequencies. The vibrations can become very large if the wind vortices rotate at the natural frequency of the structure. This is rather like pushing a child on a swing: if you push at the same frequency at which the swing oscillates, the swing goes high. If you push at the wrong frequency, the swing stays low. Large oscillations have caused bridges and chimneys to collapse. Ships have sunk when regular buffeting from waves have set up huge oscillations in the decks. Engineers test structures carefully in wind tunnels so that any design faults can be found before the structure is built.

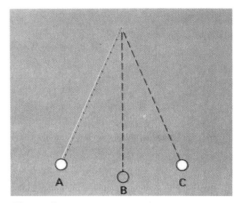

Figure 1
One complete pendulum swing or oscillation is from A to C and back to A. The amplitude of the swing is AB

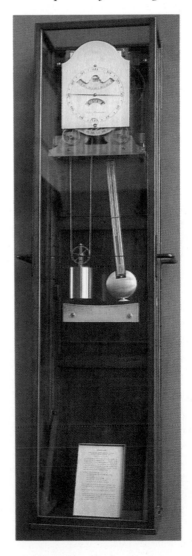

A pendulum clock

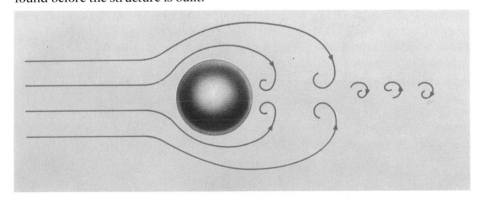

Figure 2
Vortex shedding. The arrows show the direction of wind flow

Large vibrations can be a danger in large structures

Stationary waves

A wave on the string of a musical instrument is an example of a **stationary wave**. Figure 3 shows how you can demonstrate a stationary wave in a laboratory. A string is attached to a mechanical vibrator, so that one end of the string moves up and down rapidly. If the frequency is exactly right, a stationary wave results. In this example, the wavelength of the wave is precisely 2.0 m. So when the wave reaches B, it is reflected as shown in Figure 3.

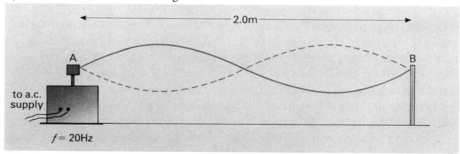

Figure 3
Stationary wave on a stretched string

Other stationary wave patterns may result. The speed of the waves does not depend on the frequency. Using the equation $v = f\lambda$, you can see that if the frequency is halved to 10 Hz, the wavelength doubles to 4.0 m. We now see just one loop between A and B (Figure 4). This frequency is the fundamental frequency of vibration of the string. A musical instrument usually vibrates at its fundamental frequency.

How do you make different notes on these instruments?

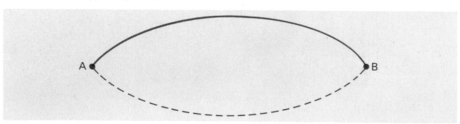

Figure 4

Medical vibrations

You can hear because sound waves make your ear drum vibrate backwards and forwards (Figure 5). If the pressure waves are small, no damage is done to the ear. But very loud or persistent noises can cause deafness. **Explosions** are shock waves of very large amplitude which can destroy body tissue.

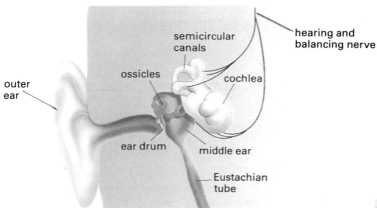

Figure 5
Sound waves set the ear drum vibrating These movements are transmitted to the inner ear by the ossicles

In medicine, controlled shock (or sound) waves can be put to good use. The picture shows how a **lithotripter** can remove kidney stones without surgery. Sound waves are focused by a reflector onto the kidney stone. Pulses of sound energy can shatter the stones and break them up into small pieces. The body can then remove the fragments by passing them out in the urine.

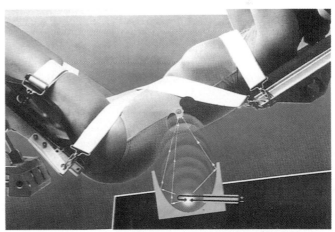

The lithotripter focuses sound waves onto the kidney. The patient is in water to ensure good transmission of the waves into her body. Without the water most of the energy would be reflected off her skin

Questions

1 The table shows measurements that Gita took using a pendulum. She wanted to investigate how the length of the pendulum affected the time taken for a swing.

Length (m)	Time for 10 swings (s)
0.1	6.3
0.2	9.0
0.5	14.2
1.0	20.0
1.5	24.6
2.0	28.4

(a) Plot a graph of 'time for one swing' (y-axis) against length of pendulum (x-axis).
(b) How long was the pendulum when the time for a complete swing was 1.0 s?
(c) Why did Gita time 10 swings?
(d) Rodney says that Gita's results show that the time for a swing is proportional to the pendulum's length. Gita disagrees. Who is right? Explain your reasoning.
2 Explain how a string can produce a musical note. How can you change the pitch of the note?

3 (a) What is 'vortex shedding'?
(b) What is meant by 'the natural frequency of vibration of a structure'?
4 (a) Explain what is meant by a stationary wave.
(b) Give two examples of stationary waves.
(c) Look at Figure 3. Draw what you would see between A and B, if the vibrator frequency is changed to (i) 30 Hz, (ii) 40 Hz.
(d) What is the speed of the waves travelling between A and B?
5 Explain why the lithotripter in the picture above needs careful focusing.

11 Light as a Wave

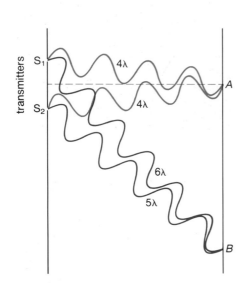

Figure 2
Constructive interference occurs at both points, A and B

The brilliant colours of the peacock are due to interference effects. Light is reflected from fine ribbings on the feathers. At certain angles constructive interference occurs for one colour of light

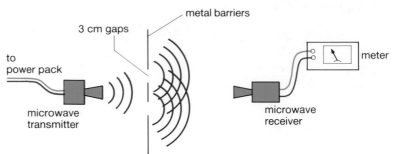

Figure 1
An experiment to detect the interference of microwaves

⚠ Microwave transmitters may be a hazard to anyone fitted with a heart pacemaker.

Microwave interference

Electromagnetic waves with a wavelength of 3 cm (**microwaves**) can be used to demonstrate interference effects in a laboratory. Figure 1 shows the sort of arrangement you can use. Microwaves from a transmitter are directed towards two small gaps in a metal barrier. The gaps should be about 3 cm wide, then the microwaves diffract out through these gaps. On the other side of the barrier the microwaves will overlap. This allows them to interfere in the same way that water waves do in a ripple tank. There will be places where the microwave receiver detects *constructive interference*. In other places the receiver detects little energy due to *destructive interference*.

Figure 2 shows you how constructive interference can occur in more than one place. Waves travel the same distance to A. So they arrive in phase there, and A is a point where you detect constructive interference. You also detect constructive interference at B. This time the waves from S_1 travel further than those from S_2 by one extra wavelength. The *path difference* between the two sets of waves is one wavelength.

You can use this idea to work out the wavelength of microwaves. For example you might measure with a ruler that $S_1B = 18$ cm and $S_2B = 15$ cm.

Path difference $S_1B - S_2B = 1$ wavelength

So 1 wavelength $= 18$ cm $- 15$ cm

$= 3$ cm

In general, constructive interference occurs if path difference $= n\lambda$

destructive interference occurs if path difference $n\lambda + \dfrac{\lambda}{2}$

(n is a whole number, 0, 1, 2, 3 etc.)

Interference of light

Figure 3 shows you how you can study the interference of light. It is the same sort of idea you used to study microwave interference. To make interference happen you must produce two beams of light that overlap. Light has a very small wavelength. So you need to use two very small slits, as shown in the diagram. The slits must be very narrow (about 0.1 mm wide) so that light is diffracted by them. The slits must also be placed close together so that the beams of light from each slit overlap.

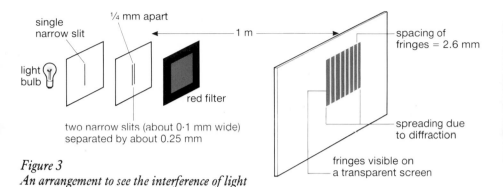

Figure 3
An arrangement to see the interference of light

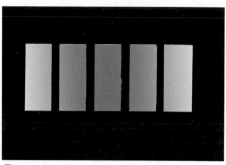

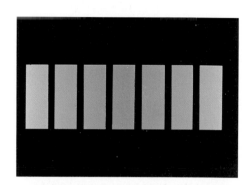

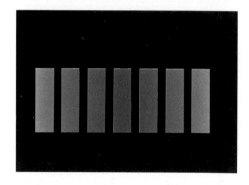

Figure 4

In a slightly darkened room it is quite easy to see **interference fringes**. With a red filter in place, you will see alternate red and dark lines. The red lines correspond to places of constructive interference; here the red light waves arrive in phase. At the dark places the light waves arrive out of phase; this is destructive interference.

The light wavelengths are too small for you to measure with a ruler as you did with microwaves. However, there is a formula which will help you.

$$\text{wavelength} = \frac{\text{fringe spacing} \times \text{slit separation}}{\text{distance from slits to screen}}$$

Using the information in Figure 3, the wavelength for red light is:

$$= \frac{(2.6 \times 10^{-3}\,\text{m}) \times (2.5 \times 10^{-4}\,\text{m})}{1\,\text{m}}$$

$$= 6.5 \times 10^{-7}\,\text{m}$$

Look at Figure 4 which shows fringes using red, green and blue light. From the formula above you can see that the wavelength is proportional to the fringe spacing. So red light has the longest wavelength and blue light the shortest.

When white light is used, you see a whole series of colours. This is because white light is made up of all colours. Each colour has a different wavelength. At some angles red light interferes constructively, while at other angles blue light does, and so on.

Questions

1 This question refers to Figure 4. The red, green and blue fringes were obtained using exactly the same apparatus but different filters. The wavelength of the red light is 6.5×10^{-7} m. Use the diagrams to work out the wavelength of green and blue light.
2 This question is about the interference fringes shown in Figure 3.
(a) The screen is moved further away from the slits. What difference does that make to: (i) the spacing of the fringes, (ii) the brightness of the fringes?
(b) The slits are moved further apart. What difference does that make to the spacing of the fringes?
(c) The slits are made slightly narrower. What difference does that make to: (i) the spacing of the fringes, (ii) the brightness of the fringes, (iii) the number of fringes that you can see?
3 Opposite, you can see a thin layer of oil floating on a puddle of water. You can see a ray of light that is partly reflected from the surface of the oil,
and then partly reflected from the water surface. Explain why if you look at oil on water you can see patches of colour.

4 All waves show interference effects. Design an experiment to show that sound waves interfere. Explain how you are going to observe this interference.

SECTION G: QUESTIONS

1 (a) Diagram (1) shows a ship 800 m from a cliff. A gun is fired on the ship. After 5 seconds the people at the front of the ship hear the sound of the gun again.

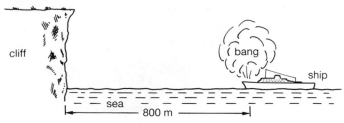

(1)

(i)   What is the name of this effect?
(ii) What happens to the sound at the cliff?
(iii) How far does the sound travel in 5 seconds?
(iv) Use the equation below to calculate the speed of sound:

$$\text{Speed} = \frac{\text{distance travelled}}{\text{time taken}}$$

(b) Diagram (2) shows three people standing around a house.

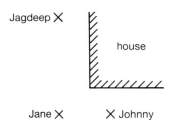

(i)   Who could see Johnny?
(ii) Who could hear Johnny?
(iii) To answer (i) you have assumed that light travels in straight lines. What did you assume about sound when you answered (ii)?

(c) Diagrams (3) and (4) show experiments which could be done in a ripple tank.

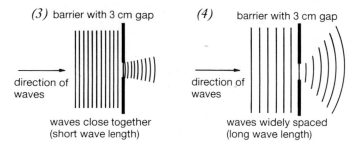

You will see that the waves spread out after going through the gap in the barrier. This effect is called diffraction. In which experiment did waves spread out less?

(d) Diagram (5) shows a ship searching for a submarine. It sends out narrow beams of sound such as AB and AC. Ordinary sound is not satisfactory. Ultrasound has to be used.

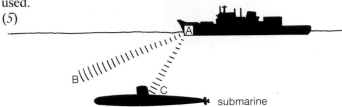

(i)   What is ultrasound?
(ii) Why is ultrasound necessary?

(e) Diagram (6) shows an oscilloscope on the ship.

(6)

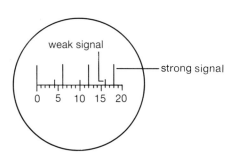

The numbers on the scale (diagram (6)) show times in tenths ($\frac{1}{10}$th) of a second. Pulses of ultrasound are sent out every 6 tenths of a second. These are shown on the oscilloscope as strong signals. Weaker signals are received when ultrasound is directed on path AC (diagram (5)).

(i)   How much time passes between the strong signal going out and the weak signal coming in?
(ii) Ultrasound travels with a speed of 1500 m/s in water. How far is the submarine from the ship?

(f) Diagram (7) shows ultrasound being used to obtain an image of an unborn baby.

(7)

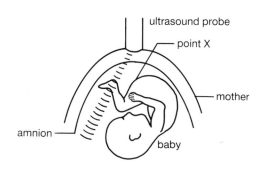

(i)   Why is ultrasound used, rather than X-rays?
(ii) When ultrasound goes from one type of material or tissue to another it may be reflected. Explain why ultrasound is reflected at X.

ULEAC

2 In the diagram below S_1 and S_2 are two loudspeakers. The speakers are both supplied by the same voltage source. The speakers produce a note of frequency 165 Hz.

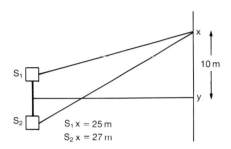

(a) The speed of sound is 330 m/s. Calculate the wavelength of the sound waves.
(b) Edward walks from y to x. At y, the sound seems loud; and, at x, it seems loud. But in between x and y the loudness decreases. Explain Edward's observations clearly.
(c) The connections to one of the speakers are reversed. Explain what Edward hears now as he walks from x to y.
(d) The frequency of the note is changed to 660 Hz. Describe what Edward hears as he walks between x and y.

3 An earthquake produces seismic waves which travel around the surface of the Earth at a speed of about 6 km/s. The graph shows how the ground moves near to the centre of the earthquake as the waves pass.
(a) What is the time period of the waves?
(b) What is the frequency of the waves?
(c) Calculate the wavelength of the seismic waves.
(d) Explain why the ground is moving most rapidly at times C and E.

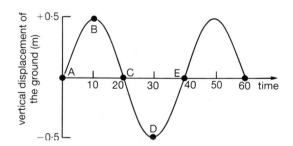

(e) When is the ground accelerating at its greatest rate?
(f) Use the graph to estimate the vertical speed of the ground at the time marked C.
(g) Make a sketched copy of the graph. Add to it a second graph, to show the ground displacement caused by a second seismic wave of the same amplitude but twice the frequency.
(h) Discuss whether high frequency or low frequency seismic waves will cause more damage to buildings.

(i) The diagram shows seismic waves passing a house. The waves produce ground displacements that have a vertical component YY[1] and a horizontal component XX[1] Which component is more likely to make the house fall down?

Explain your answer.

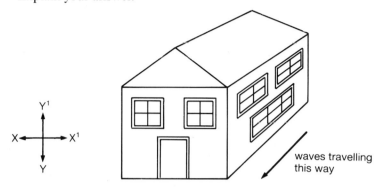

4 (a) The diagram shows a simple musical instrument made by James in his science class. To make a note he gently blows over the top of the test-tubes which contain water.

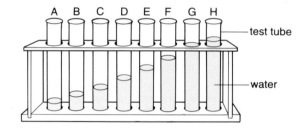

(i) James gently blows over the tubes. What happens to the air in the test-tubes?
(ii) Which test-tube will make the note with the lowest frequency?
(iii) The depth of the water in the test-tubes changes the frequency of the note. Name ONE thing which James could do that would change the loudness of the note.
(iv) Linda, instead of blowing over the test-tubes, taps them with a pencil. Explain why she will not produce notes of the same frequency as James
(b) The diagram shows a violin. It consists of four strings, **P, Q, R** and **S**. String **P** produces the highest frequency notes and string **S** the lowest frequency notes.

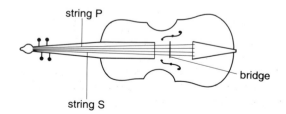

(i) What is it that vibrates to make the notes.

(ii) The notes produced from a particular string can be changed by sliding your finger up towards the bridge. What happens to the frequency of the note when this is done?

(iii) Each string is exactly the same length but each string produces a different note when bowed.

Give THREE things that could be different for the strings which would cause them to make different notes even though they are the same length. **ULEAC**

5 Read this article about whales, then answer the questions that follow.

Whale Song

Not only are blue whales the largest animals in the world, but they are also the noisiest. They give out low frequency sounds which allow them to communicate over distances of thousands of kilometres.

The whales are helped in their long distance communications because the sounds they give out are trapped in the upper surface layers of the ocean. Sounds, which hit the surface of the sea at a shallow angle, are reflected back. Sounds which travel downwards are turned back upwards. At greater depths sound travels faster, because the water is more compressed. This causes sound waves to be refracted as shown in the diagram. This is similar to the refraction of light waves on a hot day, which allows us to see a mirage. The sounds travel only in the top layers of the ocean. The sounds are trapped, rather like light waves in an optical fibre.

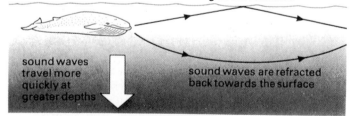

sound waves which hit the surface at shallow angles are reflected

sound waves travel more quickly at greater depths

sound waves are refracted back towards the surface

(a) Explain why the reflection and refraction of waves show in the diagram help the whales to communicate over large distances.

(b) Explain why creatures living at the bottom of the sea cannot communicate over long distances.

(c) Wallis, an amorous bull whale, is prepared to swim for a day to find a mate. After singing for half an hour, he gets a

response from Wendy 7 minutes after he stops. How far away is she, and will Wallis bother to make the journey? (Assume Wendy replies as soon as Wallis finishes his serenade.)

- Wallis can swim at 15 km/h.
- Sound travels at about 1500 m/s in water.

(d) Why are there long pauses in whale conversations?

(e) Explain carefully how the refraction of sound waves shown in the diagram, tells us that sound travels faster at greater depths. Draw a diagram to show what would happen to whale sounds, if sound travelled more *slowly* at greater depths.

(f) On a hot day, it is common to see a mirage on a road as you drive along. The mirage is an image of the sky. Draw a diagram to explain how mirages occur. (Hint: light travels slightly faster in the hot air just above the road surface.)

6 (a) Two marine biologists, Dr Fritz Muller and his research student Floella are discussing whale sounds.

Floella: We are 50 km away from a whale now. If we were 100 km away the intensity of his sound would be halved. This is because the sound has spread out into twice the area.

Fritz: That is a good idea, but I think it will be less due to energy losses.

(i) Comment on this conversation.

(ii) Analyse the data in the table to see who is right.

Intensity of sound (arbitrary units)	Distance from whales (km)
80	50
47	80
29	125
17	200
14	240

(b) Dolphins also use sound waves to communicate under water. They use high frequencies, which do not travel very far.

(i) High frequency sound has the advantage of being able to transmit more information. Explain why. What sort of information might dolphins want to exchange?

(ii) Suggest why dolphins do not need to communicate over long distances.

SECTION H

Light and Optics

The Sun emits more energy as light than it does in any other wavelengths. Animal eyes are specially adapted to take advantage of this

1 Rays and Shadows

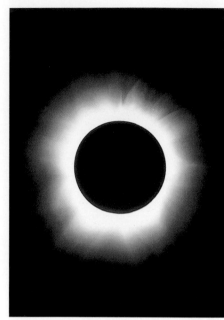

An eclipse of the Sun. In this photograph you can see the luminous atmosphere of hot gases that surround the Sun, called the corona. Normally the Sun is so bright that the corona is invisible

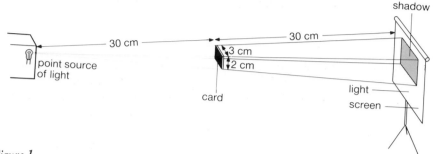

Figure 1

Shadows and eclipses

Shadows are formed when something blocks the path of light. In Figure 1 you can see a piece of card held in front of a small source of light. Some light misses the card and travels on, in a straight line, to the screen. A sharp shadow is formed on the screen behind the card.

Not all shadows are so sharp. If the source of light is large then the shadows have two parts. In the middle of the shadow there is a dark part called the **umbra**. Around the edges of the shadow there is a lighter region called the **penumbra**. A good example of shadow formation is provided by eclipses of the Sun and Moon.

An eclipse of the Sun occurs when the Moon passes between the Earth and the Sun. The Moon is a lot smaller than the Sun but it is closer to us. It is just possible for the Moon to cover the Sun completely. When this happens there is a **total eclipse** of the Sun. During a total eclipse of the Sun the sky goes black and it is possible to see stars. It is only possible to see a total eclipse of the Sun if you are in the umbra of the shadow (Figure 2). If you are inside the penumbra of the

A partial lunar eclipse. The upper left portion of the Moon has entered the umbra part of the shadow cast by the Earth

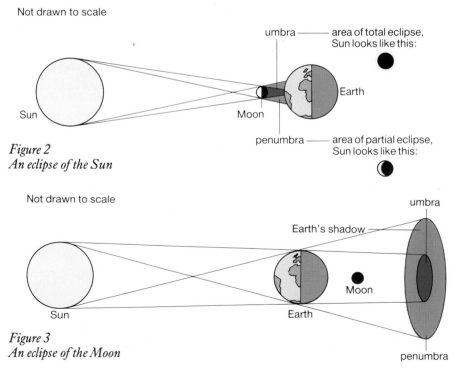

Figure 2
An eclipse of the Sun

Figure 3
An eclipse of the Moon

You can
mirror as t
You will ha
right hand,
inverted:
- An imag
  object.
Figure 4 sh
technique
diagonally
darkened s
illuminatc
When the
walking thi

shadow you will only see a **partial eclipse** of the Sun. Only part of the Sun is covered during a partial eclipse. An eclipse of the Moon happens when the Moon passes behind the Earth and into the Earth's shadow (Figure 3).

The pinhole camera

You can make a simple pinhole camera out of a cardboard shoe box. A small hole is put in one end. The other end of the box should be removed and a piece of tissue paper put in its place. If you now take the box outside and point it at some trees, you will see an image of them on the tissue paper. If you want to take a photograph of the trees, you must use a light-proof box. The photograph is made by allowing the light to fall on to a piece of photographic paper instead of the tissue paper (Figure 4).

The light from the tree travels through the pinhole in a straight line. This causes the image to be upside down. Provided the hole is small the image of the tree will be sharply defined. If the hole is too big the tree will look blurred.

The size of the image lets you work out the height of the tree, which is 30 m from the pinhole.

⚠ Never look at the Sun through a magnifying glass, binoculars or telescope. You could damage your eyes badly, or even cause blindness.

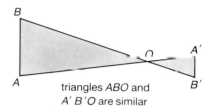

triangles *ABO* and
A′ B′O are similar

So $\dfrac{AB}{AO} = \dfrac{A'B'}{A'O}$

$AB = \dfrac{A'B'}{A'O} \times AO$

$\quad = \dfrac{10\ \text{cm}}{30\ \text{cm}} \times 30\ \text{m}$

$\quad = 10\ \text{m}$

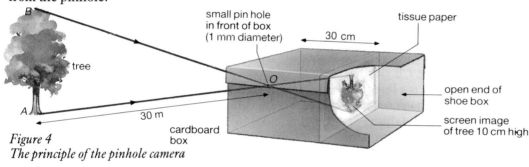

Figure 4
The principle of the pinhole camera

small pin hole
in front of box
(1 mm diameter)

30 cm

tissue paper

tree

30 m

O

open end of
shoe box

screen image
of tree 10 cm high

cardboard
box

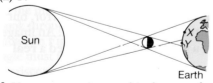

actor
brigh

Figure 4
Pepper's G

Questi

1 (a) Des
see when y
periscope
(b) At wha
fitted into
2 John rui
At what sp
him?
3 (a) Ho
see reflect
(b) Draw
of the ima
4 Show h

Questions

1 Work out the area of the shadow in Figure 1.

2 The diagram below shows how an annular eclipse of the Sun can happen. During an annular eclipse the Moon is further away from the Earth than in a total eclipse. Sketch how the Sun would appear when viewed from: (i) *X* and, (ii) *Y*.

Sun

Earth

3 Venus moves in an orbit closer to the Sun than the Earth's orbit. In the diagram you can see Venus in three positions marked V_1, V_2, V_3. On the right, X, Y and Z show how Venus looked when seen on three occasions through a telescope. Match X, Y and Z to the positions of Venus.

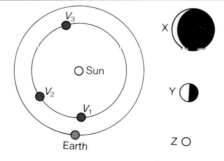

V_3

X

O Sun

V_2

Y

V_1

Earth

Z O

4(a) Explain why making the hole larger in a pinhole camera makes the image more blurred. Illustrate your answer with a diagram.

A student used a pinhole camera to form an image of the Sun. She investigated how the size of the Sun's image depended on the size of the pinhole. The table shows her results.

(b) Plot a graph of image size (*y*-axis) against hole diameter (*x*-axis).

(c) The student has made an incorrect measurement of the diameter of the hole. Which measurement is wrong and what should it have been?

(d) Use the graph to predict what the diameter of the Sun's image would be for a very small hole.

(e) Use your answer to part (d) and the extra data provided to calculate the Sun's diameter.

- Distance of Earth to Sun: 150 million km
- Length of pinhole camera: 500 mm

Diameter of Sun's image (mm)	6.5	8.0	9.5	11.0	12.5
Diameter of hole (mm)	2.0	3.5	4.0	6.5	8.0

4 Total Internal Reflection

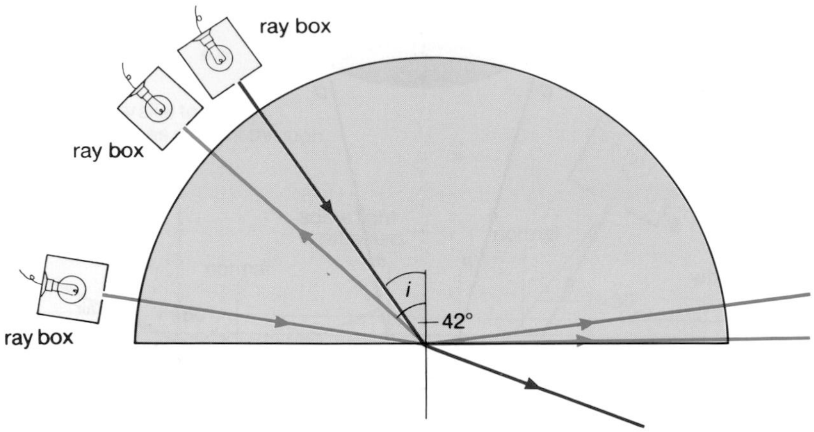

Figure 1
Refraction and reflection in a glass block

silvered surface

45°

normal

multiple
reflections

Figure 2
(a) Reflection from a mirror

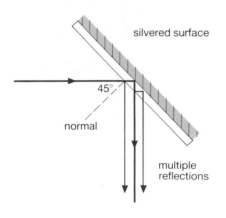

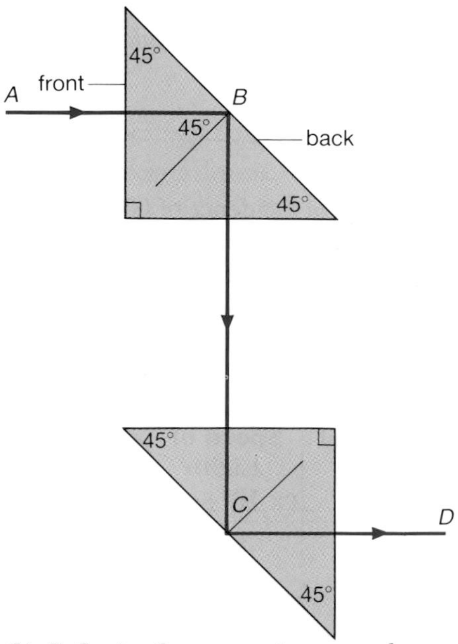

*(b) Reflection from two prisms to make a
periscope*

In the last unit you learnt that when a light ray crosses from glass into air, it bends away from the normal. However, this only happens if the angle of incidence is small. If the angle of incidence is too large all of the light is reflected back into the glass. This is called **total internal reflection**. Figure 1 shows how you can see this effect for yourself in the lab. Three rays of light are directed towards the centre of a semicircular glass block. Each ray crosses the circular part of the block along the normal, so it does not change direction. However, when a ray meets the plane surface there is a direction change. For small angles of incidence, the ray is refracted. Some light is also reflected back into the block. At an angle of 42° the ray is refracted along the surface of the block. This is called the **critical angle**. If the angle of incidence is greater than this critical angle then all of the light is reflected back into the glass. The critical angle varies from material to material. While it is 42° for glass, for water it is about 49°.

Total internal reflection in prisms

An ordinary mirror has one main disadvantage: the silver reflecting surface is at the back of the mirror. So light has to pass through glass before it is reflected by the mirror surface. This can cause several weaker reflections to be seen in the mirror, because some light is reflected back off from the glass/air surface (Figure 2(a)).

These multiple reflections can be a nuisance, for example in a periscope. We can avoid the extra reflections by using prisms. In Figure 2(b) the light ray *AB* meets the back of the glass prism at an angle of incidence of 45°. This angle is *greater* than the critical angle for glass, so the light is totally reflected. There is only one reflection because there is one surface. Total internal reflection by prisms is also put to use inside binoculars and cameras.

Refraction and cars

Refraction and reflection are put to use in your car. It is important that your rear lights are clearly visible to the car behind you. At the same time they must not dazzle the driver of a following car. Figure 3 shows how this is achieved. The cover of the rear light is made with a series of points. Any light that is travelling directly backwards is refracted to the side. A similar shape of plastic is used in the reflectors on the back of the cars and bicycles. This time the light from the headlights of a car passes straight through a plane plastic surface (Figure 4). Then total internal reflection occurs at the inside surfaces of the pointed plastic.

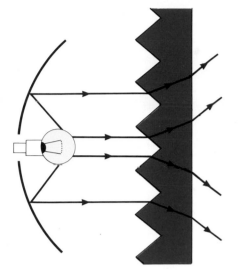

Figure 3
A car rear-light cover

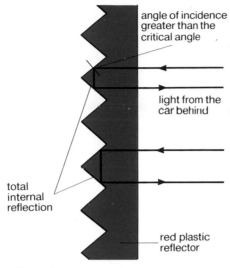

Figure 4
A reflector for a car or a bicycle

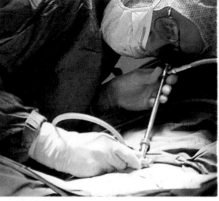

This photo shows 'laparoscopic' investigation of the abdomen. You can see the fibre optic bundle attached to the top right of the 'canula' allowing the doctor to view inside the patient

Optical fibres

Glass fibres are now used for carrying beams of light (Figure 5). The fibres usually consist of two parts. The inner part (core) carries the light beam. The outer part provides protection for the inner fibre. It is important that light travels more slowly in the outer part. Then the light inside the core is trapped due to total internal reflection.

Surgeons use a device called an *endoscope* to examine the inside of patients' bodies. This is made of two hollow light tubes. One carries light down inside the patient, and the other tube allows the surgeon to see what is there. Optical fibres are also in use by British Telecom. A small glass fibre, only about 0.01 mm in diameter, is capable of carrying hundreds of telephone calls at the same time. These fibres will soon replace the old copper cables in telephone systems.

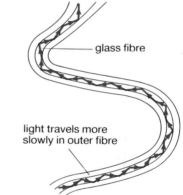

Figure 5
Internal reflection traps light inside the glass fibres of an endoscope

Questions

1 Turn back to the last unit. Explain how the graphs in Figure 3 can help you work out critical angles. What are the critical angles of glass, water and diamond?

2 Explain, with the help of a diagram, what the fish in the diagram below will see as he looks upwards.

3 Below, you can see a ray of light entering a five-sided prism. Copy the diagram and draw the path of the ray through the prism.

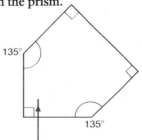

4 The diagram below shows sunlight passing through a prismatic window, that is used to light an underground public convenience.

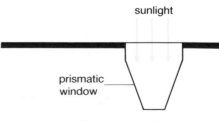

(a) Copy the diagram and show the path of the rays through the window.
(b) Explain why the window is shaped this way.

5 Converging Lenses

Every day of your life you use a converging lens; there is one in each of your eyes. There are also converging lenses in many optical instruments such as telescopes, cameras and slide projectors.

Converging lenses are usually made out of glass and they have two nearly spherical surfaces. When a light ray enters the glass it is refracted towards the normal, and then away from the normal when it leaves (Figure 1).

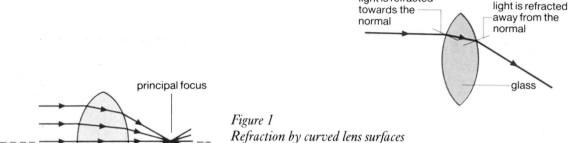

Figure 1
Refraction by curved lens surfaces

In Figure 2 you can see what happens to a lot of rays which are parallel to the **principal axis**. Each ray is refracted by a different amount, depending on where it meets the lens. After these rays have passed through the lens they converge and meet at a point. This point is called the **principal focus** of the lens. The **focal length,** f, of the lens is the distance between the lens and the principal focus. Each lens has two principal focuses. If the rays were to come from the right in Figure 2(a), they would come to a focus on the left of the lens.

Figure 3 shows how a lens can focus rays that are not parallel.

Figure 2
(a) A fat lens is a strong lens; it has a short focal length. Its curved surface refracts the light through a large angle

(b) A thin lens is a weak lens; it has a longer focal length than the strong lens

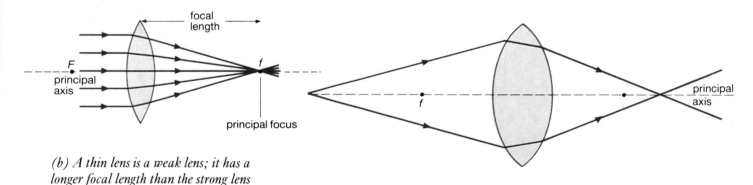

Figure 3 A lens can also focus rays that are not parallel; this time the rays meet behind the principal focus

Finding the image

If you know the focal length of a lens and the position of an object, you can work out where the lens will form an image of that object. You can construct a scale drawing using any two of the three rays shown in Figure 4(a).

- A ray parallel to the principal axis is refracted through the principal focus.
- A ray through the centre of the lens, C, does not change its direction.
- A ray through the principal focus on the first side of the lens is refracted parallel to the principal axis.

This is the method for making a scale drawing like Figure 4(a).
(1) Draw the principal axis and thin line (AB) to show the lens.
(2) Mark the position of the principal focuses. In this example the focal length is 10 cm.
(3) Mark the position of the object. In this case, it is 20 cm away from the lens.
(4) Draw the three construction rays from the top of the object. The top of the image is where these rays meet. The image can now be drawn in; the bottom of the image lies on the principal axis.

In Figures 4(b) and (c) you can see two other examples of ray diagrams. In all cases the images are real and inverted, but the sizes of the images vary. When the object is a long distance from the lens the image is small and close to the lens. When the object is just outside the focal length of the lens, the image is magnified and a large distance from the lens. A real image can be projected onto a screen.

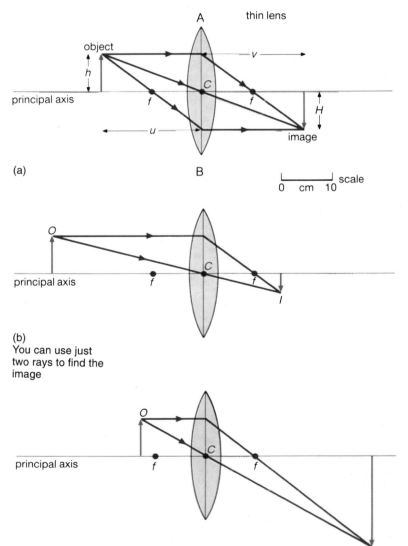

(a)

(b)
You can use just
two rays to find the
image

Figure 4
In drawing these scale diagrams we show all of the refraction occurring at the centre of the lens along line AB in Figure 4(a)

Questions

1 The diagram below shows light rays from a small object O, passing through a lens and forming an image at I.

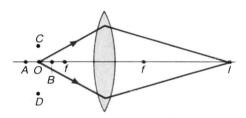

(a) Copy the diagram. Add to the diagram the position of the image, for each of the object positions, A, B, C, D. Mark your images A', B', C', D to correspond to each object position.
(b) What would happen to the image if a piece of card covered the bottom half of the lens?
2 In Figure 4 you can see that the image is magnified if the object is close to the lens. We define the magnification as:

$$M = \frac{\text{height of image}}{\text{height of object}}$$

(a) Work out the magnification for each of the images in Figure 4.
(b) Take careful measurements to prove that this formula is also true:

$$\frac{\text{image height}}{\text{object height}} = \frac{\text{distance: lens–image distance:}}{\text{distance: lens–object distance:}}$$

6 Optical Instruments

The camera

A diagram of a simple camera is shown in Figure 1. The purpose of the lens is to project an image of a distant object (a mountain, for example) on to a film. When you want to take the photograph, pressing a button opens up the shutter to allow light to fall on to the film.

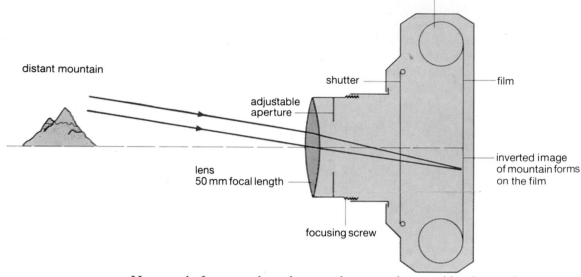

Figure 1
A camera

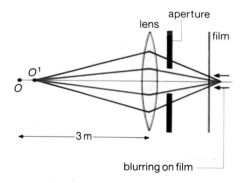

Figure 2
The depth of focus is larger if an aperture is used. In this diagram the camera is correctly focused on an object 3 m away. Rays from O¹ will be more out-of-focus if they pass through all of the lens

However, before you take a photograph you need to consider these points.
- **Choice of film.** A 'fast' film is more sensitive to light than a 'slow' film. The fast film needs less time to take a photograph than a slow film. The advantage of faster film is that you can take a photograph when the light is poor, or when something is moving quickly. The disadvantage of a fast film is that is produces poorer quality pictures than a slow film.
- **Focusing.** Your picture must be in focus. To focus, move the lens backwards and forwards with the focusing screw. For distant objects the lens is moved back towards the film. For closer objects you move the lens forwards.
- The **shutter speed** controls the amount of light coming into the camera. On a dark day you might choose a shutter speed of 1/30 s, while on a bright day you can use a faster speed of 1/60 s.
- **f-number.** This refers to the diameter of the aperture. If you set your aperture at f/8, it means that its diameter is 1/8 of the focal length of the lens. The f-number, like the shutter speed, controls the amount of light coming into the camera. A wide gap allows more light in. The f-number also determines the depth of focus of your photograph (Figure 2). A small gap (a small f-number, such as f/22) will give a large depth of focus. This means that things both near and far to the camera will be in focus.

The slide projector

Figure 3 shows how a slide projector works. A brightly illuminated slide (A) is used as an object for the projector lens (B). This lens projects an image of your slide on to a screen a few metres away. A 500 W light bulb (C) is used to make the slide bright. A concave mirror (D) behind the bulb reflects light forwards. The two condenser lenses (E) then converge the light towards the slide. A heat filter (F) is used to prevent the slide being damaged.

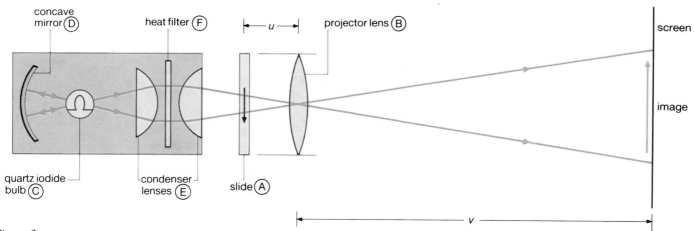

Figure 3
The principle of the slide projector

Light bulbs produce infra-red waves as well as light. The heat filter absorbs the infra-red waves, which would heat up the slide. The filter does let light through. The slide projector is cooled by a fan which blows air through it. Vents on top of the projector allow the warm air to escape.

In this slide projector, the distance, u, between the lens and slide can be adjusted between 15 cm and 20 cm. What lens should you choose for the projector? It is important that the slide is just outside the focal length of the projector lens. This makes sure that you see a large image on the screen. So a lens with a focal length of about 15 cm will be the best. Figure 4 gives a series of graphs for different lenses, to show how the distance, u, affects the distance between the image and lens, v.

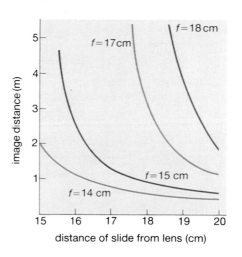

Figure 4
Graphs to show how the distance of the slide from the lens affects the image distance. Graphs for four different lenses are shown

Questions

1 Look at the three photographs. Explain in each case what factors affected the photographer's choice of:
(i) speed of film, (ii) shutter speed, (iii) f-number.

Below: f/2; 1/15 second; fast film
Right: f/4; 1/500 second fast film
Bottom right: f/16; 1/30 second; slow film

2 A lens of focal length 15 cm is used in the slide projector in Figure 3.
(a) Use Figure 4 to work out how far the slide is from the lens to project an image: (i) 3 m from the lens, (ii) 5 m from the lens.
(b) A slide measures 35 mm × 23 mm. What is the size of the picture on the screen, when the distance between the projector and the screen is 3 m? This formula may help:

$$\frac{\text{Image height}}{\text{Object height}} = \frac{\text{Image distance }(v)}{\text{Object distance }(u)}$$

(c) The lens in the projector breaks. The shop only has lenses of focal length 14 cm, 16 cm and 19 cm. Which one would you choose? Why?

7 The Eye

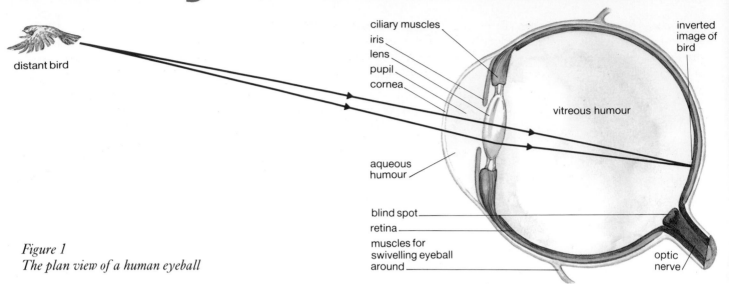

Figure 1
The plan view of a human eyeball

Labels: ciliary muscles, iris, lens, pupil, cornea, aqueous humour, blind spot, retina, muscles for swivelling eyeball around, vitreous humour, inverted image of bird, optic nerve, distant bird

Figure 1 shows you what an eyeball would look like if you could see it from the top. Below are listed the important points about the working of the eye:

- The eyeball is roughly spherical and keeps its **shape due to** the liquids inside it, the **vitreous humour** and the **aqueous humour**.
- The eye **lens** makes an image of a distant object **on the retina**.
- The retina contains cells which are sensitive to light. Some cells (**cones**) detect different colours, and other cells (**rods**) respond to the brightness of light.
- The **optic nerve** carries signals from the retina to the brain. Although the image on the retina is upside down, the brain sorts this out for us so we can see things the right way up.
- The amount of light that enters the eye is determined by the size of the **pupil**. The **iris** acts like the aperture of a camera. In bright light it closes down to protect the eye. In the dark the iris opens up to allow the eye to gather more light.

Focusing the eye

Your eye is most relaxed when you are looking at distant objects. The eye then focuses parallel rays onto the retina (Figure 2(a)). When you are looking at distant objects, your eyes cannot focus on something that is close to you at the same time. The lens is not strong enough to converge rays coming from nearby on to the retina. To look at something close to you, the eye lens has to change shape. The **ciliary muscles** make the lens fatter (Figure 2(b)). The light is now bent more when it goes through the lens and can be focused on the retina. This focusing process is called accommodation.

Wearing spectacles

A normal eye is able to see both near and distant objects clearly. However, it is quite common for people to suffer from either short or long sight:

- Someone who is short-sighted can see things nearby, but cannot focus on distant objects. The problem is that the eye lens is too powerful. Parallel rays from distant objects are focused in front of the retina. This can be corrected by using a **diverging lens**. A diverging lens will spread the rays out, so the eye can now bring them to focus on the retina (Figure 3).

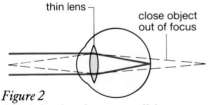

Figure 2
(a) A thin lens focuses parallel rays onto the retina

Labels: thin lens, close object out of focus

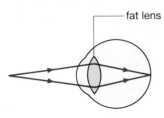

(b) A fat lens is needed to look at objects close to the eye

Label: fat lens

178

• People who are long-sighted cannot focus on objects that are close to the eye. However, they may be able to see clearly things far away. This time the problem is that the eye lens is too weak. So rays from objects close to converge at a point behind the retina. This sight defect can be corrected by using spectacles with converging lenses (Figure 4).

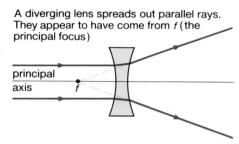

A diverging lens spreads out parallel rays. They appear to have come from *f* (the principal focus)

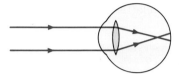

Figure 3
(a) A short-sighted eye cannot see distant objects

(b) A diverging lens corrects short sight

(c) A diverging lens

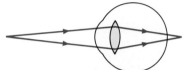

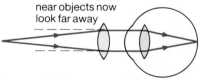

Figure 4
(a) A long-sighted eye cannot see objects close to

(b) A converging lens corrects long sight

Binocular vision

An animal that has both eyes at the front of the head is usually a hunter. Those animals that have eyes at the side of the head make good meals for the hunters. Having eyes at the side of the head gives an animal a wide range of vision. That is important if you do not want to get eaten. With two eyes at the front of your head, your brain gets two slightly different views of everything. This is very helpful when it comes to judging distances. If you are hunting your supper like the lioness in the photograph, you need to know how far away it is. Try putting a pencil in each hand and touching the points together; you will find it easy with both eyes open but difficult if you close one eye.

Fast food, Kenyan style!

Questions

1 (a) Suggest three ways in which the eye and the camera are similar.
(b) What is different about the way a camera and an eye focus light?
2 Instead of spectacles some people wear contact lenses to solve their eyesight problems. Contact lenses are curved pieces of plastic that fit directly on to the cornea. Below you can see three shapes of contact lens.

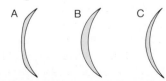

(a) Which side of each lens sticks on the eye? (the left or the right side)

(b) Which lens(es) could be used to correct for: (i) long sight, (ii) short sight?
3 (a) The eye in Figure 4 is long-sighted. How can you tell from the diagram 4(b) that the eye can see distant objects normally?
(b) Peggy wears *bifocal* spectacles. Without her bifocals Peggy can only see things about 4 m away from her eyes clearly. Explain how her bifocals help.

diverging lens for looking at distant objects

converging lens for reading

4 A surgeon in Russia has suggested that it is possible to cure short sight with an operation. His idea is to cut the muscles that control the shape of the eye lens. What are the advantages and disadvantages of this idea?
5 Close your right eye and look at the blue dot. Move your eye closer to the page until the red dot disappears. What causes this blind spot?

8 Colour

Raindrops split sunlight into separate colours to produce a rainbow

If you look into the sky when it has been raining (and if it is sunny or reasonably bright), you are quite likely to see a rainbow. White light from the sun is a mixture of many different colours. When sunlight passes through raindrops in the sky it is split up into its separate colours (Figure 1).

You can produce your own 'rainbow' or **spectrum** of colours by usng a *prism*. In Figure 2 a ray of white light is split into its separate colours. Blue light is bent more by the prism than the red light. This tells us that blue light is slowed down more by the glass than the red light.

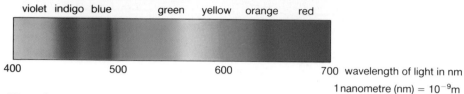

Figure 1
The colours of the rainbow. Different colours have different wavelengths

Making and mixing coloured lights

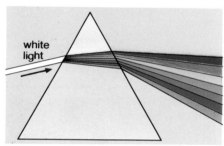

Figure 2
A prism can produce a spectrum from white light

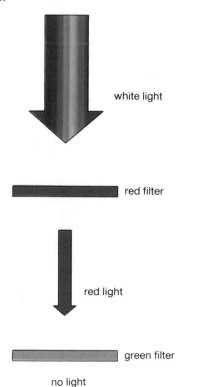

Figure 3

In this disco the mixing of four coloured lights has produced a white spot on the dancefloor

In your eyes there are cells called cones that are sensitive to colours. There are three types of cone cell; one that detects red light, one that detects green light and one that detects blue light. Red, green and blue are called the **three primary colours**. They cannot be made by mixing together other colours. Most other colours *can* be made by mixing together various amounts of the primary colours.

You can make red, green or blue light with a filter (Figure 3). A red filter, for example, allows red light to pass through it but absorbs all the other colours of light. The red light would be absorbed by a green filter. So if you looked at white light through a green and a red filter, you would see nothing. Figure 4 shows you the sort of effect that you will see if you mix together light of the three primary colours.

You can do this by using three slide projectors, one with a red filter, one with a green filter and one with a blue filter. Where the three colours overlap in the middle, white light is produced. At other places the **secondary colours** are produced. Cyan is a mixture of green and blue; magenta is a mixture of red and blue; yellow is a mixture of red and green.

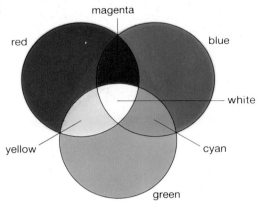

Figure 4

Making colours by reflection

Things appear coloured because they absorb some colours of light and reflect others. For example a shirt may look blue because it reflects blue light and absorbs the other colours. In fact, it is not quite as simple as that. The blue shirt will probably reflect a little green and violet light as well, but the shirt looks blue to the eye. Other colours behave in the same way (Figure 5). The blue shirt will look blue as long as you look at it in white or blue light. However, if you shine red light on the shirt it will look black, because there is no blue light for it to reflect.

What colour shirt would you get if you dyed it in a mixture of yellow and blue dyes? The answer is green. The yellow dye absorbs blue and violet, but reflects green, yellow and red. The blue dye absorbs yellow, orange and red, but reflects violet and green. The only colour that is *not* absorbed is green. Mixing paints and dyes gives a different effect from mixing light. If you mix green and red light you get yellow. If you mix green and red paint you get a nasty dark mess!

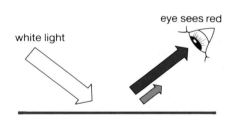

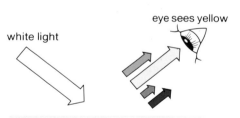

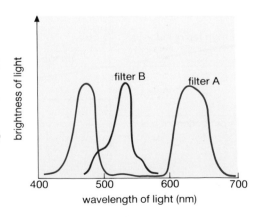

Figure 5

Questions

1 Inspector Grappler of the vice-squad is on patrol in the red light district of New York. He is in disguise wearing a red cap, blue shirt and green trousers. Copy and fill in the table below to show what colour he looks in red and yellow light.

Colour of			
light	**cap**	**shirt**	**trousers**
Red			
Yellow			

2 What colour of paint will you get if you mix yellow and cyan paints?

3 The graph (right) shows how the brightness of light, passing through two filters A and B, depends on the wavelength of light.
(a) What colour will you see if you look at a white light bulb through (i) filter A (ii) filter B? (You will find Figure 3 useful)
(b) What colour will you see if you make the light pass through both filters?
(c) What colour will a red car look if you view it through: (i) filter A, (ii) filter B?

1 Quasars are some of the most distant objects in the universe. Astronomers have observed quasars at distances of about 10 000 million light years. However, some quasars appear brighter than expected. One theory to explain this is the idea of a gravitational lens. According to Einstein's general theory of relativity, light is bent when it passes close to a massive object. A few galaxies have masses equal to that of 10 000 billion (10^{13}) suns. Such a large galaxy acts as a gravitational lens. The diagram shows the idea. Light from the quasar is bent, when it passes close to the large galaxy. The light is focused by the gravitational lens. An image of the quasar is now formed nearer to the Earth. The lens makes the quasar look brighter than it would do without it.

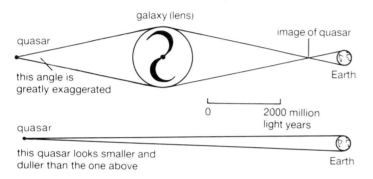

(a) Explain why the quasar would look duller without the gravitational lens.
(b) Use the diagram to measure the distance from
(i) the quasar to the galaxy, (ii) the galaxy to the image.
(c) Draw a ray diagram to show that the focal length of the gravitational lens is 2000 million light years.
(d) Explain why the quasar and its image are the same size.
(e) Now explain why the lens also makes the quasar look bigger than other quasars seen without the lens.

2 The diagram below shows the path of two rays from the top of the Moon passing through an astronomical telescope.

(a) Copy the diagram and show the passage of the two rays from the bottom of the Moon. It will make the diagram clearer to use different coloured pencils or pens.
(b) Describe the image that the eye sees.
(c) Explain why the telescope magnifies more if:
(i) the objective lens has a long focal length
(ii) the eye lens has a short focal length.

3 The diagram shows light from the Sun falling onto a lens.

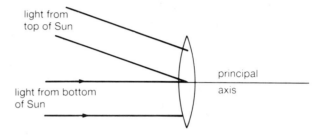

(a) Explain why light rays coming from any point on the Sun are parallel.
(b) Sketch the drawing below and complete it to show the position of the Sun's image. The focal length of the lens is 10 cm.
(c) Using the same scale make a second sketch to show the size and position of the Sun's image produced by a lens of focal length 20 cm.
(d) Use the data below to select the best lens for each of these jobs: (i) burning a piece of paper with the Sun's rays, (ii) protecting an image of the Sun to look for sun spots.

Lens	Diameter (cm)	Focal length (cm)
1	5	5
2	5	20
3	5	10
4	10	50

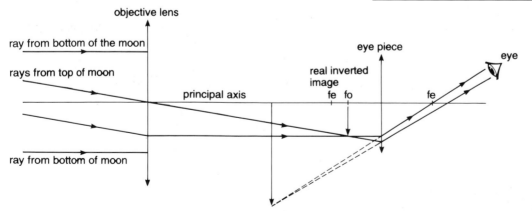

⚠ Never view the Sun directly through a telescope or lens

4 Red, green and blue lamps are arranged as shown; each lamp emits light in all directions. A metal plate is placed between the lamps and a white screen.
(a) What colours will be seen at A, B, C, D and E?
(b) Where will the screen appear white?

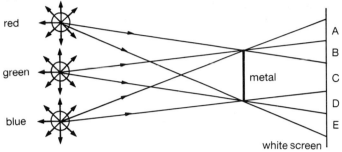

red

green

blue

metal

A
B
C
D
E

white screen

5 The diagram shows five different shapes of lens.
(a) Which is the strongest converging lens?
(b) Which is the weakest converging lens?
(c) Which lens could correct for myopia?

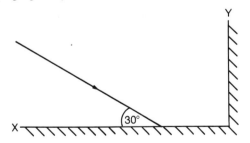

A   B   C   D   E

6 Copy the diagram and complete it to show how the incoming light ray is reflected from mirrors X and Y.

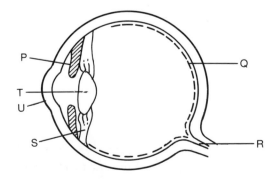

Y

30°

X

7 The diagram shows a simple plan view of the eye.
(a) Name the parts of the eye that each label points to.

P
Q
T
U
S
R

(b) Stefan leaves a dark room and goes outside into bright sunlight. What effect does this have on the pupil of the eye?

In the retina there are millions of light-sensitive cells. There are two kinds, called rods and cones. The cones enable us to distinguish colours. In a normal eye there are three types of cone. One responds to red light, one to green and one to blue.
(c) How can an eye detect: (i) yellow light, (ii) magenta light, (iii) white light?
(d) A person who suffers from red-green colour blindness cannot distinguish between red and green. Explain what might be wrong with his cone cells.

8 Copy the diagrams below, and complete them to show how the light rays pass through each block of glass.

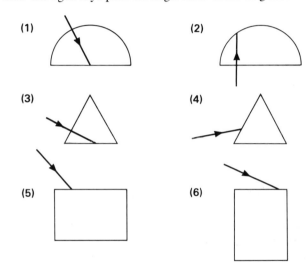

(1)   (2)

(3)   (4)

(5)   (6)

9 (a) Copy and complete diagram (1) to show refraction and dispersion by a prism. Explain why dispersion occurs.
(b) Light is also dispersed by droplets of water. Copy and complete diagram (2) and use it to explain why we can see rainbows.

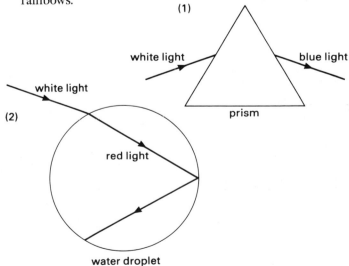

(1)

white light   blue light

prism

(2)

white light

red light

water droplet

QUESTIONS

10 The diagram shows a simple camera. The lens has to be screwed in or out to get the image in focus on the film. This movement depends on the position of the object being photographed. The graph shows how the separation of the lens and film depends on the distance of the object from the lens. The focal length of the lens is 50 mm. The distance between the lens and the film can be varied between 50 mm and 60 mm.

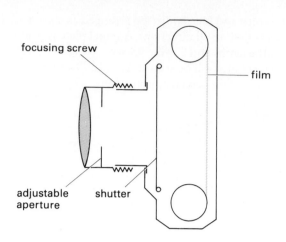

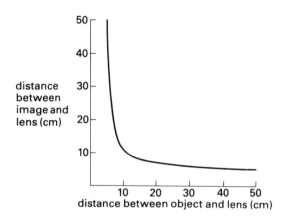

(a) Draw ray diagrams to show how the lens forms an image of an object (i) a long way away, and (ii) 30 cm from the lens. In each case state the distance between the lens and the film.
(b) Explain why the camera cannot focus on something 20 cm from its lens. How would you adapt the camera so that it could take a photograph this close up?

11 This question is about how a magnifying glass works. Look at diagram 1 below; here an eye is looking at a ladybird at a distance of about 25 cm. This distance is about as close as an adult eye can focus on an object (young eyes can often focus closer than this). This is called the **near point of vision.** Now look at diagram 2. The same eye is viewing the ladybird through a converging lens, placed so that the ladybird lies inside the focal length of the lens. You can see how the rays from the ladybird's head are diverging when they reach the eye. So the rays appear to come from behind the eye.
(a) Copy diagram 2 and extend the rays 1 and 2 to show where the image appears to be.
(b) The image should be close to the near point of vision; now explain why the image appears magnified. (It might be helpful to compare the angles α and β.)

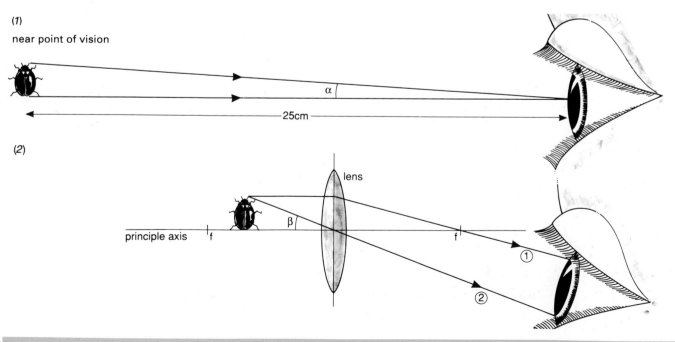

SECTION I
Electricity

An engineer repairing some overhead lines. In the background you can see the huge cooling towers of the Connah's Quay power station, and the tall pylons which carry electricity to the national grid

1 Introducing Electricity

Birmingham at night. Major cities use an enormous amount of electricity to provide light

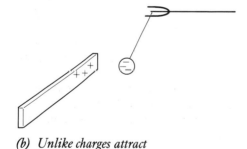

Figure 1
Attraction or repulsion?
(a) Like charges repel

(b) Unlike charges attract

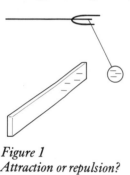

Figure 2
This beryllium atom is neutral. Four negatively-charged electrons balance four positively-charged protons

We all take electricity for granted. At home you can turn on a light or a fire at the flick of a switch. You may very well take a snack out of the fridge or deep freeze and cook it in your microwave oven, before sitting down in front of your favourite television programme. Without electricity, our lives would be completely different and less comfortable.

Electrical charge

Electricity was discovered a long time ago when the effects of rubbing materials together was noticed. You will have seen these effects for yourself. If you take a shirt off in a dark room you can hear the shirt crackle and you may also see some sparks. A well-known trick at children's parties is to rub a balloon and stick it on to the ceiling. You have probably felt an electrical shock after walking across a nylon carpet. In these examples, you, or the balloon, have become charged as a result of **friction** (rubbing).

There are two types of electrical charge, positive and negative. A **positive** charge is produced on a perspex ruler when it is rubbed with a woollen duster. You can put **negative** charge onto a plastic comb by combing it through your hair.

Some simple experiments show us that like charges repel each other, and unlike charges attract each other. These experiments are shown in Figure 1.

Where do charges come from?

There are three types of small particles inside atoms. There is a very small centre of the atom called the **nucleus**. Inside the nucleus there are **protons** and **neutrons**. Protons have a positive charge but neutrons have no charge. The **electrons** carry a negative charge and they move around the nucleus. The size of the charge on an electron and on a proton is the same. There are as many electrons as protons. This means that the positive charge of the protons is balanced by the negative charge of the electrons. So the atom is neutral or uncharged (Figure 2).

When you rub a perspex ruler with a duster some electrons are removed from the atoms in the ruler and are put onto the atoms in the duster. As a result the ruler has fewer electrons than protons and so it is positively charged. But the duster has more electrons than protons and is negatively charged (Figure 3).

Picking up litter?

You may know that you can use your comb to pick up small pieces of paper, but it is quite difficult to explain it. Figure 4 shows the idea. The comb is negatively charged, after combing your hair, so when it is placed close to the paper, **electrons** in the paper are pushed to the bottom or **repelled**. The top of the paper becomes positively charged and the bottom negatively charged. The negative charges on the comb attract the top of the paper upwards. The same charges repel the bottom, but the positive charges at the top of the paper are closer to the comb. As a result the upwards force is bigger than the repulsive downwards force and the piece of paper is picked up.

(a)
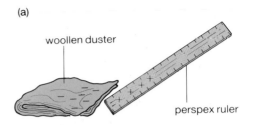
woollen duster
perspex ruler

(b)
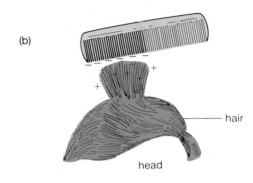
hair
head

Figure 3
Charging by friction

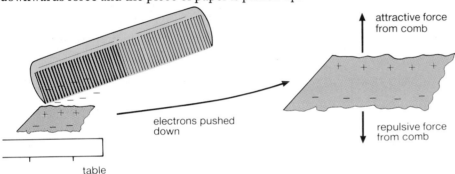
attractive force from comb
electrons pushed down
repulsive force from comb
table

Figure 4
Your comb can pick up small pieces of paper

In some parts of the world, there is no electricity supply, and people do without many of the things that we take for granted

Questions

1 (a) List 4 machines, at home, that work on electricity.
(b) How would you manage without these machines in a power cut that lasts for 3 days?
2 The diagram below shows 2 small plastic balls. The balls are charged. What can you say about these charges on the balls?

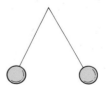

2 Electrostatics at Work

(a) *Jack is positively charged. His charge attracts electrons on to the parts of the ship near to him*

(b) *Jack touches the ship with his hand. Electrons flow on to him to cancel out his positive charge. The spark could cause an explosion*

Figure 1

An accidental spark on an oil tanker could be fatal. Great care is taken on these ships to avoid the build up of electrical charge

The conducting tail on this car prevents static electricity from building up. When could such a build-up be dangerous?

Spark hazards

Jack is a sailor who works on board an oil tanker. Jack has to wear shoes that conduct electricity. We say a material conducts electricity, if charge can flow through it. Metals are conductors of electricity. Materials like plastic and rubber do not conduct electricity; they are called insulators. Wearing shoes with rubber soles on board a tanker could be extremely dangerous. When Jack moves around the ship in rubber shoes, charges can build up on him as he works. Then when he touches the ship, there will be a small spark as charge flows away from his body. Such a spark could ignite oil fumes and cause an explosion. Some very large explosions have destroyed tankers in the past. So Jack wears shoes with soles that conduct. Now any charge on him flows away and he cannot make a spark.

Sparks are most likely to ignite the oil when it is being unloaded. To avoid this, the surface of the oil is covered with a 'blanket' of nitrogen. This gas does not burn, so a spark will not cause an explosion.

Electrostatic precipitation

Most of our power stations still burn coal to produce electrical energy. When coal is burnt a lot of soot is produced. It is important to remove this soot before it gets into the atmosphere. One way of doing this is to use an **electrostatic precipitator**. Inside the precipitator there are some wires which carry a large negative charge. As the soot passes close to these wires the soot particles become negatively charged. Now these particles are repelled away from the negative wires and are attached to some positively charged plates. The soot sticks to the plates, and can be removed later. Some large precipitators, in power stations, remove 30 or 40 tonnes of soot per hour.

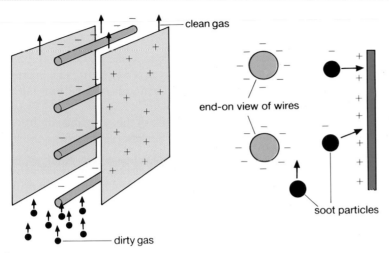

Figure 2
Soot particles in dirty fumes are removed in an electrostatic precipitator

Photocopying

Nowadays, most offices have a photocopying machine. The key to photocopying is a plate that is affected by light. When the plate is in the dark its surface is positively charged. When the plate is in the light it is uncharged.

An image of the document to be copied is projected on to the plate (Figure 3(a)). The dark parts of the plate become charged. Now the plate is covered with a dark powder, called **toner**. The particles in the toner have been negatively charged (Figure 3(b)). So the toner sticks to the dark parts of the plate, leaving a dark image. Next, a piece of paper is pressed on to the plate. This paper is positively charged, so the toner is attracted to it (Figure 3(c)).

Finally, the paper is heated. The toner melts and sticks to the paper, making the photocopy of the document (Figure 3(d)). In modern photocopiers the whole process takes two or three seconds.

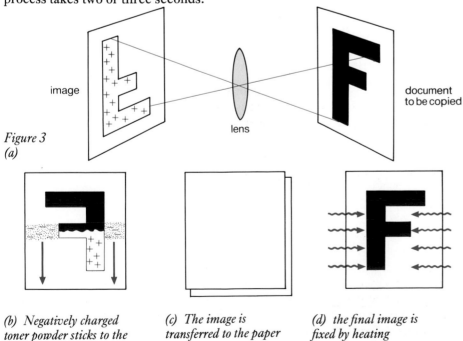

Figure 3
(a)

(b) Negatively charged toner powder sticks to the image

(c) The image is transferred to the paper

(d) the final image is fixed by heating

Questions

1 The tyres on aircraft are made from special rubber that conducts electricity. Explain why.

2 (a) Explain why a photocopier needs toner powder. Why does the powder need to be charged?
(b) When you get your photocopy out of the copier it is usually warm. Why?
(c) The lens in Figure 3(a) produces an image of the document, which is upside-down and back-to-front. Explain why the final image is the right way round.

3 (a) Explain this: When you polish a window using a dry cloth on a dry day it soon becomes dusty. Why does this not happen on wet days?
(b) Cling film is a thin plastic material that is used for wrapping up food. When you peel the film off the roll it sticks to itself. Can you suggest why this happens?

3 Electric Current

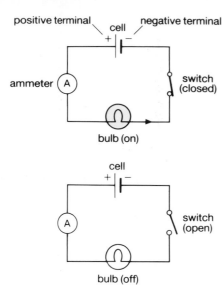

Figure 1
An electrical circuit

Figure 2
A current is a flow of charge

Conductors	Insulators
Good	rubber
metals e.g.	plastics c.g.
copper	polythene
silver	PVC
mercury	perspex
aluminium	china
steel	air
Moderate	
carbon	
silicon	
germanium	
Poor	
water	
humans	

Circuits

Figure 1 shows an electrical **circuit**. The bulb is connected to an **electric cell** by copper wires. When the lamp lights we say there is a current. The **ammeter** measures the current. The needle on the meter moves to show how big the current is.

A current is a flow of charge. In copper some of the electrons are free to move. When a current goes through the wire, electrons are repelled from the negative terminal (Figure 2).

A current flows only when there is a complete circuit without any gaps. The same idea works when water flows around your central heating system. You must have a complete pathway (or circuit), so the water goes from the boiler out to the radiators, and back to the boiler to be warmed up again.

In Figure 1 there is a switch. The simplest switch can be made out of a springy piece of copper. When the switch is open there is no conducting path for the electrons to flow round. Then the bulb is off.

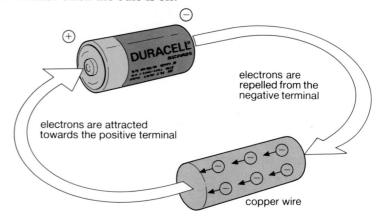

Which materials conduct?

The table (left) shows good and bad conductors, and insulators. The best conductors are metals; they contain electrons that are free to move. A current can also be carried by **ions**. The word ion is used to describe an atom (or molecule) that has lost or gained an electron. When an atom loses an electron it is a positive ion. When it gains an electron it is a negative ion.

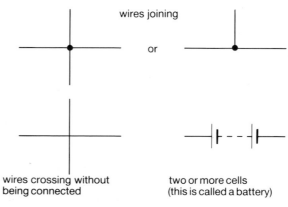

Figure 3
Circuit symbols for wires and cells

Your body is full of ions, so you conduct electricity. Electrical signals from your brain travel along nerves to give instructions to your muscles. Because you are a conductor, you can get a shock from the electricity mains supply.

Which way does current flow?

Figure 4 shows three examples of charge flowing. In all three cases the current is flowing to the right. The current is either carried by positive particles moving to the right, or negative particles moving to the left (or by both). It is easiest to say that current flows in the direction that positive charge moves in. So we say that current flows from the positive terminal of a cell to the negative terminal.

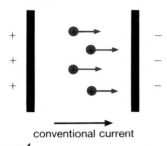

conventional current

Figure 4
(a) Positive particles in a semiconductor

Measuring charge and current

We measure the current in a circuit using an ammeter. The unit of current is the **ampere** (A), though most of us call it an amp.

When the current is big (10 A), the charge moves round the circuit quickly. When the current is small (0.001 A) the charge moves round the circuit slowly. Current is the rate at which charge flows round a circuit.

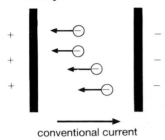

conventional current

(b) Electrons in a metal

$$\text{Current } (I) = \frac{\text{charge flowing } (Q)}{\text{time } (t)}$$

$$I = \frac{Q}{t}$$

We could measure charge by counting the number of electrons flowing. However, electrons have only a very small charge. Our unit of charge is the **coulomb** (C). 1 coulomb is equivalent to the charge on six million million million (6×10^{18}) electrons.

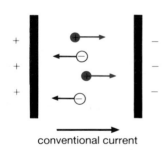

conventional current

(c) Positive and negative ions in a solution

Questions

1 (a) In circuit (a) (below) which of the switches must be closed for the bulb to light?

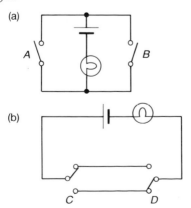

(a)

(b)

(b) Copy this table and fill in the third column to show if the bulb is on or off.

Switch C	Switch D	Lamp (on or off)
Up	Up	
Up	Down	
Down	Up	
Down	Down	

(c) What is the difference between the switches in the two circuits?
(d) Which circuit could be used for turning a light on or off at the top or bottom of a staircase?
2 Copy the circuit in the next column and mark in the direction of the current. Also show the direction in which the electrons are moving round the circuit.

3 In this circuit the ammeter A_1 reads 0.5 A. Which of the following statements is true: (i) A_2 reads 0.5 A, (ii) the current through the cell is zero, (iii) in 10 minutes, 5 C of charge flow through the bulb?

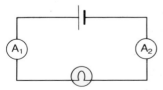

4 The bottom of a thundercloud stores about 10 C of charge. A flash of lightning lasts for about 0.001 s. Work out the current that flows between the Earth and the cloud during a flash of lightning.

4 More about Circuits

The thick cables on overhead pylons are made from aluminium. This transmission line uses a voltage of 440 kilovolts. To minimise heat loss, high voltages such as this are always used for the long distance transmission of electricity. In this case only 3% of the energy carried by these wires is wasted as heat

Splitting the current

We cannot lose current as it goes round a circuit (see Figure 1). The same current flows through bulbs B and C (and through the cell). The bulbs are **in series**. You can measure the current going into the bulbs and out of the bulbs with two ammeters. Both ammeters read 1 A. If they did not read the same, we would have lost some electrons.

Exactly the same idea applies if we have a circuit with branches in it. In Figure 2 a current of 2 A goes through bulb D. Then the current splits to go through bulbs E and F; each bulb gets a current of 1 A. But then 2 A flows back to the negative terminal of the cell.

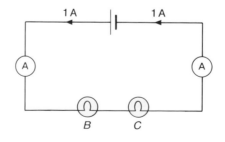

Figure 1
An electrical circuit

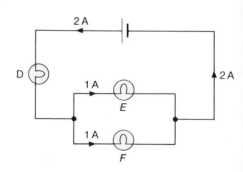

Figure 2
A current is a flow of charge

The currents do not always split equally. More current goes through the easier path. In Figure 3 it is easier for the current to go through bulb *I*, than it is to go through bulbs *G* and *H*. So 2 A goes through bulb *I*, and 1 A through the other two bulbs.

Figure 4 shows a **short circuit**. Bulb *J* has been shorted out by some copper wire. The current takes the easy path through the wire. No current goes through the bulb *J* so it is off.

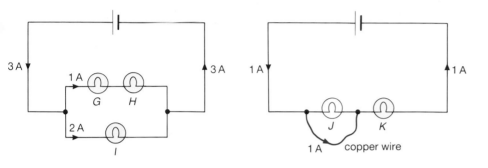

Figure 3   *Figure 4*

Voltage

Look at the circuit in Figure 5. Will the bulb work if you take the cell away? Of course it would not. The cell provides the energy for the bulb to light. The cell turns chemical energy (from the substances inside it) into electrical energy. The cell gives energy to the electrons to move around the circuit. When the electrons reach the bulb their energy is turned into heat and light energy.

There is an electrical energy difference between the positive and negative terminals in the cell. This energy difference is measured by the cell's **voltage**. (Voltage is also known as **potential difference** or **p.d.**) We use a voltmeter to measure voltages. Notice that the voltmeter is connected across the cell (**in parallel**), not in series with it.

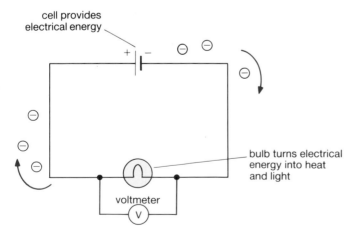

Figure 5

CURRENT RULES
- All parts of a series circuit receive the same current
- The same current goes in as comes out
- More current goes through the easier path

Many cells down the backbone on this electric eel act to produce a large voltage between its nose and tail. This voltage can be used to kill prey

High voltages can kill!

More about Circuits

VOLTAGE RULES
- Bulbs in parallel receive the same voltage
- Adding the voltage across the bulbs gives the battery voltage
- More volts means more energy

The more energy a cell or battery can provide, the larger its voltage. A transistor radio uses a 9 volt (9 V) battery. If you put your tongue across the terminals of a 9 V battery you can feel a little tingle. (DON'T use a battery of more than 9 V). Mains electricity is supplied at 240 V. A shock from the mains gives your body a lot of energy. A mains shock is very painful and could kill you.

Figure 6 shows a circuit to supply two headlamps from a car battery. The bulbs are in parallel and each gets the 12 V from the battery.

If you put the two bulbs in series, they are dimmer. Each one only gets half the battery voltage (Figure 7). The battery voltage does not always split equally. A large bulb uses more energy. So it gets more of the voltage (Figure 8). Notice that the sum of the voltages across the bulbs is always equal to the battery voltage. This is because the energy provided by the battery equals the energy used by the bulbs.

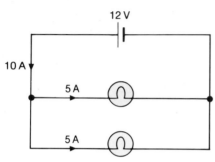

Figure 6
A circuit to supply two headlamps from a car battery

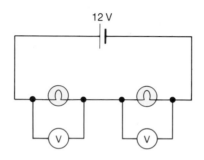

Figure 7
Each voltmeter reads 6 V

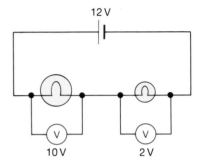

Figure 8
The larger bulb gets the larger voltage

Questions

1 Which current in the diagram below is the largest, *p*, *q* or *r*? Which current is the smallest?

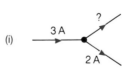

2 Bulbs *X* and *Y* (below) are identical. A_3 reads 0.2 A
What do ammeters A_1 and A_2 read?

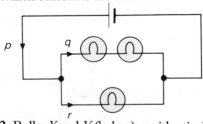

3 Copy the diagrams below and mark in the missing values of current.

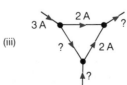

4 Some industrial machines use a 415 V supply. Explain why no machines use a 415 V supply in your home.

5 Explain carefully the energy changes that occur when a cell is connected to light a bulb.

6 In the circuit below, a voltmeter connected across *AB* measures 6 V. What does the voltmeter measure across:
(i) *CD*, (ii) *GH*, (iii) *FI*, (iv) *EI*, (v) *DI*, (vi) *DH* (quite hard)?

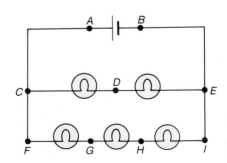

5 Resistance

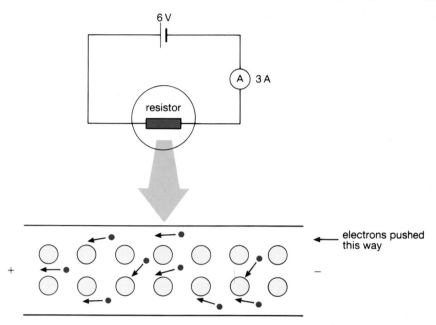

Figure 1
Moving through a resistor, electrons keep bumping into atoms. This makes the resistor hotter

Porcelain is an excellent electrical insulator, so it is used to isolate electrical cables from metal pylons

Figure 1 shows a resistor in a circuit. If we want to calculate the current flowing through it we need to know how much **resistance** it has. A resistor that has a large resistance only allows a small current through it. A small resistance allows a large current through.

$$\text{Resistance } (R) = \frac{\text{p.d. across resistor, in volts } (V)}{\text{current flowing through it, in amperes } (I)}$$

$$R = \frac{V}{I}$$

The unit of resistance is the ohm, symbol Ω. Large resistances are measured in thousands or millions of ohms. So, $1000\,\Omega = 1\,k\Omega$ (1 kilohm); and $1\,000\,000\,\Omega = 1\,M\Omega$ (1 megohm).

In an electrical circuit it is the resistors, not the wires connecting them to the battery, that get hot. The enlarged picture of a resistor in Figure 1 helps to show you why. Electrons are being pushed around the circuit by the battery and they collide with the atoms in the resistor. Energy is given to the atoms so that they vibrate faster – this means the resistor is getting hotter.

Measuring resistance

The simplest way for you to measure resistance is shown in Figure 2. The voltage across the resistor is measured using the voltmeter. The current flowing through the resistor can be read from the ammeter. Then the resistance can be calculated using the formula: $R = V/I$. A voltmeter has a large resistance so hardly any current flows through it. If current does flow through the voltmeter, the ammeter measures the extra current, and this will give you the wrong answer.

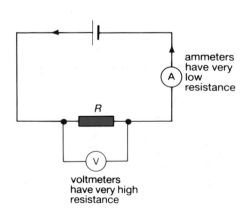

ammeters have very low resistance

voltmeters have very high resistance

Figure 2

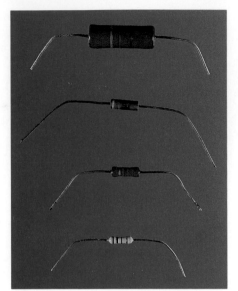

Some practical resistors. The various coloured bands show the value of the resistor and the accuracy with which it is made

You can increase the current flowing through your resistor by increasing the number of cells. You can then plot a graph of the current against the voltage. If your resistor is made from a metal or from carbon, your graph will look like Figure 3. The graph shows us that the resistance of the resistor does not change as the current increases. Such a resistor is said to be **ohmic** – one that obeys **Ohm's Law**.

Ohm's Law states that for some conductors, the current flowing is proportional to the voltage, provided the temperature does not change.

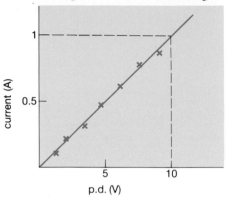

Figure 3

You can use the circuit in Figure 4 to see how the resistance of a wire is proportional to its length. A resistance wire is stretched between *AB* and a steady current of 1A flows through it. A voltmeter is attached to one end of the wire at *A*, and the other end of the voltmeter can be connected to any point on the wire. Table 1 shows some typical results. You can see that double the length of wire gives double the resistance.

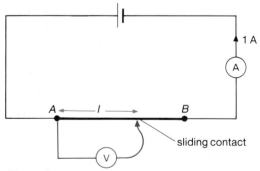

Figure 4
The resistance of a wire becomes bigger, the longer it is

Voltage (V)	Current (A)	Resistance (Ω)	Wire length (cm)
2.0	1.0	2.0	15
4.0	1.0	4.0	30
8.0	1.0	8.0	60
12.0	1.0	12.0	90

Table 1

Combination of resistors

- **Series**. The last experiment with a wire makes it easy to make a rule to work out the total resistance of two or more resistors in series. If the length of the wire is doubled the resistance is doubled. So when resistors are in series we simply **add** them to work out the total resistance. You can see in Figure 5 that the resistance between *X* and *Y* is 20 Ω.

Figure 5
Resistors in series: $R = R_1 + R_2 + R_3$

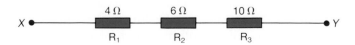

- **Parallel.** Working out how to combine resistors in parallel is harder. Figure 6 shows you an example. The most important point you should realise is that the resistance between X and Y must be *less* than either of the 3 Ω or 6 Ω resistors. When you put a resistor in parallel with another, you increase the *total* current, therefore you have made the resistance *smaller*.

  Using the equation $I = V/R$ you can see that a 6 V battery will drive 1 A through the 6 Ω resistor, and 2 A through the 3 Ω. So the *total* current flowing out of the battery is 3 A. You can now calculate the combined resistance of the two resistors.

$$R = \frac{V}{I} = \frac{6\,V}{3\,A} = 2\,\Omega$$

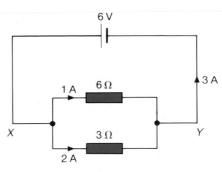

(a) Resistors in parallel

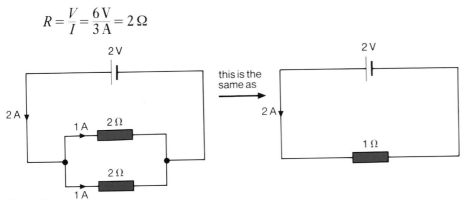

Figure 7
Putting two resistors of the same value parallel halves the resistance

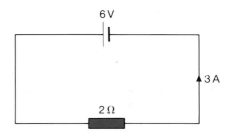

(b) The total resistance is smaller; in this example the two resistors in (a) are the same as one 2 Ω resistor

Figure 6

Questions

1 Use the data in Table 1 to calculate the resistance of 1 m of the wire.
2 Figure 3 shows an I/V graph for a resistor. Calculate its resistance.
3 Calculate the resistance of the following combination of resistors.

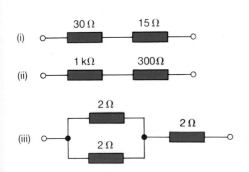

4 (a) Work out the current flowing through: (i) the 12 Ω resistor, (ii) the 4 Ω resistor, (iii) the battery.
(b) What single resistor has the same value as these two resistors which are connected in parallel?

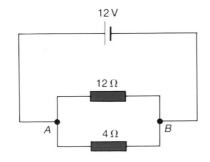

5 How does the resistance of a wire depend on its thickness? Bob chose five different thicknesses of resistance wire, each of length 1 m. Here are the measurements he made.

Diameter (mm)	Cross-sectional area (mm²)	Resistance (Ω)
0.374	0.110	10
0.315	0.078	14
0.264	0.055	20
0.189	0.028	40
0.156	0.019	56

(a) Why did he cut each wire to the same length?
(b) What conclusions can you draw from his experiment?
(c) Bob cuts a length of 3 m of this type of wire which has a cross-sectional area of 0.22 mm². What is its resistance?

6 Changing Resistance

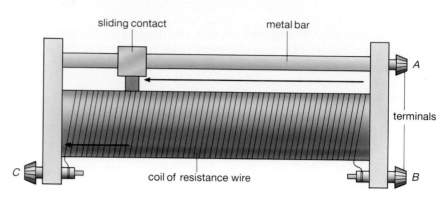

Figure 1
A variable resistor

Variable resistors

In Figure 1 you can see one type of **variable resistor**. It can be used as a fixed resistor by using terminals B and C, when the current goes through the full length of resistance wire. Or, it can be used as a variable resistor, when you use terminals C and A. The current now flows along the thick metal bar, then down through the sliding contact and along the coil of resistance wire to C. The metal bar has a very small resistance, so the further the contact slides towards C, the less the resistance becomes. When the sliding contact is half way down the bar, the resistance between A and C is about half of the total resistance of the coil between B and C. Figure 2 shows the circuit symbols for a variable resistor.

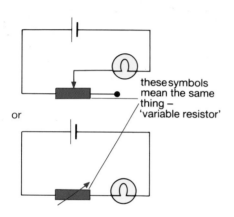

these symbols mean the same thing – 'variable resistor'

or

Figure 2
These circuits show the symbols for a variable resistor

The thermistor

A **thermistor** is a resistor whose resistance changes considerably with temperature (Figure 3). At low temperatures a thermistor's resistance is high, but as the temperature rises the resistance becomes less. At low temperatures, electrons are fixed on to atoms and so cannot move. As the electrons get hotter they receive enough energy to escape from their atoms, so the thermistor becomes a better conductor. Materials whose resistances change in this way are known as semiconductors. Carbon and silicon are two materials whose resistances decrease as they get hotter.

Resistance of a light filament

The filament of a light bulb also changes with temperature, but its resistance gets bigger as the temperature rises. In this case the number of electrons carrying the current remains constant as the temperature rises. However, the increased vibrations of the atoms as they get hotter makes it harder for the electrons to pass through the filament, so the resistance increases.

You can investigate the resistance of the light filament using the circuit in Figure 4(a). The variable resistor allows you to control the brightness of the bulb. Moving the slider towards Y reduces the resistance, R, so more current flows, making the bulb brighter. Moving the slide towards X dims the bulb. You can show your results by plotting a graph of the current against voltage across the bulb (see Figure 4(b)).

You can see that the current is *not* proportional to the voltage, so the light filament does not obey Ohm's Law. Some simple calculations shown on the graph show you how the resistance changes. At point A, where the current is only 0.2 A and the bulb is quite cool, the resistance is 5 Ω. At B the bulb is hot, having a current of 0.4 A going through it, and the resistance is now 15 Ω.

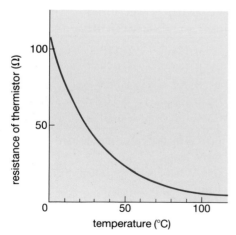

Figure 3
The resistance of a thermistor changes with temperature

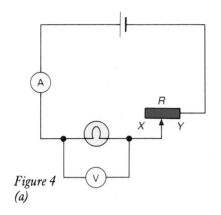

Figure 4
(a)

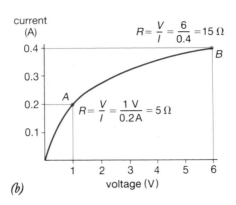

(b)

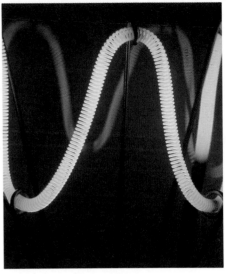

A close-up of the tightly coiled tungsten filament in a light bulb

Manufacturing light bulbs

Making a light bulb is quite a difficult problem. The wire used to make the filament is *tungsten*, which has a melting point of 3400°C. When the filament is working, its temperature is about 2500°C. At that temperature most other metals have melted. For the light bulb to work on the mains voltage of 240 V, the filament needs a high resistance. This is usually about 1000 Ω. As you know, metals are good conductors, so metal wires have low resistances. 1 m of tungsen wire with a diameter of 1 mm has a resistance of 90 Ω (at 2500°C). To make a filament have a larger resistance it must be made of very fine wire (thinner than your hair). The 1 m length of filament is coiled tightly, and then coiled again to fit into the bulb, as shown in the photograph.

You might have thought that the problems were over now. The next difficulty is that tungsten at high temperatures would oxidise in air, so the filament is placed in a mixture of argon and nitrogen gases. Even so, the filament's life is limited because at such high working temperatures it tends to evaporate. Next time a light bulb blows at home, look at it. You will see that the glass bulb has blackened due to the evaporated tungsten. Nevertheless the average light bulb has a lifetime of about 1000 hours. That is excellent value for money, at a cost of about 40p!

Questions

1 Figure 1 shows a variable resistor. The resistance of the coil between B and C is about 100 Ω. Estimate what resistance you would find if you used terminals: (i) A and B, (ii) A and C.
2 Use the graph in Figure 3 to estimate the resistance of the thermistor at: (i) 30°C, (ii) −10°C.
3 (a) The graph shows I/V graphs for three electrical components X, Y, Z.
(i) Which one obeys Ohm's law?
(ii) Which one could be a light bulb? Explain your answers.

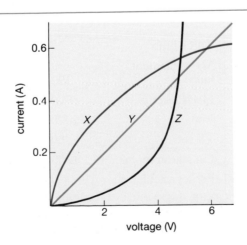

(b) The electrical components X, Y, Z are now all put in series. The current flowing through them is 0.2 A. (i) What is the voltage across each component of the circuit? (ii) Now work out the resistance between p and q. (iii) Will the resistance be the same when the current is 0.4 A? Explain your answer.

7 Electrical Power

Milk floats such as this one can deliver milk within an 11 kilometre radius of the depot. They have a top speed of just over 20 kilometres per hour. Some 15 batteries are used to provide power and after a typical day's use it takes around 10 hours to recharge them. Why are there so few electric cars around?

Measuring the power

When you switch on a fire to warm up a room, electrical energy is changed into heat energy. If you want to know how quickly the room is going to warm up you need to know the rate at which electrical energy is used, or the power used by the fire. **Power used = Energy used/second** (see page 82).

You can work out the electrical power, P, used in a circuit, if you know the voltage of the supply, V, and the current, I, by using the formula:

$$P = V \times I$$

Power is measured in **joules per second** (J/s) or **watts** (W).

Often power used comes in rather large units. For example 1 bar of an electric fire uses about 1000 W; this is called a **kilowatt (1 kW)**. If you want to talk about power stations you will need to use **megawatts**. 1 megawatt is 1 000 000 W. A large power station can produce 2000 MW of electrical power. In the cold spell of January 1987, Britain's peak power usage was 48 300 MW. Table 1 shows you the power used, and the operating currents and voltages of several electrical devices.

This photograph shows a battery charger. These 1.5 V nickel-cadmium cells need a charging current of 120 mA for 15 hours to charge them fully.
How much energy do they store?
Which way round must they be connected to the voltage supply?

Device	Power (W)	Operating voltage (V)	Current (A)
Torch bulb	0.9	3	0.3
Mains filament bulb	100	240	0.4
Electric kettle	3000	240	12.5
Iron	1000	240	4
TV set	60	240	0.25
Car starter motor	1200	12	100
Pocket calculator	0.0003	3	0.0001
Milk float	3600	72	50

Table 1.

Electricity bills

What is it that you pay the electricity board for? You pay for the energy that you use and it costs you about 7p for 1 unit. The unit that the electricity board uses is the kilowatt hour (kWh). If you leave a 1 kW fire on for 1 hour then you have used 1 kWh of energy.

The kilowatt hour does not look like a unit of energy; but it is. We can turn a kilowatt hour into joules like this:

$$\text{Energy used} = \text{Power} \times \text{time}$$
$$= 1000\,\text{W} \times 3600\,\text{s}$$
$$= 3\,600\,000\,\text{J}$$

Example. What does it cost to cook a turkey for 6 hours using a 2 kW electric oven?

$$\text{Energy used} = 2\,\text{kW} \times 6\,\text{h}$$
$$= 12\,\text{kWh}$$

Each kWh costs 7p, so total cost = $12 \times 7\text{p} = 84\text{p}$

Running a large company can be an expensive business

Questions

1 (a) A mains bulb is marked 240 V, 40 W and a torch bulb is marked 6 V, 3 W. Explain what the markings mean.
(b) Calculate the current that flows through each bulb when they are working normally.

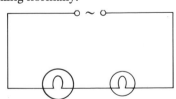

(c) Calculate the normal working resistance of each bulb.

(d) Explain what will happen if both bulbs are put in series across the mains as shown.

2 (a) Calculate the current flowing in this circuit.

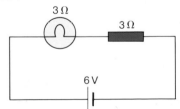

(b) What power does the battery deliver to the circuit?

(c) How much power is the light bulb using?

3 An electric kettle has an element of resistance 24 Ω. Calculate how much power it uses when it is connected to the mains (supply voltage 240 V).

4 A houseowner goes away for two weeks holiday. To deter burglars she leaves on two 100 W light bulbs in the house while she is away. If 1 kWh of electricity costs 7p, how much do these bulbs cost her over the two weeks?

5 What does it cost to use a 1 kW iron for 30 minutes?

8 Electricity in the Home

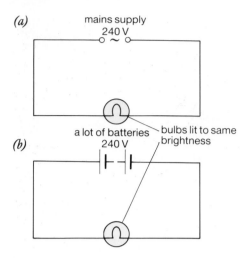

(a)
mains supply
240 V

(b)
a lot of batteries
240 V
bulbs lit to same
brightness

(c)

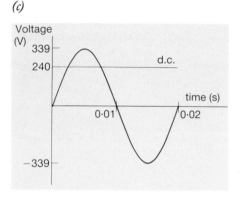

Voltage
(V)
339
240 — d.c.
time (s)
0·01   0·02
−339

Figure 1

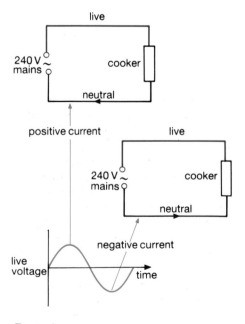

live
240 V
mains
cooker
neutral

positive current

live
240 V
mains
cooker
neutral

negative current

live
voltage
time

Figure 2

Alternating current

The electricity supply to your home is an **alternating current** supply (**a.c.**). The voltage of the supply is 240 V. Figures 1(a) and (b) show two light bulbs, one on a 240 V mains a.c. supply and the other on a 240 V d.c. supply. They both have the same brightness, so each supply delivers the same power to the light. Figure 1(c) shows how the voltage across each lamp varies with time. For the d.c. supply the voltage is constant at 240 V. However, for the light powered by the a.c. supply, the voltage across it changes from +339 V to −339 V. When the voltage becomes negative the lamp still works. It simply means that the current is flowing the other way round – just like turning the battery around in a circuit. 339 V is the peak voltage of the mains supply. The reason that the peak voltage is bigger than 240 V is that for part of the time the mains voltage is near to zero, and at those times little power is given to the lamp. So the peak voltage has to be more than 240 V, to make up for those times when little power is given to the light bulb.

We call 240 V the **root mean square** (*r.m.s.*) **value** of the mains supply. In general

$$r.m.s.\ voltage = \frac{peak\ voltage}{\sqrt{2}}$$

The frequency of the mains supply is 50 Hz, which means that there are 50 complete voltage cycles per second.

Mains supply to your house

The mains electricity supply comes into your house on two cables called the **live** and **neutral**. In any circuit, you must have two wires. The current comes into the house on one wire and returns to your local substation on the other. The live wire is the dangerous wire – the voltage on this wire changes between +339 V and −339 V. The voltage on the neutral wire is close to zero. Figure 2 illustrates how your cooker is wired to the mains. Notice how the direction of current flow reverses during one voltage cycle.

Figure 3 shows a plan of your house wiring. The supply comes through a main fuse and the electricity meter, and then to the fuse box. The fuse box is the distribution point for your house's electricity supply. In the box there are about six fuses which lead to different circuits in the house. The size of fuse depends on the current that flows in the circuit. Cookers, immersion heaters or electric shower heaters use large currents and so have their own circuit.

- Your house will have two or three **ring main circuits**, which supply all the wall sockets. On each ring main there are usually about 10 sockets. Notice that all the sockets are in parallel, so the full mains voltage is supplied to everything that is plugged into a socket. The advantage of using a ring main circuit is that current can flow two ways into a particular socket. So the connecting wires can be thinner, because they carry a smaller current than they would do otherwise. In addition to the live and neutral wires, the ring main circuit carries an **earth wire**. The earth wire is there for reasons of safety (see page 205).

- The lights for your house have their own circuit. Again each light fitting is in parallel, so that each light bulb receives the mains voltage of 240 V. Light bulbs draw a small current (about 0.4 A for a 100 W bulb) so about 10 lights can be safely run through a 5 A fuse.

Circuits in buildings are usually protected by fuses, which 'blow' if too much current flows through them. This fuse box in an office building has fuses for a large number of circuits. Circuit breakers are now often used instead of fuses.

⚠️ **It is important that the supply is turned off before fuses are changed.**

Questions

1 Explain what is meant by an alternating current.

2 Why do we say that the mains voltage is 240 V, when its maximum value is 339 V?

3 What is the advantage of using a ring main circuit?

4 What is the greatest number of 60 W bulbs that can be run off the mains, if you are not going to overload a 5 A fuse?

5 A hot water tank containing 0.2 m³ of water (200 kg) is to be heated from 15°C to 40°C, by a heater drawing a 15 A current from the mains supply (240 V). (i) What is the power of the heater? (ii) What is the resistance of the heater? (iii) How much energy is used to warm up the tank?. (1 kg of water needs 4200 J to warm it up by 1°C), (iv) How long will it take to warm up the tank, (v) How much does it cost to warm up the tank, if 1 kWh costs 7p?

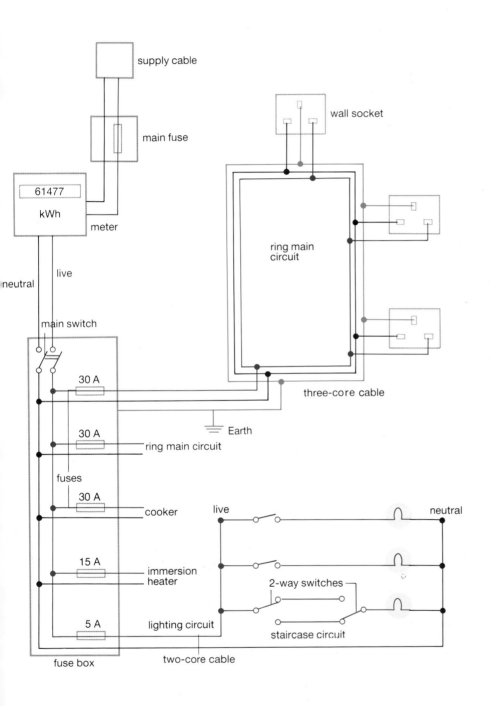

Figure 3
A house electricity supply

supply cable

main fuse

61477
kWh

meter

neutral   live

main switch

wall socket

ring main circuit

three-core cable

Earth

30 A

30 A   ring main circuit

fuses

30 A   cooker   live   neutral

15 A   immersion heater   2-way switches

5 A   lighting circuit   staircase circuit

fuse box   two-core cable

9 Electrical Safety

Although electricity is very useful to us it can also be very dangerous. Electrical faults can cause fires which damage property and sometimes result in injury or death. You can also give yourself a very painful electric shock from the mains, and electric shocks do occasionally kill people. The purpose of this unit is to teach you how to keep your home safe.

This adaptor is being used dangerously; far too many devices are plugged into it. Question 3 opposite shows the problems that can occur

Fuses

The main purpose of a fuse is to prevent a fire, although a fuse can also prevent damage to expensive electrical equipment. In electrical cables that are bent and twisted it is quite common for the electrical insulation to crack and break. For example this could easily happen with your electric lawn mower or an iron. As a result of the broken insulation, the live wire could come directly into contact with the neutral wire. This would short-circuit the electrical device (Figure 1).

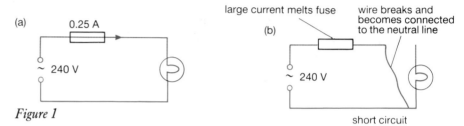

Figure 1

Normally, the current flowing through the connecting wires is quite small: 0.25 A in our example. By contrast, if there is a short circuit, the resistance in the circuit is so small that the current becomes very large: 100 A or more. Such a large current would cause enormous heating and a fire is the likely result. However, a fuse protects against a fire. A fuse contains a thin piece of resistance wire which melts easily if the current becomes too large. So when a fault happens, the fuse melts and the circuit is broken (Figure 1(b)).

The fuse is put into the *live* wire. Then if a fault occurs and the fuse melts, the live (dangerous) wire is disconnected. Switches are also put into the live wire for the same reason. If a switch is put into the neutral wire, you could get a shock from an electric fire element even though it is switched off (Figure 2).

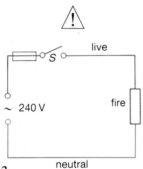

Figure 2
Switches, as well as fuses, are put into the live wire

All devices are protected by a fuse in the plug, and if that fails, the main fuse box in the house also has a fuse for each circuit. Nowadays, it is quite common to find **circuit breakers** in houses. If the current becomes too big the circuit breaker throws open a switch which can be reset when the fault has been corrected.

Choosing a fuse

When wiring up a plug you should always choose the right fuse for the device you are going to use. For example, what fuse should you use for a 1000 W iron?

$$P = V \times I$$

$$I = \frac{P}{V} = \frac{1000\,\text{W}}{240\,\text{V}} = 4\,\text{A}$$

The most commonly used fuses in the home are 3 A, 5 A, and 13 A. So you should choose the 5 A for your iron, then the fuse will blow if any small extra current flows due to a fault. Table 1 shows some appliances and their recommended fuses.

This extension lead has a 13 A fuse. This is a safe way of plugging many things into one socket

Appliance	Power (W)	Normal Current (A)	Recommended Fuse (A)
Lamp	100	0.4	3
TV set	70	0.3	3
Hair dryer	500	2.1	5
Toaster	1200	5.0	13
Kettle	2750	11.5	13

Table 1

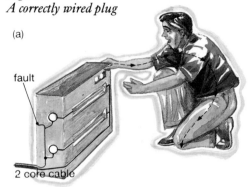

Figure 3
A correctly wired plug

Wiring a plug

You will often want to wire up a new plug to some electrical device. Don't rush at it; it will take you at least 15 minutes to do it well. Figure 3 shows you what it should look like when finished. These are the points to watch for:

- Make sure the wires are in the right place; earth is green/yellow, live is brown and neutral is blue.
- Do not strip the insulation back too far.
- Make sure the main thick cable is held in the cord grip.
- Do the screws up well.
- Choose the right fuse.

Earthing appliances

The pictures in Figure 4 tell a story. In (a) a fault has developed in an electric fire and the live wire has come into contact with the metal casing. The man gets a shock. In (b) the casing was attached to earth, so as soon as the live wire touched the case, a large current flowed and the fuse blew. In the second case the man does not get a shock.

A lot of modern appliances are made out of plastic, such as the bases of electric lamps. In these cases there is no need for an earth wire since plastic does not conduct electricity. This method of protection is called **double insulation** and it is shown by the sign ▣

(a)

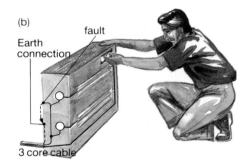
(b)

Figure 4

Questions

1 (a) Explain why we need fuses and how they work.
(b) Why is it dangerous to put a fuse into the neutral wire?
2 The diagram shows a thermal circuit breaker. Explain how it works.

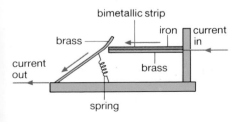

3 (a) Why is it dangerous to touch an electrical supply with wet hands?
(b) Explain three special precautions taken over the use of mains electricity in the bathroom.
(c) What is an earth leakage trip? Why are some electric lawnmowers fitted with one?

4 The devices below are all drawing current through an adaptor plugged into a single socket.
(a) Copy the table and fill in the column to show how much current each device takes.

Device	Power	Voltage	Current
Kettle	3 kW	240 V	
Fan heater	2 kW	240 V	
Iron	1 kW	240 V	

10 Electron Beam Tubes

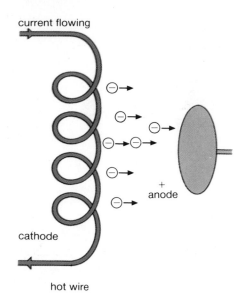

current flowing

⊖→

⊖→
⊖→
⊖→⊖→

⊖→

+
anode

⊖→

cathode

hot wire

Figure 1
Thermionic emission

⚠ There are high voltages inside a
CRO or TV set

Thermionic emission

When a current goes through a wire, the wire gets hot. At very high temperatures
some electrons have enough energy to escape from the wire. If the wire is charged
negatively the electrons are repelled and can be collected by a positive plate
(Figure 1). The positive plate is called an **anode** and the negative wire the
cathode. When this effect (thermionic emission) was first discovered, the emitted
electrons were called **cathode rays**, because they were emitted from the cathode.

The cathode ray oscilloscope (CRO)

Figure 2 shows a **cathode ray oscilloscope (CRO)**. Electrons are accelerated
away from a heated filament by the pull of a positive anode. Then the electrons
travel through a vacuum until they hit a fluorescent screen. The CRO produces a
very fine beam of electrons, which produces a small spot of light on the screen.
The beam is focused by the two anodes. The brightness of the spot is controlled by
the grid. If the grid is charged slightly negative, electrons from the filament are
repelled so that fewer of them hit the screen. This makes a smaller, duller spot.

The electron beam can be deflected vertically or horizontally by charging the
Y-plates or the X-plates. Figure 3 illustrates how the X and Y plates may be
charged to deflect the beam. In this diagram you are looking down the length of
the oscilloscope, through the screen towards the deflection plates. Notice that
when alternating current is applied to the Y-plates a vertical line is produced. As a
result of the alternating current, the charge on the Y-plates is switching around
very rapidly. Sometimes the top plate is positive, but a fraction of a second later it
is negative. This makes the electron beam switch quickly from being deflected
upwards to being deflected downwards. The switching occurs so rapidly that the
spot on the screen has no time to fade, and we therefore see a continuous straight
line.

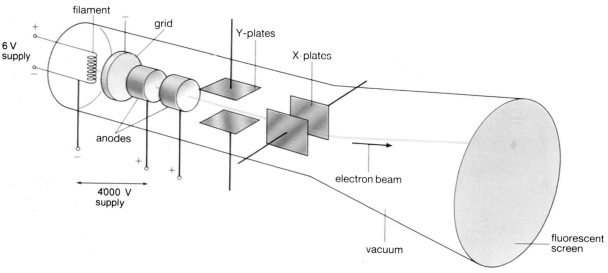

filament

grid

Y-plates

X-plates

+
6 V
supply
−

anodes

+
+

−

4000 V
supply

electron beam

vacuum

fluorescent
screen

Figure 2
A cathode ray oscilloscope (CRO)

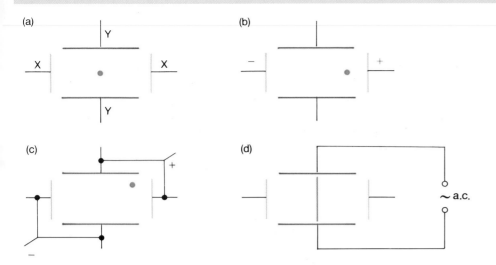

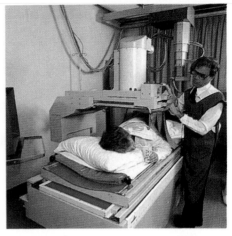

The radiographer (wearing a protective apron) positions the X-ray source over the patient's back prior to making an exposure

⚠ Overexposure to X-ray radiation is dangerous.

Figure 3
Negatively-charged electrons are attracted towards the positive plate and repelled from the negative plate

X-ray production

Thermionic emission is also used to produce X-rays. Figure 4 shows the principle of an X-ray tube. Electrons are emitted from a hot filament. A high voltage (10 kV or so) is applied between this filament and an anode. The anode is made from a dense metal such as tungsten. When the electrons strike the tungsten, X-rays are produced. The higher the voltage, the shorter the wavelength of the X-rays. Short wavelength X-rays are more penetrating than longer wavelength rays. A greater intensity of X-rays can be produced by increasing the current in the filament. More electrons are emitted as the filament's temperature rises. You can find out about the uses and hazards of X-rays in sections G6, L7 and L9.

Questions

1 Look carefully at Figure 2.
(a) (i) Does it matter which way round the 6 V battery is connected to the filament? (ii) Does it matter which way round the 4000 V supply is connected? Explain your answers.
(b) What change would you see if you reduced the voltage between the cathode and the anode to 3000 V?
(c) What would happen if the vacuum tube developed a leak?
2 Draw diagrams similar to those in Figure 3 to show the deflection of the spot in an oscilloscope in the following cases:
(a) top Y-plate negatively charged, bottom Y-plate positively charged;
(b) alternating current supplied to both Y and X plates together.
3 Look at Figure 4. What changes will you notice in the X-rays if these two changes are made:
(i) the anode–cathode voltage is reduced to 9 kV;
(ii) the filament voltage is increased to 8 V?

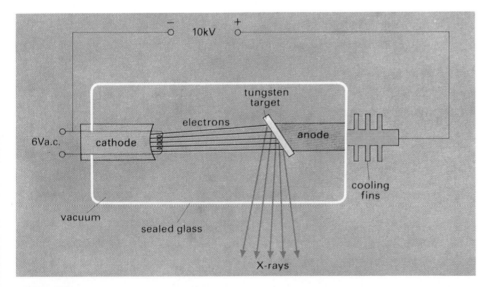

Figure 4
An X-ray tube

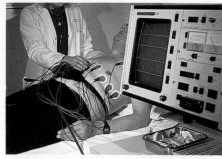

Cathode ray tubes are used in electrocardiograph (ECG) machines which monitor the electrical activity of the heart. This photograph shows an ECG machine being used to investigate whether the patient has suffered a heart attack.

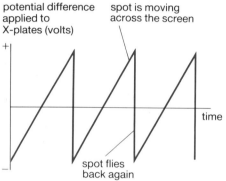

potential difference applied to X-plates (volts)

spot is moving across the screen

+

time

spot flies back again

Figure 1
How the voltage applied to the X-plates varies with time

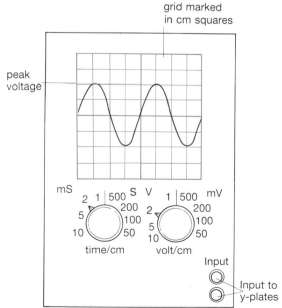

grid marked in cm squares

peak voltage

mS

2 1 500 S V 1 500 mV
200 200
5 100 2 100
10 50 5 50
10

time/cm   volt/cm

Input

Input to y-plates

Figure 2
A typical CRO signal

TV studios use Cathode ray tubes as monitors. This picture shows the presentation control room during transmission. From here, the transmission controller keeps a close watch on the schedule and on the technical quality of the programmes

The time base

In the last unit you read how the X and Y plates in an oscilloscope can be used to deflect the electron beam. What really makes the oscilloscope useful to us is its time base. The **time base circuit** works like this. A changing voltage is applied to the X-plates of the CRO, so that the spot moves across the screen from left to right. The moment the spot reaches the right-hand side of the screen it returns immediately to the other side to start its journey across the screen again. Figure 1 shows how the voltage applied to the X plates varies with time.

The time base allows us to look at changing voltages (*signals*). The CRO plots a graph of voltage (*y*-axis) against time (*x*-axis). We can apply a signal to the Y-plates so that the spot is moving up and down. At the same time the time base makes the spot move sideways. A typical signal is shown in Figure 2. In this diagram the right-hand dial shows you that each centimetre in a vertical direction means a voltage of 2 V. So the peak voltage is 4 V. The left-hand dial shows you that the beam crosses the screen horizontally, taking 2 ms (0.002 s) for each centimetre. One complete cycle of the voltage waveform takes 4 squares of the grid, or 8 ms. The time period of the cycle is 8/1000 s. Therefore the frequency is 1000/8 = 125 cycles per second or 125 Hz.

Oscilloscopes are widely used in industry to monitor and display changing signals. They are also useful in hospitals where they are used to show the strength and frequency of a patient's heart beat, or to build up a picture after an X-ray or ultrasound scan.

Televisions

Getting our television pictures is a complicated process, but the television set itself is rather like a CRO. The photograph opposite shows a magnified picture of a TV screen. You can see that it is made up of lots of small dots which are blue, green or red. These are the three primary colours, and if they are mixed together they make white. Various combinations of the colours can make the other colours. Table 1 gives some examples.

To make a colour picture the TV set needs three electron beams, as shown in Figure 3. A beam of electrons leaves the 'green' gun and passes through a hole in the shadow mask. Then the beam hits a small fluorescent spot that gives out green light. Similarly electron beams leave the other two guns and hit fluorescent spots that give out blue and red light. The purpose of the shadow mask is to make sure the beams hit the right part of the screen. The small holes in it only allow the beam to hit a small part of the screen.

Like an oscilloscope, the television has a time base. The whole television picture is built up by the beams sweeping backwards and forwards across the screen very rapidly indeed. The beams go from top to bottom of the screen in about 0.025 seconds. During this time they have crossed the screen about 625 times.

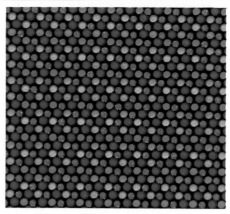

A close-up of a TV screen, showing the tiny dots that seem to blend together at a distance to make up a smooth picture

Blue	Green	Red	Mixture	
●	●	●	○	white
●	●	●	●	cyan
●	●	●		yellow
●	●	●	●	magenta

Table 1

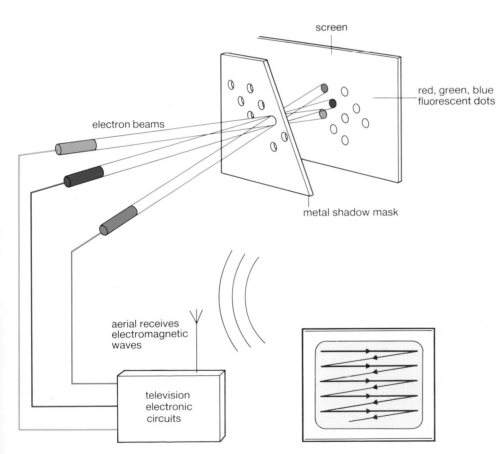

Figure 3
The principle of the colour television

Figure 4
The time taken for the beam to travel from top to bottom of a screen is 0.025 seconds

Questions

1 Look at the signal in Figure 2. Draw another diagram to show how the signal looks when you make these two separate changes: (i) increase the Y-gain to 1 V/cm, (ii) increase the time base speed to 1 ms/cm.

2 The diagram shows a waveform on an oscilloscope screen; the time base is set to 1 ms/cm and the Y-gain control to 2 V/cm. Calculate the peak voltage and frequency of the a.c. supply.

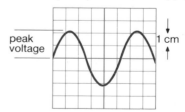

3 Explain why a colour television needs three electron beams. What colour do you see on the screen when two red spots are illuminated for every blue and green spot?

SECTION I: QUESTIONS

1 Diagram (1) below shows the inside of a mains-operated hair drier. The fan can blow either hot or cold air. Diagram (2) is a circuit diagram of the same drier showing how it is wired up.

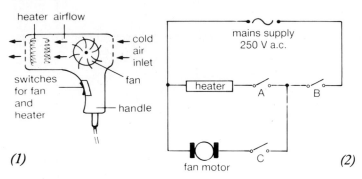

(1)

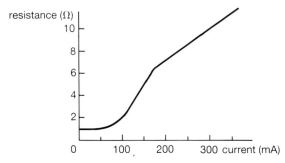

(2)

(a) Copy the table below and show, by placing ticks in the boxes, which switches need to be ON to get the result shown.
(You may use each switch more than once, once, or not at all.)

Result	Switch A	Switch B	Switch C
a blow of cold air	☐	☐	☐
a blow of hot air	☐	☐	☐

(b) The heater must not be on without the fan.
(i) Which of the switches, A, B, C, must always be on to achieve this?
(ii) Explain carefully what you would expect to happen if the fan failed to work when the heater was on.
(c) The manufacturer wishes to include a two-speed fan. This could be done by connecting a suitable resistor across one of the switches as follows:

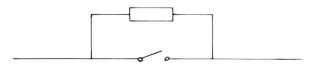

(i) Copy diagram (2) and draw a resistor across the correct switch in order to make a two-speed fan.
(ii) When this switch is open, will this give a fast or slow speed? Explain your answer.
(d) The details on the hair drier are 250 V, 500 W. Calculate the current from the mains supply when the drier is working at the stated power.
(e) Fuses for the mains of 3 A, 5 A and 13 A are available.
(i) Which fuse would you choose for use in the plug attached to the drier?
(ii) Which wire in the mains cable should be connected to the fuse?

(f) A girl needs to use the drier for 10 minutes. Calculate the energy converted during this time.
(g) The manufacturer makes a different hair drier which will work from a 12 V car battery. You are required to find the power taken by the 12 V drier. Copy and complete the circuit diagram below to show how you would connect up an ammeter and voltmeter to do this.

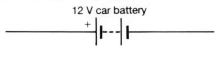

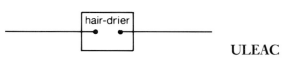

ULEAC

2 The graph shows how the resistance of a bulb varies with the current through it.

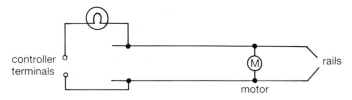

(a) What is the value of the resistance of the bulb when it is not switched on?
(b) The bulb can be used as a simple overload protection and warning device for use with model railway controllers. The bulb is placed in one of the wires leading to the track, as in the circuit diagram below.

When operating normally the current through a model railway locomotive is about 250 mA.
(i) What is the potential difference across the bulb when 250 mA passes through it?
(ii) If the output voltage of the controller is 12 V, what is the potential difference across the motor when 250 mA passes? (Assume the resistance of the rails is negligible.)

The train derails and short-circuits the two rails (i.e. it joins the two rails together by a good conductor).
(iii) State and explain what happens to the bulb when this occurs.   **MEG**

3 Two light bulbs (A and B) light normally when connected to a 5 V battery. The graph shows how the current varies with the applied voltage (potential difference) for each bulb.

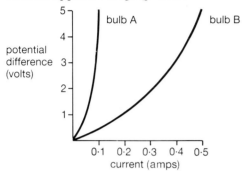

(a) When they are working normally, what is the resistance of each bulb?
(b) A 5 V battery is connected across XY in diagram (1).
(i) Which bulb will be brighter?
(ii) How much power is used by each bulb?

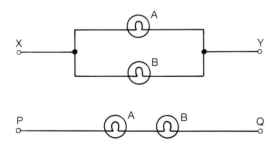

(c) The bulbs are now connected in series as shown in diagram (2). Describe carefully what happens as the voltage across PQ is increased slowly from 0 to 5 V.

4 The diagram below shows a **gold leaf electroscope**. This can be used for detecting charges. When the electroscope is uncharged the leaf hangs down. But when a charged rod is brought close to the plate at the top, the leaf rises. This is because the positive charged rod attracts the electrons to the top. The leaf and stem are now left positively charged so the leaf is pushed away from the stem.

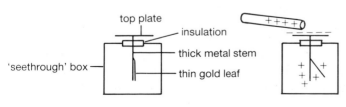

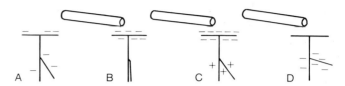

Negative charge is now placed on the electroscope. The leaf rises as shown in A.
(a) Three rods are brought close to the electroscope. What can you say about the sign of the charges on B, C and D?
(b) What can you say about the size of the charges on B and C?

5 In the circuit, R is a variable resistor. The table shows six different values of R. For two of these values, the current flowing, the voltage (potential difference) across R and the power used by R have been calculated.

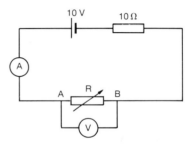

R (Ω)	Current (A)	Voltage across AB (V)	Power used in R (W)
2	$\dfrac{10}{12}$	$\dfrac{10}{12} \times 2$	1.4
5			
10			
15	$\dfrac{10}{25}$	$\dfrac{10}{25} \times 15$	2.4
20			
40			

(a) Explain how these values were calculated.
(b) Copy the table and fill in the missing values of current, voltage and power.
(c) Plot a graph of power (y-axis) against R (x-axis). For what value of R is the power used by R the greatest?
(d) What can we say about the power used by R when R is:
(i) very small – nearly zero;
(ii) very large – about 10 000 Ω?
Explain your answers.

QUESTIONS

6 (a) Calculate the readings on each of the four ammeters A_1, A_2, A_3, A_4, shown in diagrams (1) and (2).
(b) Calculate the power transformed in each of the resistors shown in the diagrams.

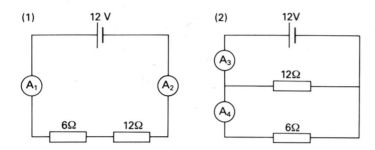

7 An electric iron is marked 240 V, 720 W.
(a) Explain the meaning of these markings.
(b) Calculate the normal working current of the iron.
(c) What is the resistance of the iron element?
(d) Which fuse would you use for the iron: 3 A, 5 A or 13 A?
(e) Explain why it is important to have a fuse.

8 The manufacturers of a new bulb claim that its long life and high efficiency will save you about £60 during its lifetime, if you use it rather than the normal bulb. It emits the same amount of light as a normal bulb, but it uses less power.

	Cost	Lifetime	Power rating
'long life' bulb	£12	12 000 h	20 W
normal bulb	50p	1 000 h	100 W

(a) Explain what is meant by 'high efficiency'.
(b) Use the data in the table to investigate the truth of the claim. (At today's prices, electricity costs 7p per kWh.)

9 In the circuit shown, a battery delivers energy to a resistor. This energy is measured by a joulemeter. After 5 minutes, 10.80 kJ of energy have been supplied to the resistor.

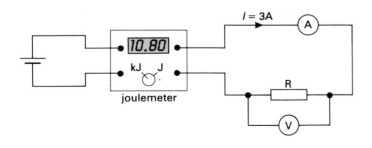

(a) Use the information in the diagram to calculate the number of coulombs that flowed round the circuit in 5 minutes.
(b) How much power was transferred to the resistor? (Use the formula: power = energy supplied/time.)
(c) Now work out the voltage across the resistor, R. (Use the formula: $P = V \times I$.)
(d) Calculate the energy carried by each coulomb of charge, using your answer to part (a). Your answer should confirm that the voltage measures the number of joules of energy carried by each coulomb of charge.
● A potential difference (voltage) of 1 volt exists across a resistor if 1 joule of energy is transferred to the resistor when 1 coulomb of charge passes through it.
(e) Use the definition of the volt to work out the speed of an electron reaching the anode of this electron gun. The mass of an electron is 9.1×10^{-31} kg and its charge is 1.6×10^{-19} C.

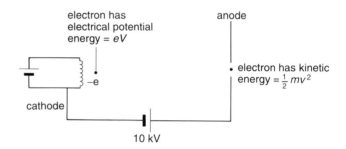

10 Below you can see a typical waveform on the screen of an oscilloscope. Also shown are the Y-gain and time base controls of the oscilloscope.
(a) What is the peak voltage of the signal?
(b) What is the time period of one oscillation of the signal?
(c) Calculate the frequency of the signal.
(d) Make a sketch to show what the trace would look like if two changes are made together: the time base changes to 1 ms per cm, and the Y-gain is changed to 2 V per cm.

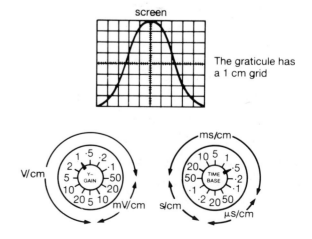

SECTION J

Magnetism and Electromagnets

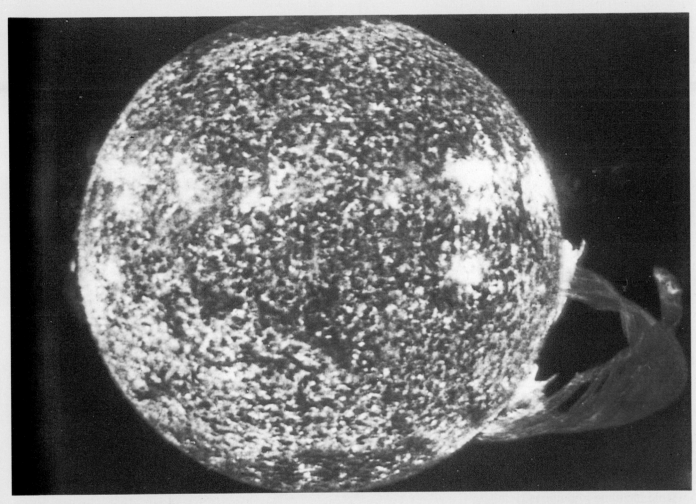

This photograph of the Sun, taken in ultraviolet light, shows a loop of material above the Sun's surface. Loops like this are called prominences. They are held up by the Sun's magnetic field, and are colossal. The Earth could fit many times into the prominence shown here

1 Magnets

Some metals, for example iron, cobalt and nickel, are **magnetic**. A magnet will attract them. If you drop a lot of pins on to the floor the easiest way to pick them all up again is to use a magnet.

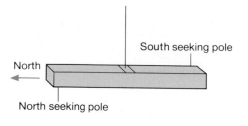

North seeking pole

South seeking pole

North

North seeking pole

Figure 1

Poles

In Figure 1 you can see a bar magnet which is hanging from a fine thread. When it is left for a while, one end always points north. This end of the magnet is called the **north-seeking pole**. The other end of the magnet is the **south-seeking pole**. We usually refer to these poles as the north and south poles of the magnet.

The forces on pins, iron filings and other magnetic objects are always greatest when they are near the poles of a magnet. Every magnet has two poles which are equally strong.

When you hold two magnets together you find that two north poles (or two south poles) repel each other, but a south pole attracts a north pole (Figure 2).

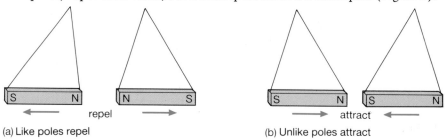

(a) Like poles repel

(b) Unlike poles attract

Figure 2

Magnetic fields

There is a magnetic field in the area around a magnet. In this area there is a force on a magnetic object. If the field is strong the force is big. In a weak field the force is small.

The direction of a magnetic field can be found by using a small plotting compass. The compass needle always lies along the direction of the field. Figure 3 shows how you can investigate the field near to a bar magnet, using a compass. We use magnetic field lines to represent a magnetic field. Magnetic field lines always start at a north pole and finish on a south pole. When the field lines are close together then the field is strong. The further apart the lines are the weaker the field is. (Magnetic field lines are not real, but they make a useful model that helps us to understand magnetic fields.)

⚠ Avoid handling iron filings. Wash your hands after use.

You can see the shape of a magnetic field by using iron filings as in this photograph

Plotting compass

Figure 3

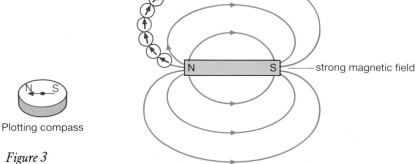

weak magnetic field

strong magnetic field

Combining magnetic fields

The pattern of magnetic field lines close to two (or more) magnets becomes complicated. The magnetic fields from the two magnets combine. The field lines from the two magnets *never* cross. If field lines did cross, it would mean that a compass would have to point in two directions at once.

Figure 4 shows the sort of pattern when two north poles are near each other. The field lines repel. At the point X there are no field lines. The two magnetic fields cancel out. X is called a neutral point.

Figure 5 shows the pattern produced by a north and a south pole. Notice that there is an area where field lines are equally spaced and all point in the same direction. This is called a uniform field.

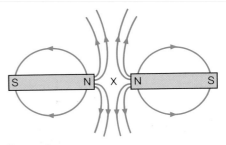

Figure 4
X is a neutral point; a compass placed here can point in any direction

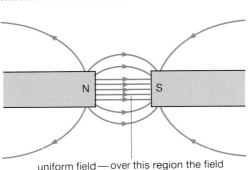

uniform field—over this region the field has the same strength and direction

Figure 5

The Earth's magnetic field

Figure 6 shows the shape of the Earth's magnetic field. Magnetic north is not in the same place as the geographic North Pole. At the moment magnetic north is in the sea north of Canada! Over a period of centuries the direction of the field alters.

The north (seeking) pole of a compass points towards magnetic north. Unlike poles attract. This means that magnetic north behaves like a south-seeking magnetic pole.

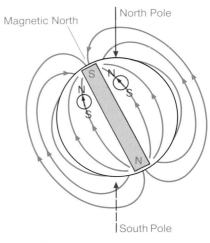

Figure 6
The Earth's magnetic field

Questions

1 (a) The diagram shows a bar magnet surrounded by four plotting compasses. Copy the diagram and mark in the direction of the compass needle for each of the cases *B, C, D*.

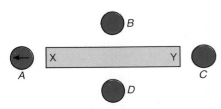

(b) Which is a north pole, X or Y?

2 Draw carefully the shape of the magnetic field surrounding these magnets. Mark in any neutral points.

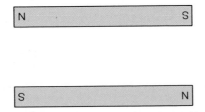

3 Two bar magnets have been hidden in a box. Use the information in the

diagram below to suggest how they have been placed inside the box.

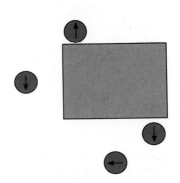

2 Currents and Magnetism

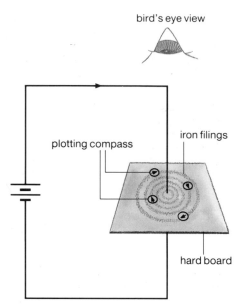

birds's eye view

plotting compass

iron filings

hard board

Figure 1
This experiment shows there is a magnetic
field around a current-carrying wire

The field near a straight wire

In Figure 1, a long straight wire, carrying an electrical current, is placed vertically so that it passes through a horizontal piece of hardboard. Iron fillings have been sprinkled onto the board to show the shape of the field. Below are summarised the important points of the experiment:

- If the current is small, the field is weak. But if a large current is used (30 A) the iron filings show a circular magnetic field pattern.
- The magnetic field gets weaker further away from the wire.
- The direction of the magnetic field can be found using a compass. If the current direction is reversed, the direction of the magnetic field is reversed.

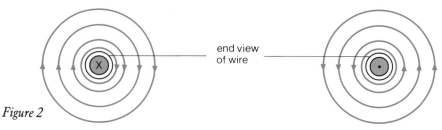

end view
of wire

Figure 2

Figure 2 shows the pattern of magnetic field lines surrounding a wire. When the current flows into the paper (shown $\otimes$) the field lines point in a clockwise direction around the wire. When the current flows out of the paper (shown $\odot$) the field lines point anticlockwise. The **right-hand grip rule** will help you to remember this. Put the thumb of your right hand along a wire in the direction of the current. Now your fingers point in the direction of the magnetic field.

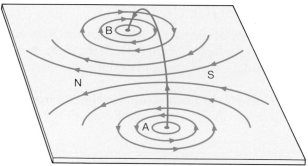

Figure 3
The magnetic field near a single loop of wire

The field near coils of wire

Figure 3 shows the magnetic field around a single loop of wire, which carries a current. You can use the right-hand grip rule to work out the field near to each part of the loop. Near A the field lines point anticlockwise as you look at them, and near B the lines point clockwise. In the middle, the fields from each part of the loop combine to produce a magnetic field running from right to left. This loop of wire is like a very short bar magnet. Magnetic field lines come out of the left-hand side (north pole) and go back into the right-hand side (south pole). Figure 4 shows the sort of magnetic field that is produced by a current flowing through a long coil or **solenoid**. The magnet field from each loop of wire adds on to the next. The result is a magnetic field that is like a long bar magnet's field.

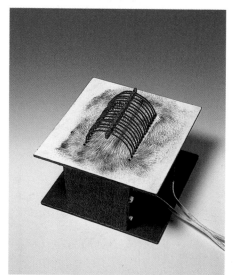

The iron filings show the shape of the
magnetic field around a solenoid

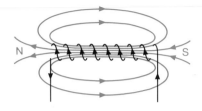

Figure 4
The magnetic field near a long solenoid

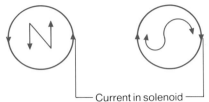

Current in solenoid

Figure 5
This is a good way to work out the polarity of the end of a solenoid

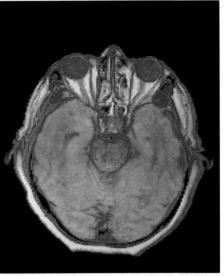

Producing large magnetic fields

The strength of the magnetic field produced by a solenoid can be increased by:
- using a larger current.
- using more turns of wire.
- putting some iron into the middle of the solenoid.

But there is a limit to how strong you can make a magnetic field. If the current is made too large the solenoid will get very hot and start to melt. For this reason large solenoids must be cooled by water. There is a limit to the number of turns of wire you can put into a space. Eventually iron becomes magnetically 'saturated' and its magnetisation gets no stronger.

Nowadays, the world's strongest magnetic fields are produced by **superconducting magnets**. At very low temperatures (about 4 K) some materials, such as nobium and lead, become superconductors. A superconductor has no electrical resistance. This means that a current can flow without causing any heating effect. So a large magnetic field can be produced by making enormous currents (5000 A) flow through a solenoid which is kept cold in liquid helium.

Strong magnetic fields are used by doctors to examine soft body tissue using a technique called nuclear magnetic resonance (NMR). This photograph shows a section through a brain. NMR is a very safe technique because unlike X-rays it does not damage the body

Questions

1 The diagram below shows two plotting compasses, one above and one below a wire. Draw diagrams to show the position of the needles when: (i) there is no current, (ii) the current is very large (30 A), (iii) the current is small (1 A).

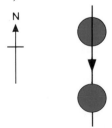

2 Diagram (i) below shows an electron moving around the nucleus of an atom.
(a) Explain why this atom is magnetic.
(b) Sketch the shape of the magnetic field near the atom.

(c) What happens to the magnetic field if the electron moves the opposite way.
(d) What can you say about the atom in diagram (ii)? Is it more or less magnetic than the other atom?

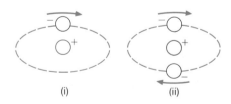

(i)   (ii)

3 The diagram shows two coils which carry the same current *I*. The graph

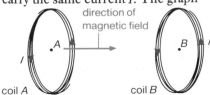

coil A   coil B

shows how the strengths of the magnetic fields, from each coil, change along the line *AB*.

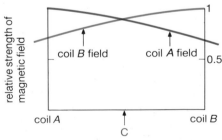

(a) Use the information in the graph to plot another graph to show the resultant strength of both fields added together.
(b) What is the strength of the magnetic field at *C* when these separate changes are made? (i) The current in coil *A* is reversed, (ii) The current in coil *B* is doubled.

3 Magnetising

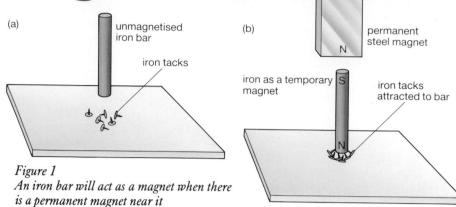

Figure 1
*An iron bar will act as a magnet when there
is a permanent magnet near it*

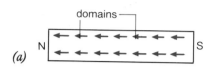

(a)

(b)

Figure 2

Jo's music is stored on magnetic tape

*Magnetic discs store information for these
computers which are in use at the London
International Financial Futures
Exchange*

Magnetic domains

Steel can be a permanent magnet; once a piece of steel has been magnetised it
remains a magnet. Iron can be a temporary magnet; iron is only magnetised when
in a magnetic field. This field can be provided by a magnet or by a solenoid
(Figure 1).

We think that the insides of magnetic materials are split up into small regions,
which we call **domains**. Each domain acts like a very small magnet. In any iron or
steel bar there are thousands of domains. This idea helps us to understand
permanent and temporary magnets.

In a steel bar, once the domains have been lined up, they stay pointing in one
direction (Figure 2(a)).

In an iron bar the domains are jumbled up when there is no magnet near
(Figure 2(b)). But as soon as a magnet is put near to an iron bar the domains are
made to line up. So an iron bar will be attracted to either a north or south pole of
a permanent magnet.

Magnetic memories

We use magnetic tapes to store information. These tapes are used for recording
music or as computer memories. The tape is coated with very small magnetic
particles. The direction of magnetisation of these particles is changed by applying
a strong magnetic field. The information is stored by the pattern in the magnetised
particles.

Magnetising

One way to make a magnet out of an unmagnetised steel bar is to stroke it with a
permanent magnet. The movement of the magnet along the steel bar is enough to
make the domains line up (Figure 3(a)).

A magnet can also be made by putting a steel bar inside a solenoid. A short but
very large pulse of current through the solenoid produces a strong magnetic field.
This magnetises the bar.

The apparatus shown in Figure 3(b) can also be used to make an
electromagnet. However an iron bar is used instead of a steel bar in the
solenoid. When a current is switched on the iron becomes magnetised; when the
current is switched off the iron is no longer magnetised. Electromagnets are
widely used in industry.

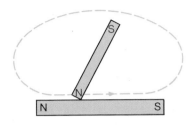

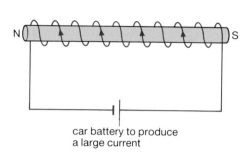

car battery to produce
a large current

Figure 3
(a) To magnetise a steel bar stroke it 20 times like this

(b) Magnetisation using a solenoid

A pair of electromagnets lifting railway lines in a stockyard

Demagnetising

To make a steel bar lose its magnetism you need to jumble up the domains inside the bar. There are three ways to do this:

- Hit the bar with a hammer.
- Put the bar inside a solenoid that has an alternating current supply. The alternating current produces a magnetic field that switches backwards and forwards rapidly. The domains are left jumbled up after the current has been reduced gradually to zero.
- Heat the bar to about 700°C. At low temperatures the atoms inside the steel line up to magnetise each domain. At a very high temperature the atoms vibrate at random so much that each domain is no longer magnetised.

Questions

1 (a) Tracey has magnetised a needle as shown below. She then cuts the needle into four smaller bits as shown. Copy the diagram and label the poles *A–H*.

(b) Lee says 'this experiment helps to show that there are magnetic domains in the needle'. Comment on this observation.

2 The diagram below shows an electromagnet. As the current increases the magnet can lift a larger load, because the domains in the iron core line up more and more.

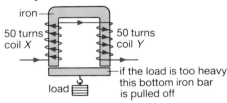

(a) Use the data provided to draw a graph of the load supported (*y*-axis) against current (*x*-axis).
(b) Use the graph to predict the load supported when the current is:
(i) 1.5 A, (ii) 6.0 A.
(c) Sketch on your graph how the load varies with current when each of the coils *X* and *Y* has: (i) 25 (ii) 100 turns.

(d) What happens when you reverse the windings on coil *X*?
(e) Make sketches to show how the domains are lined up when the current (in the original diagram) is: (i) 0, (ii) 0.5 A, (iii) 5 A.
(f) Use your answer to part (e) to explain why there is a maximum load that an electromagnet can lift.

Load (N)	Current (A)
0	0
5	0.5
8.5	1.0
12.0	2.0
14.0	3.0
14.8	4.0
15.0	5.0

4 The Motor Effect

Aluminium foil carrying a current is pushed out of a magnetic field

In the photograph you can see a piece of aluminium foil that has been fixed between the poles of a strong magnet. When there is a current through the foil it is pushed upwards away from the magnet. This is called the **motor effect**. It happens because of an interaction between the two magnetic fields, one from the magnet and one from the current.

In Figure 1 below you can see the way in which the two fields combine. By itself the field between the poles of the magnet would be nearly uniform. The current through the foil produces a circular magnetic field. The magnetic field from the current squashes the field between the poles of a magnet. It is the squashing of the field that catapults the foil upwards.

The force acting on the foil is proportional to:
- the strength of the magnetic field between the poles.
- the current.
- the length of the foil between the poles.

Fleming's left-hand rule

To predict the direction in which a wire moves in a magnetic field you can use the **left-hand rule** (Figure 1). Spread out the first two fingers and the thumb of your left hand so that they are at right angles to each other. Let your first finger point along the direction of the magnet's field, and your second finger point in the direction of the current. Your thumb then points in the direction that the wire moves.

This rule works when the field and the current are at right angles to each other. When the field and current are parallel to each other, there is no force on the wire and it stays where it is.

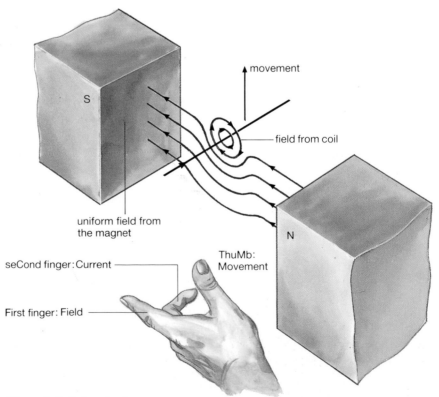

Figure 1 Left-hand rule

Ammeters

Figures 2 and 3 show the idea behind a moving coil ammeter. A loop of wire has been pivoted on an axle between the poles of a magnet. When a current is switched on, the left-hand side of the loop moves downwards and the right-hand side moves upwards. (Use the left-hand rule to check this). If nothing stops the loop it turns until side *DC* is at the top and *AB* is at the bottom. However, when a spring is attached to the loop it only turns a little way. When you pass a larger current through the loop, the force is larger. This will stretch the spring more and so the loop turns further.

The photograph in the questions section shows how a model ammeter can be made using a coil of wire, a spring and some magnets.

We say that an ammeter is sensitive if it turns a long way when a small current flows through it. The sensitivity of an ammeter will be large when:
- a large number of turns is used on the coil.
- strong magnets are used.
- weak springs are used.

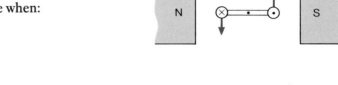

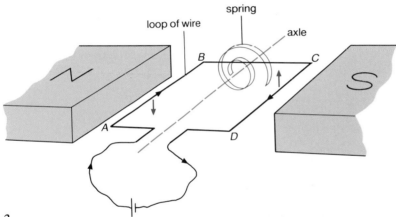

Figure 2
The principle of the moving-coil ammeter

Figure 3
(a) The end-on view of the wire loop in Figure 2. The turning effect is maximum in this position

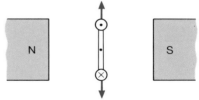

(b) In this position the forces acting on the coil produce no turning effect

Questions

1 (a) This question is about the photograph on page 220. Draw the magnetic fields near to: (i) the two magnetic poles shown, (ii) the wire carrying a current into the paper.
(b) Draw the combined field when the wire is placed between the magnetic poles.

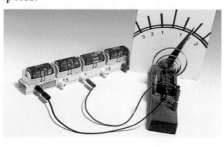

(c) Which way does the wire move when: (i) the current is reversed, (ii) the north and south poles are changed round?
2 This question is about the model ammeter shown in the photograph. The graph shows the angle that the coil turned through, against the current flowing through the coil.
(a) Use the graph to work out the current when the angle is: (i) 30°, (ii) 50°
(b) Make a sketch of the graph. Add to it further graphs to show the deflection for different currents when the ammeter is made: (i) with stronger springs, (ii) with twice the number of turns on its coil.

(c) Could this ammeter be used to measure alternating currents? Explain.
(d) (Quite hard) Why is the graph not a straight line?

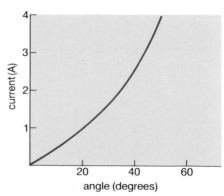

5 Electric Motors

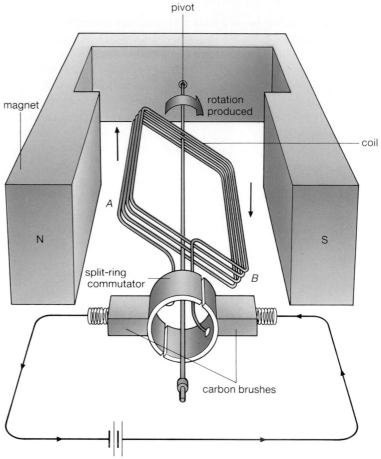

Figure 1
A design for a simple motor

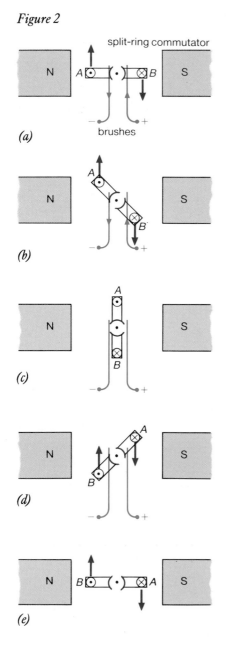

Figure 2

In the last unit you learnt that a coil carrying a current rotates, when it is in a magnetic field. However, the coil can only rotate through 90° and then it gets stuck. This is not good for making a motor. We need to make a motor rotate all the time.

Figure 1 shows the design of a simple motor that you can make for yourself. A coil carrying a current rotates between the poles of a magnet. The coil is kept rotating continuously by the use of a **split-ring commutator**. This causes the direction of the current in the coil to reverse, so that forces continue to act on the coil to keep it turning. Figure 2 explains the action of the commutator.

A current flows into the coil through the commutator so that there is an upwards push on side A, a downwards push on B (Figure 2(a)). The coil rotates in a clockwise direction (Figure 2(b)).

When the coil reaches the vertical position no current flows through the coil. The coil continues to rotate past the vertical due to its own momentum (Figure 2(c)).

Now side A is on the right-hand side. The direction of current flowing into the coil has been reversed (Figure 2(d)).

Side A is pushed down and side B is pushed upwards. The coil continues to rotate in a clockwise direction (Figure 2(e)).

Commercial motors

You use electric motors every day. Every time you put your washing into the washing machine or use a vacuum cleaner, you are using an electric motor. Your car also uses an electric motor to get it started. For these sort of uses, motors need to be made as powerful as possible.

Here are some of the ways that can be used to make a motor more powerful (Figure 3):

- A large current should be used.
- As many turns as possible should be put on the coil.
- More than one coil can be put on the rotor. Each coil then experiences a force, so the total force turning the motor is larger.
- The coils can be wound on an iron core to increase the magnetic field.
- Curved pole pieces make sure that the magnetic field is always at right angles to the coils. This gives the greatest turning effect.

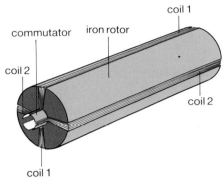

Figure 3
(a)

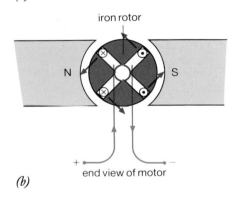

(b)

A cut-away view of a Black and Decker drill, showing the electric motor and gears inside it

Questions

1 Name three machines (other than those mentioned in the text) that use electric motors.

2 What two properties of carbon make it a good material to use for motor brushes?

3 (a) Look at Figure 2. Which position is the motor likely to stick in?

(b) Explain why the motor shown in Figure 3 is less likely to stick.

4 Copy Figure 3(b) and draw in some field lines to show the shape of the magnetic field.

5 The motor shown to the right is used to lift up a load.

(a) Use the information in the diagram to calculate the following: (i) the work done in lifting the load, (ii) the power output of the motor while lifting the

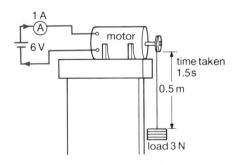

load, (iii) the electrical power input to the motor.

(b) Calculate the efficiency of the motor.

(c) Explain carefully the energy changes that occur during the lifting process.

6 The diagram below shows how an electric motor can be made using an electromagnet. You can see the forces acting on the coil when a current flows.

(a) The battery is now turned round. What effect does that have on: (i) the polarity of the magnet? (ii) the direction of current in the coil? (iii) the forces acting on the coil?

(b) Would this motor work with an a.c. supply? Explain.

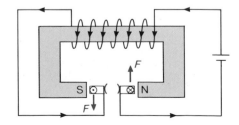

6 Electromagnets in Action

Reed switches

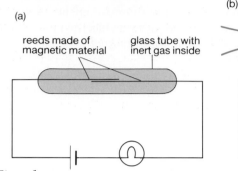

(a)

reeds made of magnetic material

glass tube with inert gas inside

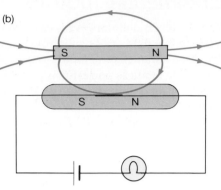

(b)

Figure 1

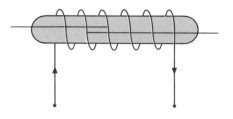

Figure 2
Reed switches can also be opened and closed using the magnetic field from a solenoid. This device is sometimes called a reed relay

Two reeds made of magnetic materials are enclosed inside a glass tube. Normally these reeds are not in contact so the switch is open (Figure 1(a)). When a magnet is brought close to the reeds they become magnetised so that they attract each other. The switch is now closed (Figure 1(b)). These switches can be used to work circuits using a magnetic field. This field can be supplied using a permanent magnet or a solenoid (Figure 2).

Relays

A car starter motor needs a very large current of about 100 A to make it turn round. Switching large currents on and off needs a special heavy-duty switch. If you had such a large switch inside the car it would be a nuisance since it would take up a lot of space. The switch would spark and it would be unpleasant and dangerous. A way round this problem is to use a relay.

Figures 3(a) and 3(b) show how a car starter relay works.

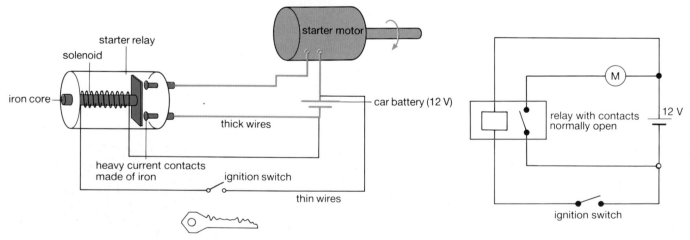

Figure 3
(a) A car starter relay

(b) The circuit diagram for the starter motor

Inside the relay a solenoid is wound round an iron core. When the car ignition is turned, a small current magnetises the solenoid and its iron core. The solenoid is attracted towards the heavy-duty electrical contacts, which are also made of iron. Now current can flow from the battery to the starter motor. The advantage of this system is that the car engine can be started by turning a key at a safe distance!

Electric bells

Figure 4 shows the idea behind an electric bell. When the bell push is pressed the circuit is completed and a current flows. The electromagnet becomes magnetised and the soft iron is attracted to the electromagnet. The movement of the iron causes the striker to hit the bell. As the bell is hit, A moves away from S and the circuit is broken. The electromagnet is now demagnetised and a spring pulls the iron back to its original position. The circuit is remade and the process starts over again. The striker can be made to hit the bell a few times every second.

The moving-coil loudspeaker

The loudspeaker in your radio and television sets is probably made like the one shown in Figure 5. A loudspeaker is an example of a transducer. Transducers turn one form of energy into another.

Electrical signals from the radio cause the current flowing in the coil to change. A change in the current causes a change in the force acting on the coil. In this way the paper cone is made to move in and out. The vibrations of the cone produce sound waves.

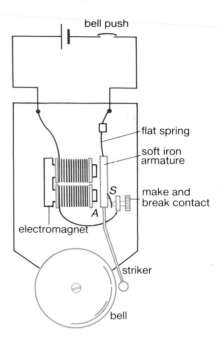

Figure 4
An electric bell

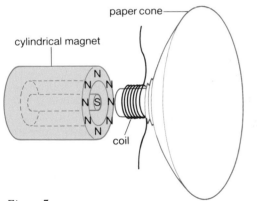

Figure 5
The moving-coil loudspeaker

Questions

1 The diagram below shows the detonator circuit for a magnetic mine. This sort of mine is placed on the seabed in shallow water. The detonator is activated when it is connected across the battery. Explain how a ship can trigger the mine.

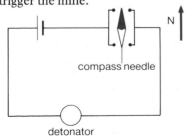

2 (a) In Figure 3(a) the resistance of the relay's solenoid is 48 Ω. Calculate the current that flows through it when it is connected to the battery.
(b) Now explain why two different thicknesses of wire have been used in the circuit.

3 This question refers to the loudspeaker in Figure 5.
(a) Explain carefully why an alternating current will make the coil vibrate backwards and forwards.
(b) What factors affect the amount of noise that the loudspeaker makes?

Minesweepers have glass-fibre hulls. If their hulls were metallic, they would interfere with the sensitive electrical equipment that they use to detect the mines. Non-metallic hulls also make the ship less likely to cause mines to explode

7 Electromagnetic Induction

The devices shown here are used by security guards at airports to check that passengers are not carrying any guns or bombs. Metal objects cause changes in an electromagnetic field when they pass below the archway or near the hand-held device. A circuit detects the changes and sets off an alarm

Electric motors work because forces act on current-carrying wires that are placed in magnetic fields. A motor converts electrical energy into mechanical energy. The process can be put into reverse. If you turn a small motor by hand you produce a current. Then mechanical energy has been turned into electrical energy. This is called **electromagnetic induction**.

Size and direction of current

Figure 1 shows an experiment to find out what affects the making of a current.

(1) **Direction of movement**. To get a current the wire must cut across lines of magnetic field. The wire must be moved up and down along the direction XX'. There is no current if the wire moves along ZZ' or YY'. Reversing the direction of movement reverses the current. If moving the wire up makes the meter move to the right, moving the wire down will make the meter go to the left.

(2) **Size of current**. You can make a larger current flow in the following ways:
- Moving the wire more quickly. When the wire is stationary between the poles of the magnet, there is no current.
- Using stronger and bigger magnets.
- Looping the wire so that several turns of wire pass through the poles.

These facts about electromagnetic induction were first discovered by Faraday.

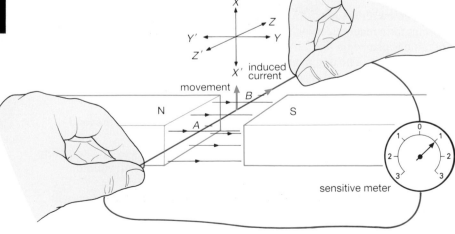

Figure 1

Energy changes

In Figure 2(a) a current is being produced by moving a magnet into a solenoid. When the current flows a compass needle is attracted towards end Y. So end Y behaves as a south pole and end X as a north pole. This means that as the magnet moves towards the solenoid there is a magnetic force that repels it. There is a force acting against the magnet, so you have to do some work to push it into the solenoid. The work done pushing the magnet produces the electrical energy.

In Figure 2(b) the magnet is being pulled out of the solenoid. The direction of the current is reversed and now there is an attractive force acting on the magnet. The hand pulling the magnet still does work to produce electrical energy. When the switch S is opened, there is no current. So the magnet can move in and out of the coil without any repulsion or attraction.

When a current is produced by electromagnetic induction, energy is always used to create the electrical energy. In the example described in Figure 2 the energy originally came from the muscles pushing the magnet.

Michael Faraday, the father of electromagnetism. In 1831 Faraday discovered that a changing magnetic field produces electric current

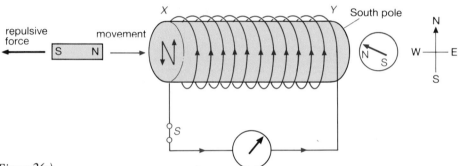

Figure 2(a)
Lenz's law of electromagnetic induction says: 'When a current is induced it always opposes the change in magnetic field that caused it.' The end X of the solenoid is a north pole. So the effect of the induced current is to push the magnet back. This opposes the motion of the magnet, and agrees with Lenz's law

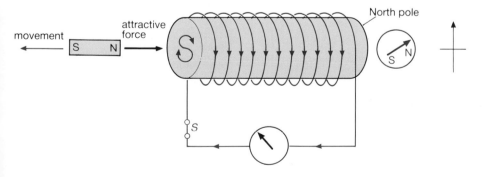

Figure 2(b)
Now the magnet moves away. X is now a south pole so the magnet is attracted to the solenoid. This opposes the motion and so agrees with Lenz's law

An electromagnetic flow meter

Figure 3 shows a way to measure the rate of flow of oil through an oil pipeline. A small turbine is placed in the pipe, so that the oil flow turns the blades round. Some magnets have been placed in the rim of the turbine, so that they move past a solenoid. These moving magnets induce a voltage in the solenoid which can be measured on an oscilloscope (Figure 3(b)). The faster the turbine rotates, the larger is the voltage induced in the solenoid. By measuring this voltage an engineer can tell at what rate the oil is flowing.

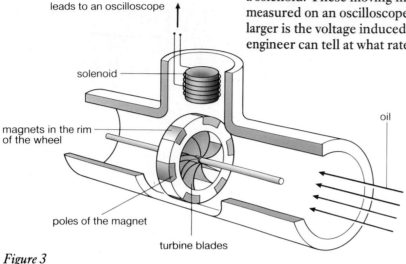

Figure 3
(a) An electromagnetic flow meter

(b) The oscilloscope trace. The time base is set so that the dot crosses the screen in 0.1 s

Questions

1 (a) In Figure 1 the wire moves up at a speed of 2 m/s. This makes the ammeter kick one division to the right. Say what will happen to the meter in each of the following cases: (i) the wire moves up at 4 m/s, (ii) the wire moves down at 6 m/s, (iii) the wire moves along YY' at 3 m/s, (iv) the magnets are turned round and the wire moved up at 3 m/s, (v) a larger pair of magnets is used so that the length of the wire AB in the magnetic field is doubled. The wire is now moved up at 1 m/s with the direction of the field, as shown in Figure 1.
(b) In the diagram (right), explain which arrangement will produce a larger deflection on the meter.

2 This question refers to the flow meter shown in Figure 3.
(a) What does the oscilloscope trace tell you about the way the magnets are arranged in the rim of the turbine? Are they all arranged with their north poles facing outwards or what?

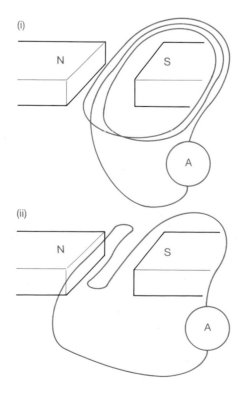

(i)

(ii)

(b) Sketch the trace on the oscilloscope for the following (separate) changes:
(i) the number of turns on the solenoid is made 3 times larger, (ii) the flow of oil is increased, so that the turbine rotates twice as quickly.
(c) Use the information in Figures 3(a) and 3(b) to calculate how many times per second the turbine is rotating.
(d) Gita and Sarah are two apprentice engineers who are using the flowmeter for the first time. This is part of a conversation they have about the meter.
Sarah: Because a current is induced there must be a force that opposes the motion. This means that the turbine will slow down the oil flow.
Gita: The resistance of the oscilloscope is very large so only a small current can flow. So the slowing down effect can be ignored.
Comment on their conversation.

8 Generators

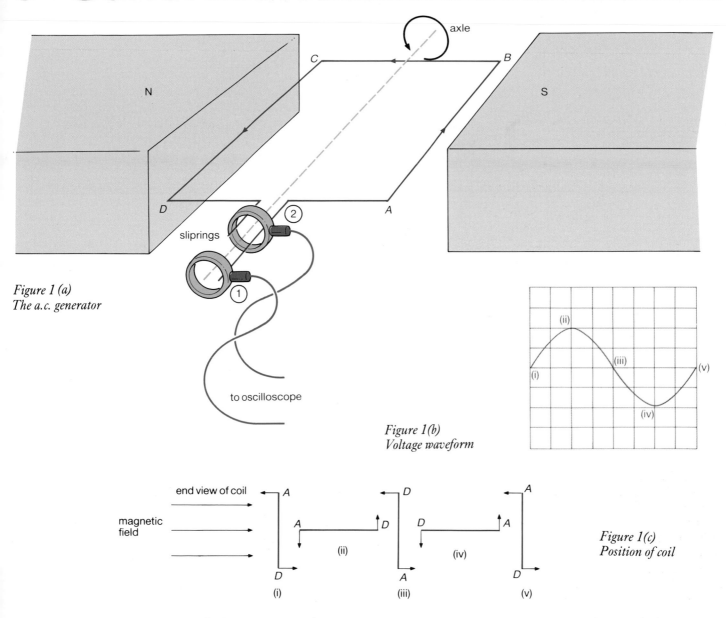

Figure 1 (a)
The a.c. generator

sliprings

to oscilloscope

Figure 1(b)
Voltage waveform

end view of coil

magnetic
field

Figure 1(c)
Position of coil

The a.c. generator (alternator)

Figure 1 shows the design of a very simple **alternating current** (a.c.) generator. By turning the axle you can make a coil of wire move through a magnetic field. This causes a voltage to be induced between the ends of the coil.

You can see how the voltage waveform, produced by this generator, looks on an oscilloscope screen. In position (i) the coil is vertical with AB above CD. In this position the sides CD and AB are moving parallel to the magnetic field. No voltage is generated since the wires are not cutting across the magnetic field lines.

When the coil has been rotated through a ¼ turn to position (ii), the coil produces its greatest voltage. Now the sides CD and AB are cutting through the magnetic field at the greatest rate.

In position (iii), the coil is again vertical and no voltage is produced. In position (iv) a maximum voltage is produced, but in the opposite direction. Side AB is now moving upwards and side CD downwards.

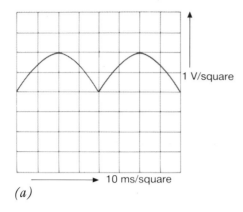

(a)

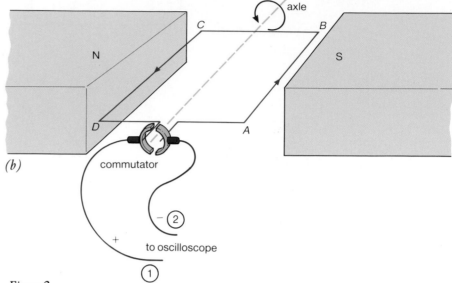

(b)

Figure 2
The d.c. generator

The d.c. generator (dynamo)

Figure 2 shows how **direct current** (d.c.) can be generated. The design of the dynamo is very similar to that of the alternator in Figure 1. The difference is that the ends of the coil are now fixed to a split-ring commutator rather than two separate slip rings. Now it does not matter which side, *AB* or *CD*, moves upwards, current will always flow out of one side of the commutator. So direct current is produced.

The d.c. generator is identical to the d.c. motor in design. However, a motor is used to turn electrical energy into kinetic energy; whereas the dynamo turns kinetic energy into electrical energy.

Producing power on a large scale

The electricity that you use in your home is produced by very large generators in power stations. These generators work in a slightly different way from the ones you have seen so far.

Instead of having a rotating coil and a stationary magnet, large generators have rotating magnets and stationary coils (see Figure 3). The advantage of this set-up is that no moving parts are needed to collect the large electrical current that is produced.

The important steps in the generation of electricity in a power station (Figure 4) are these:
(1) Coal is burnt to boil water.
(2) High pressure steam from the boiler is used to turn a turbine.
(3) The drive shaft from the turbine is connected to the generator magnets, which rotate near to the stationary coils. The output from the coils has a voltage of about 25 000 V.
(4) The turbine's drive shaft also powers the **exciter**. The exciter is a direct current generator which produces current for the rotating magnets, which are in fact electromagnets.

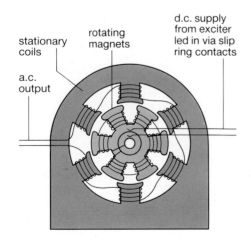

Figure 3
A section through a large generator

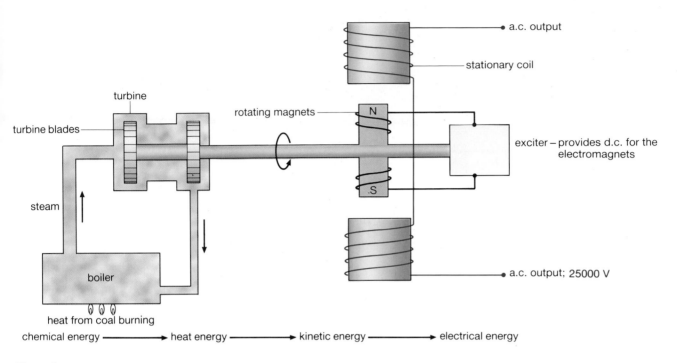

chemical energy ⟶ heat energy ⟶ kinetic energy ⟶ electrical energy

Figure 4
The layout of a power station

Questions

1 (a) Copy Figure 2(b) and mark on the graph the places where the coil is: (i) horizontal, (ii) vertical.
(b) On the same graph show how the voltage appears when the coil is rotated in the opposite direction: (i) at the same speed, (ii) at twice the speed.
2 In the diagram (right) the dynamo is connected to a flywheel by a drive belt. The dynamo is operated by winding the handle. When Surindra turns the handle as fast as she can, the dynamo produces a voltage of 6 V. The graph shows how long the dynamo takes to stop after Surindra stops winding the handle. The time taken to stop depends on how many bulbs are connected to the dynamo.
(a) Explain the energy changes that occur after Surindra stops winding the handle: (i) when the switches S_1, S_2 and S_3 are open (as shown), (ii) when the switches are closed.

(b) Why does the flywheel take longer to stop when no light bulbs are being lit by the dynamo?
(c) Make a copy of the graph. Use your graph to predict how long the flywheel turns when you make it light 4 bulbs.
(d) The experiment is repeated using the bulbs that use a smaller current. They use 0.15 A rather than 0.3 A. Using the same axes, sketch a graph to show how the running time of the dynamo depends on the number of bulbs, in this second case.

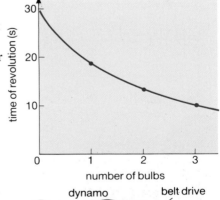

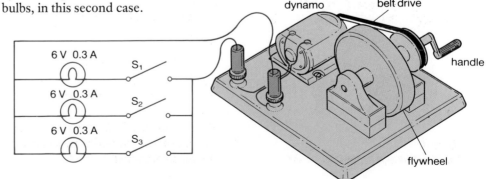

9 Introduction to Transformers

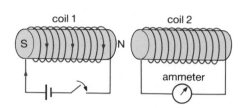

Figure 1

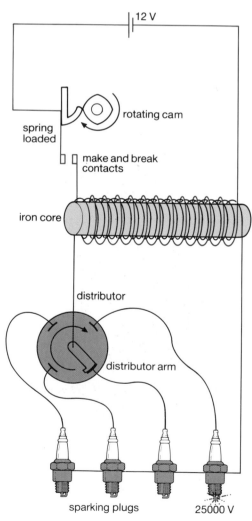

Figure 2
An induction coil in a car

This transformer delivers power to the national grid

Mutual inductance

In Figure 1 (above left) when the switch is closed in the first circuit, the ammeter in the second circuit kicks to the right. For a moment a current flows through coil 2. When the switch is opened again the ammeter kicks to the left.

Closing the switch makes the current through coil 1 grow quickly. This makes the coil's magnetic field grow quickly. For coil 2 this is like pushing the north pole of a magnet towards it, so a current is induced in coil 2. When the switch is opened the magnetic field near coil 1 falls rapidly. This is like pulling a north pole away from coil 2. Now the induced current flows the other way.

The ammeter reads zero when there is a constant current through coil 1. A current is only induced in the second coil by a changing magnetic field. This happens when the switch is opened and closed.

The induction coil

Inside the cylinders of a car's engine, the mixture of air and petrol explodes when it is sparked off by a sparking plug. To make sparks large voltages, about 25 000V, are needed. Figure 2 shows how this is done.

Two solenoids have been wound round an iron core. The red solenoid is connected to the car battery through a pair of make-and-break contacts. When the circuit is complete there is a current. The iron core becomes magnetised and a strong magnetic field is produced. As the cam rotates the circuit is broken. This causes the magnetic field to fall very rapidly. A very large voltage is now induced in the second (black) solenoid, which has a large number of turns. The rotating distributor arm feeds the high voltage to each sparking plug in turn.

Transformers

A **transformer** is made by putting two coils of wire onto a soft iron core as shown in Figure 3. The primary coil is connected to a 2 V alternating current supply. The alternating current in the primary coil makes a magnetic field that rises and then falls again. The soft iron core carries this changing magnetic field to the secondary coil. Now a changing voltage is induced in the secondary coil. In this way, energy can be transferred continuously from the primary circuit to the secondary circuit.

Transformers are useful because they allow you to change the voltage of a supply. For example, model railways have transformers that decrease the mains supply from 240 V to a safe 12 V. These are **step-down transformers**. The transformer in Figure 3 steps *up* the voltage from 2 V to 12 V.

To make a step-up transformer the secondary coil must have more turns of wire in it than the primary. In a step-down transformer the secondary has fewer turns of wire than the primary coil.

The rule for calculating voltages in a transformer is:

$$\frac{V_s}{V_p} = \frac{N_s}{N_p}$$

V_p = primary voltage; V_s = secondary voltage; N_p = number of turns on the primary coil; N_s = number of turns on the secondary.

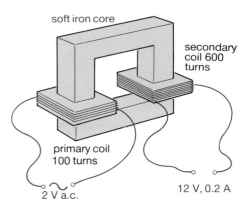

(a) A step-up transformer

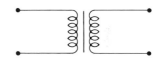

(b) The circuit symbol for a transformer

Figure 3

Questions

1 (a) In Figure 1 when the switch is opened the ammeter kicks to the left. Describe what happens to the ammeter during each of the following (i) the switch is closed and left closed so that a current flows through coil 1, (ii) the coils are now pushed towards each other, (iii) the coils are left close together, (iv) the coils are pulled apart.
(b) The battery is replaced by an a.c. voltage supply which has a frequency of 2 Hz. What will the ammeter show when S is closed?
2 (a) Explain why the black solenoid in Figure 2 must have such a large number of turns.
(b) Will the car still work if the black

circuit is connected to the battery, and the red solenoid to the distributor?
3 Explain why a transformer does not work when you plug in the primary to a battery.
4 The question refers to Figure 3.
(a) What power is used in the secondary circuit?
(b) Explain why the smallest current that can be flowing in the primary circuit is 1.2 A
(c) Why is the primary current likely to be a little larger than 1.2 A?
5 The table below gives some data about 4 transformers. Copy the table and fill the gaps.

A small transformer for use in a laboratory

Primary turns	Secondary turns	Primary voltage (V)	Secondary voltage (V)	Step-up or step-down
100	20		3	
400	10 000	10		
	50	240	12	
	5 000	33 000	11 000	

10 Transmitting Power

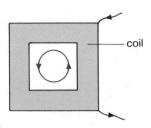

Figure 1
(a) Eddy currents flow in an unlaminated core

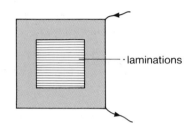

(b) Laminations help to stop eddy currents flowing

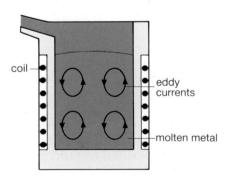

Figure 2
An induction furnace, which puts eddy currents to good use. In this furnace, scrap metal is melted down for re-use. The heating is due to eddy currents that are induced in the metal itself. These are induced by an alternating current flowing through a coil on the outside of the furnace

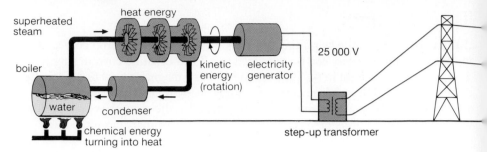

Figure 3
How the power gets to your home. The diagram shows how electricity is generated at a power station, and how it is distributed around the country through the national grid

Power in transformers

We use transformers to transfer electrical power from the primary circuit to the secondary circuit. Many transformers do this very efficiently and there is little loss of power in the transformer itself. For a transformer that is 100% efficient we can write:

> Power supplied by primary circuit = Power used in the secondary circuit
> $$V_p \times I_p = V_s \times I_s$$

In practice transformers are not 100% efficient. These are the most important reasons for transformers losing energy:
- The windings on the coils have a small resistance. So when a current flows through them they heat up a little.
- The iron core conducts electricity. Therefore the changing magnetic field from the coils can induce currents in the iron. These are called **eddy currents**. To reduce energy lost in this way the core is laminated. This means the core is made out of thin slices of iron with sheets of insulating material between each slice (Figure 1).
- Energy is lost if some of the magnetic field from the primary coil does not pass through the secondary coil.
- Energy is also used to switch the direction of the domains in the iron core, 50 times each second. This can be done easily in the iron, so energy losses are small.

The national grid

You may have seen a sign at the bottom of an electricity pylon saying 'Danger high voltage'. Power is transmitted around the country at voltages as high as 400 000 V. There is a very good reason for this – it saves a lot of energy. The following calculations explain why.

Figure 4 on the next page suggests two ways of transmitting 25 MW of power from a Yorkshire power station to the Midlands:
(a) The 25 000 V supply from the power station could be used to send 1000 A down the power cables.
(b) The voltage could be stepped up to 250 000 V and 100 A could be sent along the cables.

How much power would be wasted in heating the cables in each case, given that 200 km of cable has a resistance of 10 Ω?

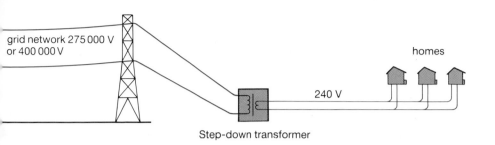

grid network 275 000 V or 400 000 V

homes

240 V

Step-down transformer

The transformers in this substation are used to change the voltage of the electricity being transmitted through the national grid

(a) Power lost = voltage drop along cable × current

$$= IR \times I$$
$$= I^2R$$
$$= (1000)^2 \times 10$$
$$= 10\,000\,000 \text{ W or } 10 \text{ MW}$$

(b) Power lost = I^2R

$$= (100)^2 \times 10$$
$$= 100\,000 \text{ W or } 0.1 \text{ MW}$$

We waste a lot less power in the second case. The power loss is proportional to the square of the current. Transmitting power at high voltages allows smaller currents to flow along our overhead power lines.

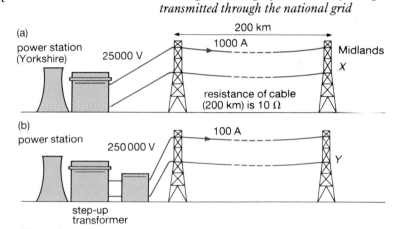

(a)
power station (Yorkshire)

200 km
1000 A
25000 V
Midlands
X

resistance of cable (200 km) is 10 Ω

(b)
power station

100 A
250 000 V
Y

step-up transformer

Figure 4

Questions

1 This is about how a transformer could be used to melt a nail. You will need to use the data provided.
(a) Calculate the voltage across the nail.

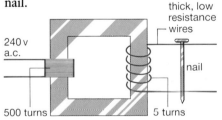

240 v a.c.

thick, low resistance wires

nail

500 turns

5 turns

- The nail has a resistance of 0.02 Ω.
- The melting point of the nail is 1540°C.
- The nail needs 10 J to warm it through 1°C.

(b) Calculate the current flowing in:
(i) the secondary circuit,
(ii) the primary circuit.
(c) Calculate the rate at which power is used in the secondary circuit to heat the nail.
(d) Estimate roughly how long it will take the nail to melt. Mention any assumptions or approximations that you make in this calculation.
(e) Explain how a transformer can be used to produce high currents for welding.

2 Explain why the electricity supply in your home is a.c. rather than d.c.
3 Use the data in Figure 4 to calculate the voltage at each of the points X and Y.
4 The data in the table below was obtained using samples of copper, aluminium and steel wires. Each wire was 100 m long and had a diameter of 2 mm. Use this data to explain why our overhead power cables are made out of aluminium with a steel core.

Material	Resistance (Ω)	Force needed to break wire	Density in kg/m³	Cost of wire
Copper	2.2 Ω	320 N	8900	£5.60
Aluminium	3.2 Ω	160 N	2700	£1.60
Steel	127 Ω	1600 N	9000	£0.14

SECTION J: QUESTIONS

1 Two students, Rajeeb and Emma have designed a new ammeter shown below. They have plotted a calibration graph to show the extension of the spring for a particular current.

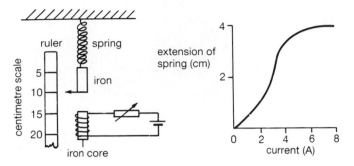

(a) Explain why their device can be used as an ammeter.
(b) Explain the shape of the graph in as much detail as you can.
(c) Here are some remarks that the students made about their ammeter. Evaluate their comments.
(i) *Rajeeb:* Our ammeter is not really any good for measuring currents above 4 A.
Emma: It would be better at measuring large currents if we used fewer turns on the solenoid.
(ii) *Rajeeb:* It does not matter which way the current flows the spring still gets pulled down.
Emma: That's useful, we can use our ammeter to measure a.c.

2 You can use magnetic tapes to store information. These tapes can be used as computer memories or for recording music. The tape is made of plastic and is coated with very small magnetic particles. The direction of magnetisation of these particles can be changed by applying a strong magnetic field.

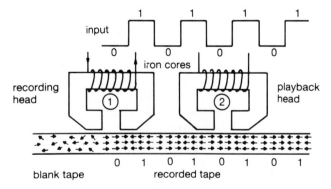

(a) In the diagram a series of pulses are sent into the recording head of a tape recorder. Explain how these produce the pattern of magnetism shown in the tape.
(b) Explain how the same tape could be used as a computer memory or for recording music.

(c) (i) When the tape is passed by the playback head, a voltage is induced in the coil. Explain why.
(ii) Draw a sketch to show the form of this induced voltage. Assume the tape moves at the same speed as it did when it was recording.
(d) To erase the tape, an erase head is used. The erase head is supplied with a very high frequency voltage (about 50 kHz). Explain why this high frequency is chosen.

3 The diagram below shows a coil connected to a sensitive meter. The meter is a 'centre zero' type. (This means that when no current is flowing the needle points to the centre of the scale.)

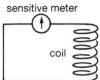

A girl performs two experiments.

First experiment
She pushes a magnet into the coil. A moment later, she removes the magnet. The meter needle deflects (moves away from zero) only when the magnet is moving.

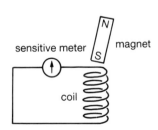

Second experiment
She brings a second coil close. She switches a current, supplied by a battery, on and off in this second coil. The needle deflects only for a very short time each time she switches on or off.

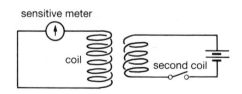

(a) In the first experiment:
(i) What is happening in the coil while the magnet is moving?
(ii) How does the deflection of the needle as the magnet is pushed towards the coil compare with the deflection of the needle as the magnet is pulled away?

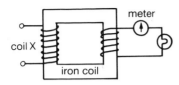

coil X   meter   iron coil

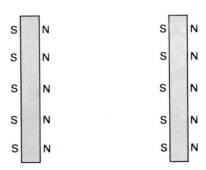

(iii) Without changing any of the apparatus, how could she make the needle deflection as large as possible?
(b) In both experiments, what is happening in the space around the coil at the times when the needle is deflected?
(c) Two coils are wound around an iron core as shown. What, if anything, will you see on the meter if:
(i) a constant direct current is passed through coil X?
(ii) the constant direct current in coil X is switched on and then off again?
(iii) an alternating current of frequency 50 Hz is passed through coil X?
(d) The device in the diagram is a transformer. Name a household device which includes a transformer.

MEG

4 The diagram shows a laboratory electromagnet, made from two iron cores and a length of wire. W and X are connected to a d.c. supply.

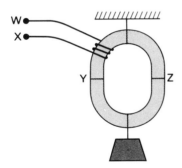

The number of turns of wire on the magnet is varied, and the largest weight that the magnet can support is shown in the table. For each measurement, the current remains at the same value of 2 A.

Number of turns in coil	5	10	15	20	25	30
Weight supported (N)	1.1		3.2	4.0	4.3	4.3

(a) What weight do you think can be supported when the magnet has 10 turns?
(b) Explain why the same load is supported by 25 or 30 turns.
(c) What happens if poor contact is made between the cores at Y and Z?

5 (a) Copy diagram (1). Then draw lines to represent the magnetic field between the two magnets.

(b) Diagram (2) shows a wire that is at right-angles to the plane of the paper. The current in the wire flows up out of the paper. Copy the diagram and add lines to represent the magnetic field.

•

(c) In diagram (3) the wire has been placed between the poles of the magnet. Draw the field lines now.

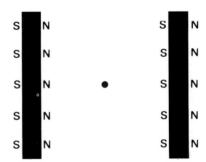

(d) Which way will the wire move in diagram (3)?
(e) Diagram (4) shows a model motor.

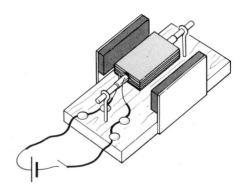

(i) Which position is the coil likely to stick in?
(ii) Explain why the coil rotates.
(iii) State and explain two changes that will make the coil turn faster.

QUESTIONS

6 In this question you are asked to investigate the cost of producing electricity by three different means. The methods are:
- a conventional coal-fired power station
- a nuclear-powered station
- a tidal barrage.

Each scheme would be financed by borrowing money from a bank. The capital investment is to be repaid at 12% per annum.

	Coal-fired power station	Nuclear power station	Severn tidal barrage
Capital cost of buildings and plant, or barrage	£1200 million	£2500 million	£11 000 million
Running costs (maintenance and salaries) per annum	£90 million	£110 million	£40 million
Cost of coal	£50 per tonne	–	–
Waste disposal per annum	–	£40 million	–
Cost of reprocessing spent fuel	–	£150 000 per tonne	–
Coal/fuel used per year	4.5 million tonnes	800 tonnes	–
Maximum power output	2000 MW	2000 MW	7000 MW
Number of kWh produced per year	12 000 million	14 000 million	24 000 million

(a) For the coal-fired station:
(i) Work out the annual interest payable.
(ii) Work out the total annual cost of the coal.
(iii) Work out the total annual cost of generating electricity.
(iv) What does it cost the station to produce 1 kWh of electricity?
(v) Explain why electricity companies will charge more than this per kWh.
(b) Calculate the cost of generating 1 kWh of electricity by using (i) nuclear power, (ii) the tidal barrage.
(c) Which of the three methods runs its generators for the greatest time each year?

(d) Comment on the environmental impact of each scheme. Which has the greatest drawback: acid rain produced by coal burning, possible damages from nuclear fallout, or changing high tide levels to remove the habitat of thousands of waders and wild fowl?
(e) Producing electricity using the Severn Tidal Barrage seems the most expensive at the present time. If it were built now, would it still be the most expensive means of producing electricity in 20 years time? (Interest repayments will be the same but all other costs will have risen by about a factor of 2.)
(f) Which of the schemes would you choose? Justify your answer.

SECTION K

Electronics

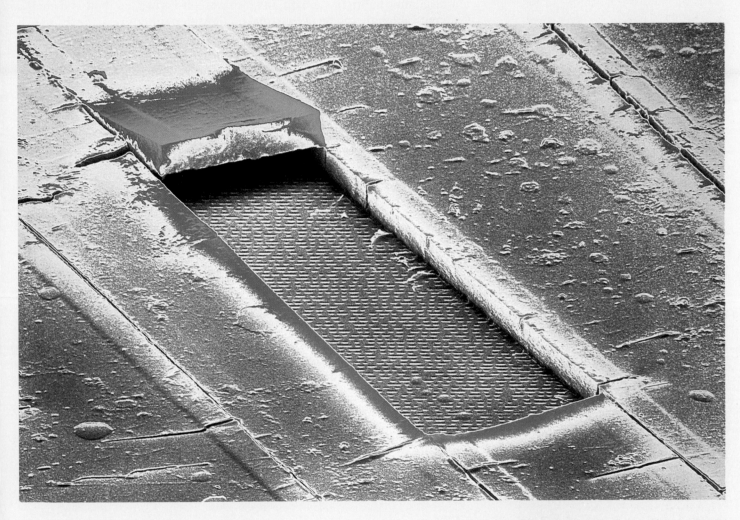

A false colour scanning electron micrograph of the surface of a compact disc. The surface of the disc, which is plastic coated, has been cracked in the shape of a rectangle, to show the music layer beneath. The music is coded in the series of tiny notches which are just visible. Light from a laser is bounced off these notches, and decoded by the player as music

1 Electronic Devices

A photographer's light meter uses light dependent resistors. The reading from the light meter enables the photographer to adjust his camera to take a properly exposed photograph

Light dependent resistor (LDR)

The electrical resistance of some materials depends on the brightness of the light around us. When it is dark the resistance of a piece of cadmium sulphide can be a few megohms. In bright sunlight, the resistance drops to a few thousand ohms. Cadmium sulphide can be used to make a **light dependent resistor** (LDR).

Figure 1 shows how you can make a simple light meter using an LDR and an ammeter. When it is bright the resistance of the LDR is low and the reading on the ammeter is high. When it is dark the resistance of the LDR is high, so the current is low. Such a light meter can help a cricket umpire decide when it is too dark to carry on playing safely.

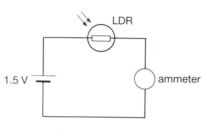

Figure 1
A simple light meter

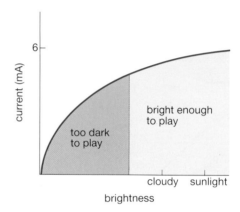

Silicon diodes

A silicon diode allows current to pass through it in one direction only. Figure 2(a) shows the circuit symbol for the diode. Current flows easily from anode to cathode. However, current cannot flow the other way. When the anode is positive and the cathode negative, we say the diode is **forward biased**. The diode is **reverse biased** when the cathode is positive and the anode negative. Figure 2(b) shows how the current flowing through a diode varies with the applied voltage. Notice that once the voltage across the diode is about 0.7 V, the current increases very rapidly.

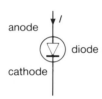

Figure 2
(a)

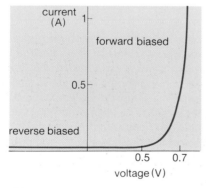

(b)

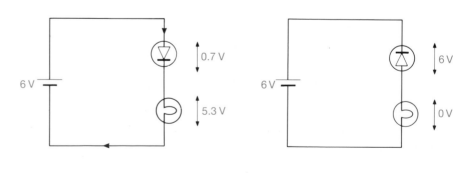

Figure 3
(a) Forward biased   *(b) Reverse biased*

Figure 3 shows a diode in action. When it is forward biased the bulb is lit. There is about 0.7 V across the diode. This means that the voltage across the bulb is 6 V − 0.7 V = 5.3 V. When the diode is reverse biased no current flows. There is no voltage across the light bulb and so 6 V across the diode.

Light-emitting diode (LED)

A **light-emitting diode (LED)** is a diode that gives out (emits) light when it is forward biased. Diodes can emit green, red or yellow light.

When an LED is working, it usually has a voltage of about 2 V across it and a current of 10 mA flowing through it. An LED can easily be damaged if too large a current (50 mA) flows through it. So we protect an LED by putting a resistance in series with it.

Example. What value of resistor do you need for the circuit in Figure 4? The voltage across R must be 4 V (6 V – 2 V). The current flowing through R is

10 mA,   or $\dfrac{10\,A}{1000}$ = 0.01 A

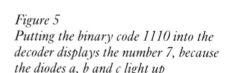

Figure 4

$$R = \frac{V}{I}$$

$$= \frac{4V}{0.01\,A}$$

$$= 400\,\Omega$$

Displaying numbers

A common use of LEDs is to make a **7 segment display**. Here 7 LEDs are arranged to make a figure-of-eight pattern (Figure 5). Each LED can be switched on separately. Depending on which bars are lit, the display shows a different number. It is possible to display all the numbers from 0 to 9. You often see 7 segment displays on calculators and clocks.

To operate, a 7 segment display needs a **decoder,** which is an example of an **integrated circuit, IC** (see unit K6). This IC has the job of turning a binary number into a form that can be displayed by the diodes. Numbers are transmitted electronically in binary. The inputs DCBA can be in two possible 'states'. The voltage to them can be high, which we represent by '1'; or the voltage can be low, which is represented by '0'. The number going into the decoder is given by:

$$8 \times D + 4 \times C + 2 \times B + 1 \times A$$

Example. If D = 0, C = 1, D = 1, A = 1, the number (or code) 0111 means 7.

The decoder now turns this into a suitable message to light the diodes. Outputs abc receive high voltages to light diodes abc. The other diodes receive low voltages, so they are off. This way the display shows the number 7.

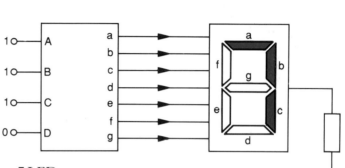

Figure 5
Putting the binary code 1110 into the decoder displays the number 7, because the diodes a, b and c light up

Binary codes

When we write the number 362, we mean: $(3 \times 100) + (6 \times 10) + (2 \times 1)$. So we are using powers of 10 to express numbers. The same number can also be written $(3 \times 10^2) + (6 \times 10^1) + (2 \times 10^0)$. We use numbers involving powers of ten, because we have ten digits on our two hands. However, computers can store numbers using either a high voltage level (1), or a low voltage level (0). So computers write

Digital score boards, like this one at Crystal Palace Sports Centre, use an integrated circuit to decode electronic binary signals and display the numbers or letters as a visual, readable message

numbers using powers of 2 or **binary**. The code going into the decoder was a binary number. The number 0111 means: $(0 \times 2^3) + (1 \times 2^2) + (1 \times 2^1) + (1 \times 2^0)$. This works out as: $4 + 2 + 1 = 7$.

Compact discs also use binary codes, to store information about music. Notches in the surface of the disc reflect a laser (Figure 6). Then a light-sensitive detector can pick up the information from the surface of the disc and turn it into an analogue signal (see unit G9). The analogue signal is then used to power loud-speakers and create the music we hear.

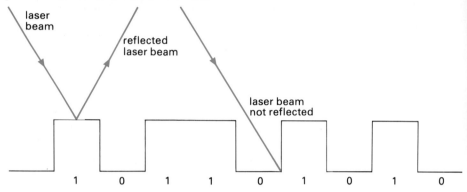

Figure 6

Questions

1 Look carefully at the light meter in Figure 1.
(a) What is the resistance of the LDR when it is just too dark to play cricket safely?
(b) Sketch a graph to show how the resistance of the LDR depends on the brightness of the light falling on it.

2 (a) Which bulbs light in the circuits below?

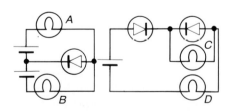

(b) Which bulbs light if you turn the batteries round?

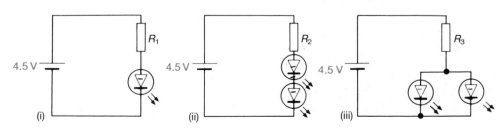

3 The graph shows how the current through an LED varies with the applied voltage. The LED works normally when there is a current of 20 mA through it.

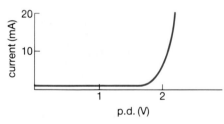

(a) What is the voltage across the diode when the current is 20 mA?
(b) For each of the circuits below choose a resistor that will make the diode(s) work correctly. You may only choose from the resistors given.
Values of available resistors (Ω): 33, 47, 56, 82, 100, 120, 150.

4 (a) This question refers to the diagram. Which diodes light for these inputs: (i) DCBA = 1001, (ii) DCBA = 0101?
(b) Why does no diode light for the input DCBA = 1100?

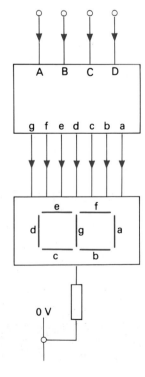

2 Rectification and Smoothing

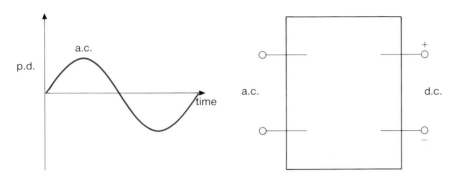

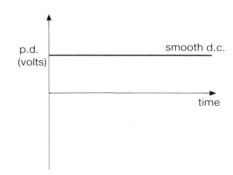

Figure 1
Rectification
(a) The voltage produced by the mains   *(b) The rectifier changes a.c. to d.c.*   *(c) The smoothed d.c. voltage*

The mains electricity supply produces alternating current (a.c.). Sometimes we want to turn this supply into a direct current (d.c.) supply. This process is called **rectification**. We also want to have a 'smooth' d.c. supply. This means that the d.c. supply must be constant (Figure 1).

Half-wave rectifiers

Figure 2 shows the simplest way of rectifying an a.c. supply, to produce d.c. for a resistor, in which a single diode is used. This is called **half-wave rectification**.

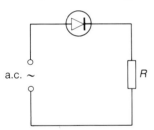

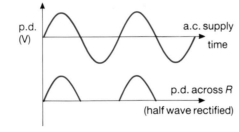

Figure 2
Half-wave rectification

The a.c. supply tries to drive current first one way, then the other, through the resistor. However, the diode only allows current to flow one way. Half of the time, current flows through the resistor. The other half of the time there is no current, because the diode is reverse biased. To make a smooth d.c. supply, we need to use a **capacitor**.

Some practical capacitors

Capacitors

A capacitor is like a rechargeable battery. It stores a small amount of charge and energy. This energy can be used to make a current flow, for a short time, through a resistance or a light bulb. The construction of capacitors is explained in Figure 3.

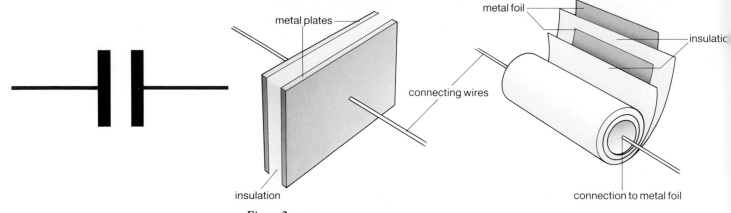

(c) Circuit symbol for a capacitor

Figure 3
(a) The simplest form of capacitor is made from two metal plates, with insulation between them

(b) Practical capacitors are made like a swiss roll. Metal foil, separated by layers of insulation, is rolled up

The size of a capacitor is measured in units called **farads**. A farad is a very large unit; so we usually use microfarads or μF to measure the size of a capacitor. 1 000 000 μF = 1 farad.

In Figure 4(a) the capacitor is charged by connecting it to the battery. When the switch is closed there is a quick pulse of current. This leaves one side of the capacitor with positive charge, and the other side with negative charge. Once the capacitor is charged, no more current passes through it.

In Figure 4(b) the capacitor is discharged by connecting it to a resistor and LED. Charge flows, until the positive charge has neutralised the negative charge. A large capacitor stores more charge than a smaller capacitor. This means that a large capacitor keeps the current going for a longer time. So the LED lights for longer. A large resistance also makes the current last longer. When R is big, I is small. So the capacitor loses its charge more slowly.

For a slow discharge, $R \times C$ must be large

Figure 4
(a) Charging the capacitor

(b) Discharging the capacitor

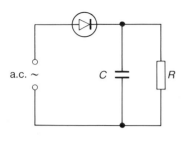

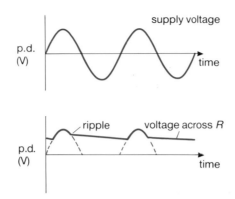

Figure 5
Smoothing the voltage from an a.c. supply

Smoothing

Figure 5 shows how to smooth the voltage from an a.c. supply, using a large capacitor. Each time the a.c. supply reaches its maximum voltage, the capacitor gets charged up to this voltage. When the a.c. voltage drops, the voltage across the resistor stays roughly constant. This is because the capacitor acts like a battery to provide the extra current. Now the d.c. voltage shows only a small 'ripple'.

Increasing the current drawn from the supply makes the ripple worse. This is because the capacitor cannot supply enough charge. You can make the ripple less by using a large capacitor.

Questions

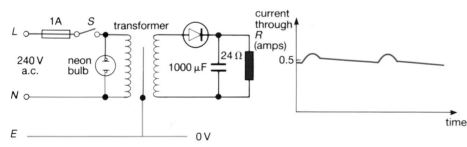

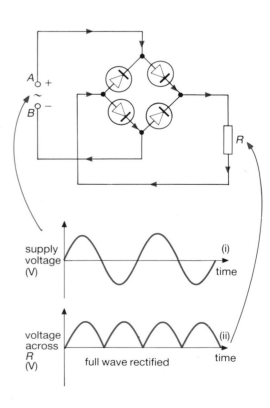

1 The circuit above shows a d.c. power supply which plugs into the mains.
(a) What does the transformer do?
(b) Why is the core of the transformer earthed?
(c) Why does the live wire have: (i) a switch, (ii) a fuse?
(d) Why is there a neon bulb?
(e) The graph shows how the current through R changes with time. Copy the graph. Add to it two further graphs to show the current when these two separate changes are made:
(i) $C = 2000\ \mu F$, (ii) $R = 12\ \Omega$

2 This diagram (right) shows a bridge rectifier. It produces a voltage which is 'full-wave rectified'.
(a) The arrows in the diagram show the current path when A is positive. Copy the diagram and show the current path when B is positive and A negative.
(b) Now explain the shape of the graph showing the voltage across R. Why is it called 'full-wave rectified'?
(c) Explain how a capacitor can be used to smooth this rectified voltage. Show on your diagram how the capacitor should be connected to the circuit.

3 Transistors

V_{BE} (V)	I_B (mA)	I_C (mA)	I_E (mA)
0	0	0	0
0.3	0	0	0
0.6	0	0	0
0.7	1	100	101

Table 1

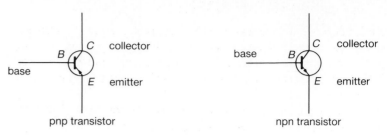

Figure 1
These are the circuit symbols for two types of transistor. The arrows show the direction of the current through the emitter.

The outside of a transistor is nothing special to look at. But the transistor is the key to all modern electronic circuits. It looks like a small piece of plastic or metal, with three long legs sticking out. These legs are connecting wires to the transistor's three terminals. These terminals are called the **base**, the **emitter** and the **collector**. Transistors are made out of specially manufactured germanium or silicon. Figure 1 shows the circuit symbols for two types of transistor.

Figure 2 shows a circuit that will help you to understand the action of a transistor. A 6 V battery has been connected across the collector and the emitter. The power supply connected between the base and the emitter can vary the voltage between 0 and 1 V. Table 1 shows how the voltage between the base and the emitter, V_{BE}, affects the currents going into and out of the transistor. The currents are called the base, collector and emitter currents, (I_B, I_C and I_E).

Turning a transistor on and off

When there is a voltage of about 0.7 V between base and emitter, the transistor is on. This means a current goes into the base and a current flows from collector to emitter. If the voltage (V_{BE}) is less than 0.7 V, the transistor is off. No current flows into or out of it.

The transistor is so useful because it can be switched on and off by a small change in base voltage. It is an electronic switch with no moving parts, so it does not wear out. In computers, transistors are switched on and off millions of times a second! That is why computers can calculate so quickly.

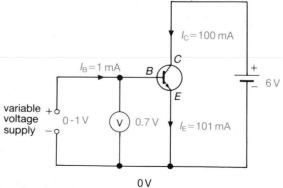

Figure 2
A circuit showing how a transistor works

Controlling the transistor

Figure 3, below, shows one way to control the base current of a transistor. A variable resistor is connected between the base and the positive side of the battery. When the resistance is very high (1 MΩ) no current flows into the base. The lamp is off. When the resistance is made lower (1 kΩ) current flows into the base. Then the lamp is on.

The extra resistance, R_B, is there to protect the transistor. If the resistance of the variable resistor is made zero, a large current would flow into the base and cause damage.

A potential divider (Figure 4) can also be used to control a transistor. The two resistors are in series, so the same current flows through them. However, resistor X is five times bigger than resistor Y. So the voltage across X is five times bigger than the voltage across Y. There is a voltage of 5 V across AB and a voltage of 1 V across BC. The voltage at A is 6 V, at B, 1 V, and at C it is 0 V.

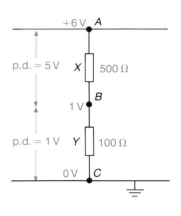

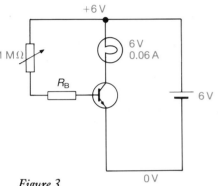

Figure 3
Controlling the base current of a transistor
(a) High resistance; current cannot flow into the base, so the lamp is off

(b) Low resistance; a small current flows into the base, so the lamp is on

$$\text{In general:} \quad \frac{V_1}{V_2} = \frac{R_1}{R_2}$$

Figure 4
A potential divider (a)

Questions

1 In this question you may assume that the voltmeters V_1 and V_2 have a very high resistance, so you can ignore any current that flows through them.
(a) What do the voltmeters read in the circuit below when: (i) $R_1 = 5$ kΩ; $R_2 = 10$ kΩ, (ii) $R_1 = 2$ kΩ; $R_2 = 2$ kΩ, (iii) $R_1 = 300$ Ω; $R_2 = 1500$ Ω?

(b) (i) what does V_1 read in the circuit below? (ii) What does V_2 now read? (iii) What is the value of R_3?

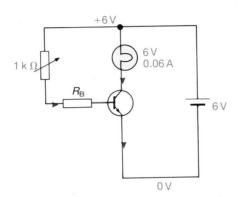

2 In the diagrams shown opposite the transistors are similar to the transistor shown in Figure 3. The light bulbs

need a current of about 100 mA (0.1 A) to make them light. Explain which of the bulbs will light in these diagrams.

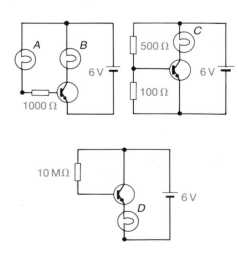

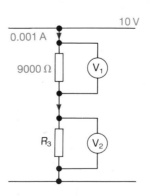

4 Using Transistors

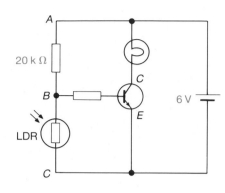

Figure 1
Making a light come on in the dark

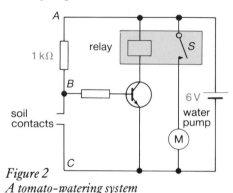

Figure 2
A tomato-watering system

Here are some circuits which use transistors to switch things on and off.

Turning on a light in the dark

In daylight an LDR has a resistance of about 500 Ω. So when it is light the voltage across BC is very small (Figure 1). This is because the resistance of an LDR is much smaller than the 20 kΩ resistor. Now the voltage at B is close to zero. The transistor is switched off and the bulb is out.

When it is dark the resistance of the LDR becomes very high, about 1 MΩ (1 000 000 Ω). The voltage across the 20 kΩ resistor is small compared with the voltage across the LDR. Now the voltage at B is nearly 6 V. The transistor is switched on and the lamp lights.

An automatic tomato waterer

The circuit in Figure 2 makes sure your tomatoes get watered when you are away on holiday. The soil contacts are placed into the soil around the tomatoes. When the soil is wet it conducts electricity well. This means that there is a short circuit between the base and the emitter of the transistor. So the transistor is switched off. When the soil dries out the resistance between B and C becomes large. This makes the voltage at B rise and the transistor is switched on. Now the current flowing through the transistor energises the relay coil. The magnetic field from the coil makes the switch S close. This switches the pump on. We need a relay to switch on the pump because it uses a current of 2 A. This is a far larger current than can flow through the transistor.

Transistors make it easy to keep these tomatoes watered while you are away on holiday

Controlling the temperature

This circuit might help you grow tomatoes in winter. The idea is to turn on a heater when your greenhouse gets too cold. The resistance of the thermistor changes as shown in the graph (Figure 3). The thermistor's resistance is highest when it is cold.

We can calculate what value the variable resistor must be set to, so that the heater is switched on when the temperature drops below freezing point (0°C) (Figure 3). From the graph you can see that the resistance of the thermistor is 700 Ω at 0°C. For the transistor to be switched on, the voltage drop across the thermistor needs to be 0.7 V. This means there must be 5.3 V (6.0 V − 0.7 V) across the variable resistor (Figure 4).

The same current goes through the resistor and the thermistor. (We ignore any small base current going into the transistor.)

$$\text{So } I = \frac{V}{R} = \frac{\text{voltage across resistor}}{\text{resistance of } R} = \frac{\text{voltage across thermistor}}{\text{resistance of thermistor}}$$

$$\text{or} \qquad \frac{5.3 \text{ V}}{R} = \frac{0.7 \text{ V}}{700}$$

$$\text{So} \qquad R = 700 \text{ Ω} \times \frac{5.3}{0.7} = 5300 \text{ Ω}$$

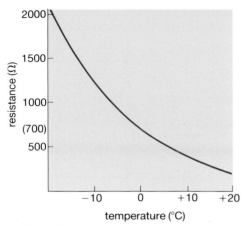

Figure 3
The thermistor resistance changes with temperature

Adjusting R allows you to control the temperature of your greenhouse. This is an example of control to an electronic system being achieved through negative feedback. The temperature is selected by adjusting the variable resistor. The thermistor senses the temperature in the greenhouse, and the heater is controlled by the action of the transistor and relay. The heater switches on and off and the temperature oscillates around 0°C.

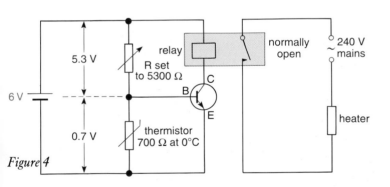

Figure 4

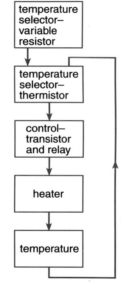

Figure 5
Achieving control through negative feedback

Questions

1 Jamal has designed the circuit shown, so that his light bulb will come on in the dark. Has he got his circuit right? Explain your answer.

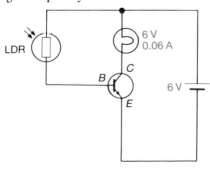

2 Look at the circuit in Figure 4. You have decided that you want the heater on the greenhouse to switch on when the temperature is 10°C.
(a) What is the resistance of the thermistor at this temperature?
(b) What is the voltage across the thermistor when the transistor is switched on?
(c) Therefore, what is the voltage across the variable resistor when the transistor is on?
(d) Now work out what value R must be set to, so that the heater comes on at a temperature of 10°C.

3 Design an electronic system to turn on a freezer motor, when the temperature inside the freezer rises above −10°C. In your system you might use a relay, a thermistor and variable resistor similar to those in Figure 4. Explain how your system works, using appropriate calculations.
4 Design an electronic system for a photographic dark room, to sound an alarm if the light intensity is too high.

5 Transistor Amplifiers

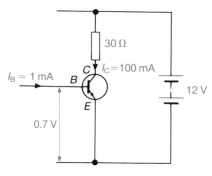

Figure 1

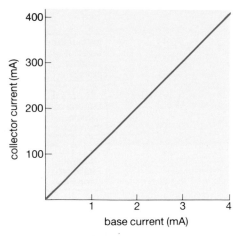

Figure 2
The collector current is proportional to the base current

(not drawn to scale)

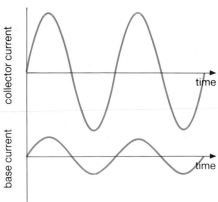

Figure 3

Current gain

In Figure 1 the voltage between the base and the emitter is 0.7 V. The current going into the base is 1 mA and the current going into the collector is 100 mA. So a small base current controls a much larger collector current. The **current gain** of this transistor is 100.

$$\text{current gain} = \frac{I_C}{I_B}$$

The gain varies for different transistors; it can be anywhere between 10 and 1000.

The transistor is very sensitive to changes in its base current. Figure 2 shows that I_C is proportional to I_B. So if we double the base current to 2 mA the collector current increases to 200 mA.

This behaviour of the transistor is very important. It allows us to build **amplifiers**. When a small alternating current goes into the base, a much larger alternating current flows into the collector (Figure 3).

Instead of the resistor in Figure 1, you can put in a loudspeaker or earphone. The amplifier in your stereo system uses lots of transistors. When you play a tape, a small current goes into the base of a transistor. A much larger current flows through the loudspeaker. Now you can hear the music.

Some amplifiers are for personal use

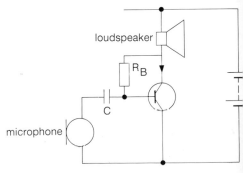

Figure 4
A circuit for a simple amplifier

Making an amplifier

Figure 4 shows how to make a simple amplifier. A microphone is used to pick up a speaker's voice. A loudspeaker relays what is said to an audience:
- The base resistor R_B is there to allow a small current to go into the base. This switches the transistor on. We call this biasing the transistor.
- When someone speaks into the microphone, a small changing voltage is produced across it. This signal passes through the capacitor, and makes changes to the base current.

Big amplifiers are needed for large outdoor concerts. This photo shows Luciano Pavarotti's 1991 concert in Hyde Park

- The small changes to the base current are amplified by the transistor. The current is now large enough to drive the loudspeaker.
- It is important to have a capacitor in the circuit. Capacitors allow alternating currents through, but not direct currents. This means that the signal can get through to the transistor. Without the capacitor the biasing of the transistor would be upset. This is because a direct current would go from the base through the microphone. Now the base would be at the wrong voltage, and the transistor would not work properly.

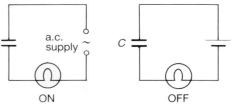

Figure 5
A capacitor passes a.c. but not d.c.

Questions

1 Below you can see a transistor circuit driving a loudspeaker. An alternating current is going into the base (graph 1). Graph 2 shows how the base current controls the collector current.

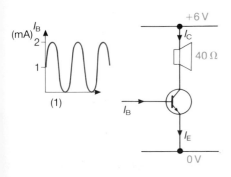

(a) What is the current gain of the transistor?
(b) What are the maximum and minimum values of: (i) I_B, (ii) I_C?
(c) Sketch a graph to show how I_C varies with time.
(d) When I_B is 1 mA, how big is the current I_E?

2 Look again at the diagram in question 1.
(a) When the collector current I_C is 50 mA, what is the voltage across the loudspeaker?
(b) Now explain why the largest collector current that can flow through this circuit is 150 mA.
(c) A very large base current is now fed into the transistor (graph 3). The music now sounds odd. Can you explain why? This is called distortion.

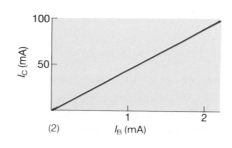

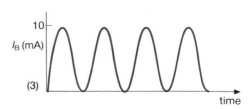

6 Operational Amplifiers

Here you can see a photograph of two operational amplifiers. Operational amplifiers like these are used in many electrical devices

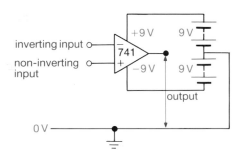

Figure 1
The 741 op-amp

Input	Output
V+ > V−	+ 9 V
V− > V+	− 9 V

Table 1

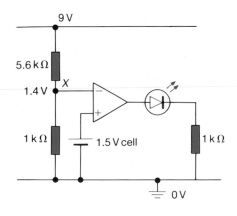

Figure 2
A cell tester

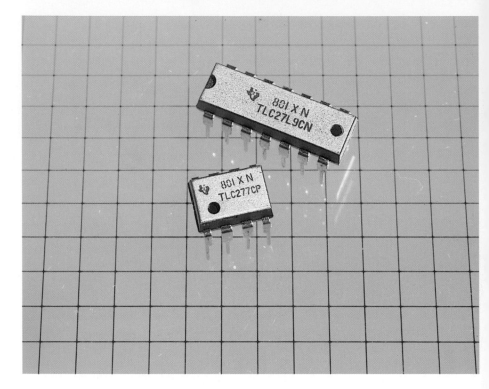

Operational amplifiers are used for comparing and amplifying voltages. One of the most widely used operational amplifiers (or **op-amps**) is the 741. This op-amp contains about 20 transistors, a capacitor and several resistors. All these components are on a single piece of silicon. This is called a **silicon chip**. The 741 op-amp is an example of an **integrated circuit**.

Figure 1 shows the various connections to the 741. It is powered by two 9 V batteries. It has two inputs and an output. When the non-inverting input voltage, V+, is greater than the inverting input voltage, V−, the output voltage is +9 V. When V− is greater than V+ the output voltage is −9 V (table 1). (The 741 must always be connected to two batteries. We do not usually show the connections on circuit diagrams. This makes the diagrams easier to follow.)

A battery tester

When you are making an electronic circuit, it is really annoying to find that you have a flat battery. A 741 can be used as a battery (or cell) tester. Figure 2 shows the idea. A potential divider fixes the non-inverting input at about 1.4 V. The cells to be tested should produce a voltage of about 1.5 V. When the voltage is greater than 1.4 V the output is +9 V. Then the LED is forward biased and it lights. If the battery is flat (voltage less than 1.4 V) the output is −9 V. Then the LED is reverse biased and it does not light.

Making an amplifier

The 741 can also be used as a voltage amplifier. The output of the 741 depends on the voltage difference between the two inputs. The 741 has a very high voltage gain. When the difference between the two inputs is about 0.001 mV, the output is about 1 V.

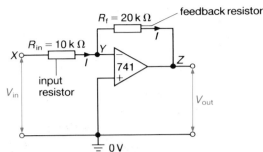

Figure 3
An inverting amplifier. (The connections of the op-amp to the power supply are not shown)

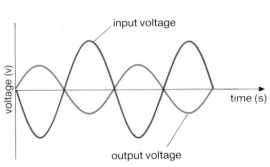

Figure 4
The input and output voltages are out of phase

To make a stable amplifier we apply some **negative feedback**. The output is connected to the inverting input through a feedback resistor. When the op-amp has negative feedback we can say:

- The two inputs have very nearly the same voltage, V+ = V−
- No current flows into the op-amp itself. (In fact a current of about 0.000 000 000 01 A goes into the op-amp, but it is so small we will ignore it!)

We will now apply these two rules to the amplifier shown in Figure 3. Since V+ = V−, the point Y must be at earth voltage (0 V). Also, since no current goes into the op-amp, the same current I goes through the input and feedback resistors.

$$I = \frac{V_{XY}}{R_{in}} = \frac{V_{YZ}}{R_f}$$

V_{XY} = voltage between X and Y
V_{YZ} = voltage between Y and Z

So $$\frac{V_{in} - 0}{R_{in}} = \frac{0 - V_{out}}{R_f}$$

or $$\frac{V_{out}}{V_{in}} = -\frac{R_f}{R_{in}}$$

In this example $\frac{R_f}{R_{in}} = 2$. So the amplifier multiplies this voltage by −2.

The minus sign means that the phase of the voltage has been changed (Figure 4). The output voltage of an op-amp can be no bigger than the supply voltage. If the supply voltage is from two 9 V batteries, you can get no more than 9 V out of the amplifier. This means that if the input voltage is too big, the output gets distorted (Figure 5).

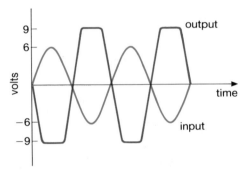

Figure 5
This amplifier tries to multiply the voltage by 2, but it cannot produce more than 9 V

Questions

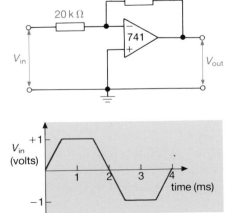

1 On the left you can see a circuit diagram for an operational amplifier. The graph shows how the input voltage varies with time. Copy the graph. Add to it a second graph to show how the output voltage varies with time. Give as much detail as possible.

2 This question is about using an op-amp to add voltages (see circuit on right).
(a) What is the voltage at C?
(b) Calculate the current I_1 and I_2.
(c) Now calculate the current I_3.

(Remember no current goes into the op-amp).
(d) What is the voltage at D?
(e) Explain how this circuit could be useful in an audio-mixer.

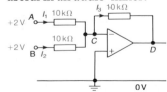

3 Explain why the voltage at point X in Figure 2 is about 1.4 V.

7 Logic Gates

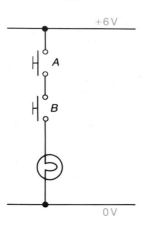

Figure 1
A two-state system: the bulb is on or off

Two states: ON and OFF

Logic gates are widely used in computers and other electronic systems. They collect and display information.

A very simple example of a logic gate is shown in Figure 1. To turn the light bulb on, you have to press both switches. It is no good just pressing one of them. The behaviour of the circuit is summarised in table 1. This simple circuit can collect information about your fingers! When two fingers press A and B the bulb lights.

Figure 2 shows a similar circuit that works with relays. This time the circuit collects information about voltages. The relay switches in the circuit are normally open. However, if the voltage at A is high (for example, above 3 V), the top switch is closed. So when A and B are both above 3 V the lamp switches on.

Truth tables

Truth tables show the behaviour of an electronic system in shorthand. A high voltage (above 3 V in our example) is defined as **logic state '1'**. A low voltage (below 3 V) is defined as **logic state '0'**.

You can describe the behaviour of the circuit in Figure 2 like this. When both A AND B are 1, Q is 1; otherwise Q is 0. Table 2 summarises this. We call this sort of table a **truth table**.

The circuit you have just been looking at makes an **AND-gate**. This gate was made with switches and relays. Logic gates are also made easily with transistors in integrated circuits. The advantages of integrated circuit logic gates are: they are cheap, small and can be switched millions of times each second. The circuit symbol for an AND-gate is shown above Table 2.

All integrated circuits need batteries to power them. We usually leave out connections to batteries in the circuit diagrams though. This helps to make the circuits easy to follow.

State of switch		State of bulb
A	**B**	
open	open	OFF
closed	open	OFF
open	closed	OFF
closed	closed	ON

Table 1

The magnet around Boris' neck will open the cat door for him. Can you explain how it works?

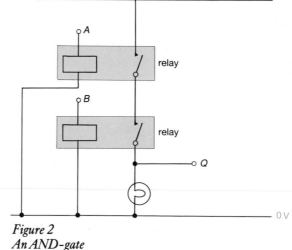

Figure 2
An AND-gate

A ———[AND]——— Q
B ———

State of inputs		State of outputs Q
A	**B**	
0	0	0
0	1	0
1	0	0
1	1	1

Table 2 Truth table for an AND-gate, with its circuit symbol.

More logic gates

- An **OR-gate** is shown on the next page. When A OR B (OR both) is 1 then Q is 1 (Table 3).
- A **NOT-gate** is shown on the next page. When A is NOT 1, Q is 1 (Table 4).

State of inputs		State of outputs
A	**B**	**Q**
0	0	0
0	1	1
1	0	1
1	1	1

State of input A	State of output, Q
0	1
1	0

Input A	Input B	Q
0	0	1
0	1	0
1	0	0
1	1	0

Table 3 Circuit symbol and truth table for an OR-gate.

Table 4 Circuit symbol and truth table for a NOT-gate.

Table 5. Circuit symbol and truth table for a NOR-gate.

- A **NOR-gate** symbol and truth table are shown on the right. When neither A NOR B is 1, Q is 1 (Table 5).
- A **NAND-gate** symbol and truth table are shown below right. When both A AND B are NOT 1, Q is 1 (Table 6).

Input A	Input B	Q
0	0	1
0	1	1
1	0	1
1	1	0

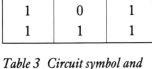

Table 6. Circuit symbol and truth table for a NAND-gate.

Questions

1 This question is about working out logic states. For example when A is 1, B is 0. What are the states of the points C and D?

2 The circuit below shows a security system for a car.
(a) Which switches have to be closed to turn on the starter motor?
(b) What happens if any other switch is closed?

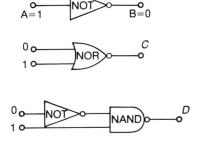

3 What single gate can be made from the AND-gate and the NOT-gate? Explain your answer.

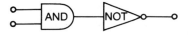

4 Design these electronic systems using the gates in this unit:
(i) a 4 input AND-gate,
(ii) a 3 input NOR-gate.

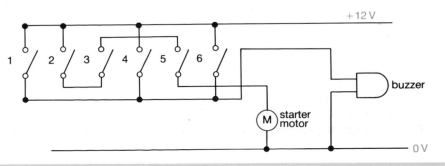

8 Using some Logic

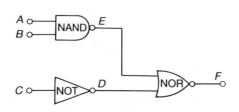

Figure 1

A	B	C	D	E	F
0	0	0	1	1	0
0	0	1	0	1	0
0	1	0	1	1	0
1	0	0	1	1	0
0	1	1	0	1	0
1	0	1	0	1	0
1	1	0	1	0	0
1	1	1	0	0	1

Table 1

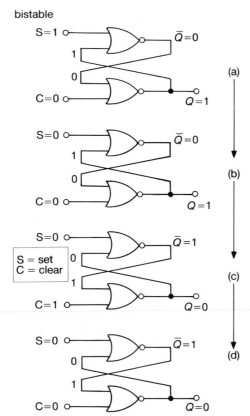

Figure 2
This shows how a bistable changes from one state to another

An Apple Macintosh microcomputer contains about half a million NOR gates

Analysing logic systems

More complicated logic systems can be made by using a combination of gates. In Figure 1, how does the output F depend on the states of A, B and C? You can work out this problem by constructing a truth table (Table 1).

First you write out the columns A, B and C, working out all the possible combinations of states. The state of E can be worked out from the truth table of a NAND-gate. E is only 0 when A and B are both 1. D is 0 when C is 1; D is 1 when C is 0. You can now fill in the columns D and E. D and E are the inputs to the NOR-gate; F is the output. F is 1 when D and E are 0.

So this circuit makes a 3-input AND gate. When A AND B AND C are 1, F is 1. Otherwise F is 0.

The bistable

The **bistable** or **flip-flop** is the basis of electronic computer memories. A bistable can be made from two NOR gates, as shown in Figure 2. The sequence of diagrams shows how the bistable works. You should remember that the output (Q or $\bar{Q}$) of a NOR gate is only 1 when both inputs are 0. If either input (or both) is 1, the output is 0.

In diagram (a) the bistable is set with S = 1 and C = 0. Since S = 1, $\bar{Q}$ = 0. Now both inputs to the lower NOR gate are 0, so Q = 1.

In diagram (b), S has changed to 0. This does not affect Q or $\bar{Q}$. So Q = 1 still.

However, in diagram (c) S = 0 and C = 1. This means that Q = 0 and $\bar{Q}$ = 1. The bistable has been cleared.

In diagram (d) C goes back to 0. This does not affect Q or $\bar{Q}$.

Notice that the bistable remembers which of S or C was *last* in the logic state 1. The output, Q, of the bistable has two states. Q is either 0 or 1. We say that the bistable remembers one **bit of information**. The word 'bit' is short for binary digit. Computers calculate only in terms of binary numbers. All binary numbers can be expressed in terms of noughts and ones.

The electronic memory for a home computer has about 250 000 bistables inside it. These can be accommodated on a few small silicon chips. When you turn your computer off, it forgets what is stored in its electronic memory. So to store information permanently, you must transfer it to a magnetic disc or tape.

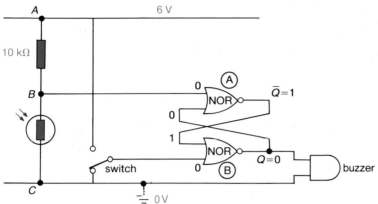

Figure 3
A latched burglar alarm

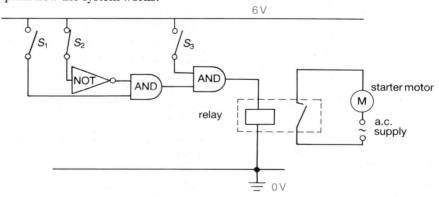

Figure 4

An electronic latch

Bistables are often used in alarm systems. This is because they work as **electronic latches**. When the alarm has been triggered, the latch keeps the alarm on.

Figure 3 shows a circuit for a 'latched' burglar alarm. An infra-red beam of light falls onto a LDR. This keeps the LDR's resistance low. B is at a low voltage. Both inputs of NOR gate A are low. This makes $\bar{Q}$ high, and Q low. The buzzer is off.

If the burglar now walks across the infra-red beam, the resistance of the LDR goes high for a moment. Now B goes high and $\bar{Q}$ goes low. Both inputs to NOR gate B are low and Q goes high. The buzzer sounds the alarm. But the alarm stays on because the bistable 'remembers' that the burglar was there.

This is a useful example of positive feedback in an electronic system.

Figure 4
Here is another example of positive feedback, but not as useful as a bistable. By placing a microphone close to a loudspeaker a loud pitched whine can be produced. The microphone picks up a small noise, which is amplified and emitted by the loudspeaker. The microphone picks this noise up and it is amplified again and so on

Questions

1 (a) Why is an infra-red beam of light used in the burglar alarm?
(b) Explain how you can turn the burglar alarm off.
(c) Think of another way that you could use an electronic latch.

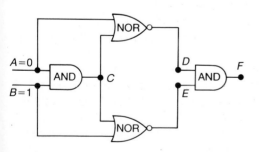

2 (a) In the diagram above, work out the logic state of the points C, D, E, F.
(b) When F is in logic state 1, what are the states of A and B?

3 The circuit below shows part of a security system for a car ferry. S_1 closes when the bow doors are shut. S_2 opens when the passenger gangway is lifted. S_3 is an ignition switch on the bridge. Explain how the system works.

9 Data Logging

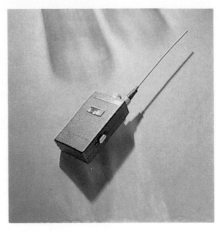

A LogIT data logger

We are all used to taking measurements. We use a ruler to measure length, a thermometer to measure temperature, a watch to measure time, and voltmeters and ammeters in electrical experiments. We usually measure two or more things at once. For example, we might want to know how the temperature of a cup of tea changes with time; so you place a thermometer in the tea and record the temperature at 1 minute intervals. This is a very simple, manual way of **recording, or logging,** data. Nowadays we can use preprogrammed instruments to do the job for us. These instruments are called **data loggers.**

LogIT

LogIT is an example of a cheap and easy-to-use data logger. At the press of a button, LogIT can be started and it will record automatically from a variety of sensors. These sensors can record such variables as temperature, light intensity and voltage, so the experimenter can carry out a wide range of experiments. The data logger can record measurements every few milliseconds, or slow its rate down to one measurement every few hours for a long experiment where changes are very slow. So, not only does LogIT make it easier to do some experiments, but it enables you to measure changes that are too fast or too slow for normal laboratory work. LogIT can even measure *three* variables simultaneously. Finally when the experiment is finished, LogIT is connected to a computer system which is programmed to receive and analyse the data. The computer can then be instructed to tabulate the data or display it graphically.

Examples of use

- **Light output from a bulb**

A bulb was pulsed on for about 1 second in a darkened room. Figure 1 shows the circuit used. LogIT recorded readings of light intensity and voltage across the bulb every 10 milliseconds (Figure 2).

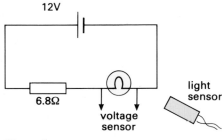

Figure 1

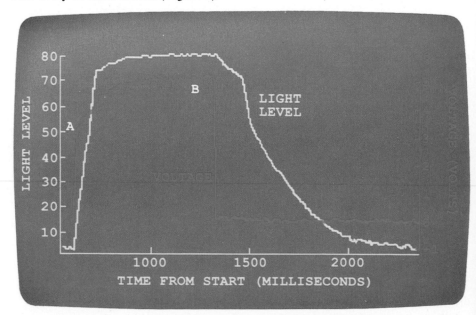

Figure 2

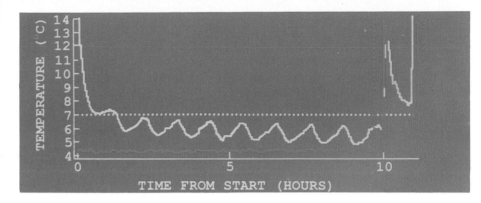

Figure 3

- **Temperature inside Steve's fridge overnight**
  LogIT records temperature and light intensity every 5 minutes (Figure 3).

- **Cooling curves**
  This is a familiar experiment you can do yourself; but LogIT can record temperatures of two beakers of water simultaneously (Figure 4). The results can be displayed immediately on a VDU. Which beaker in Figure 4 is insulated?

- **How fit are you?**
  The graph in Figure 5 shows Judy's heart rate before, during and after a run. A sensor strapped to the body detects electrical pulses caused by the heart beating. The sensor then transmits these pulses, using radiowaves, to a receiver which is plugged into LogIT. Judy carries LogIT on her belt. On returning from the run the data can be displayed on a computer again.

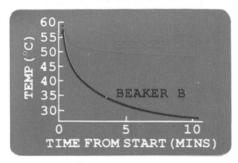

Figure 4

Judy's belt carries LogIT to record her heart rate as she runs

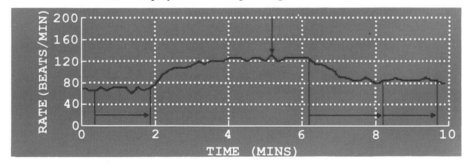

Figure 5

Questions

1 This question is about the light levels and voltage changes of the light bulb (Figure 2).
(a) What is the voltage across the light bulb at times (i) A, (ii) B?
(b) Use the answers to part (a), and the data in Figure 2, to calculate the voltage across the 6.8 Ω resistor at times (i) A, (ii) B.
(c) Now deduce the current flowing through the circuit at each time (i) A, (ii) B.

(d) Calculate the resistance of the bulb at times A and B. Explain why the changes occurred.
(e) Describe how the light level changed immediately after the bulb was turned off. Account for these changes.
2 (a) Describe how the temperature changed in Steve's fridge overnight.
(b) What happened after 10 hours?
(c) The motor on Steve's fridge is rated at 100 W. Use the graph to

calculate the cost of running his fridge overnight, given that 1 kWh costs 7p.
3 (a) How long did Judy run for?
(b) What was her average heart rate (i) before, (ii) during, (iii) after, the run?
(c) What was Judy's recovery time?
4 Design an experiment using LogIT to monitor a chemical reaction which involves a colour change in a liquid.

10 Storage and Retrieval

If you go into a music shop to buy a music centre, it is possible to buy one that allows you to play records (or vinyls), tapes or compact discs (CDs). However, a lot of companies have stopped cutting records and concentrate on producing CDs. Whichever system you choose, information has been stored on record, tape or CD, which you retrieve when you listen to the music.

Vinyl discs

The photograph opposite shows a high powered magnification of the grooves that have been cut into a vinyl disc. When you play a record, a very small stylus tracks along the groove. The indentations along this groove make the stylus vibrate, up and down, or sideways. A transducer turns these vibrations into electrical pulses, which are amplified to make the loudspeaker vibrate. You then hear a reproduction of the original music.

A stylus following the grooves on a record

Magnetic tapes

Figure 1 shows the principles behind a tape recorder. The tape is made of thin plastic, coated with a special form of ferric oxide which can easily be magnetised. When a recording is made the microphone turns the sound into audio frequency electrical oscillations. These oscillations are amplified and fed into the recording head, which is rather like a small electromagnet. As the tape passes under the recording head it is magnetised. The information about the music that you are recording is then stored in the tape.

To play the music back, the tape is fed past the playback head. The process now goes into reverse. The magnetised tape induces currents in the coil in the playback head. These currents are amplified to power the loudspeaker. Tape recorders have an erase head to wipe the tape clean, so that you can rerecord over material on a tape. A very high frequency current passes through the erase head, which jumbles up all the directions of magnetisation on the tape. The information about the music is lost. You can erase a tape for yourself using a magnet. Take a tape that you have recorded and move a magnet backwards and forwards past the exposed part. Play it back and listen to the effect.

The small currents induced in the playback head coil of a cassette deck can be amplified to drive loudspeakers

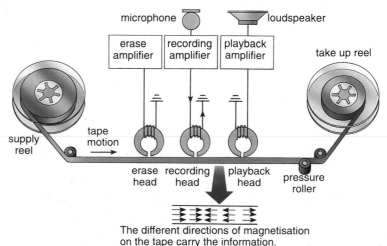

Figure 1
Layout of a tape recorder

Compact disc

The photograph at the start of this section on electronics (page 239) shows an electron micrograph of a compact disc. The music is coded in a series of tiny notches. Laser light is bounced off these notches and decoded to play the music.

A compact disc is similar to the vinyl disc in that they are both **read only memories** (ROMs). The information is stored permanently, cannot be altered, and can only be read. Magnetic tape is an example of an erasable memory. A compact disc differs from vinyls and tapes in that the music is recorded in a digital code; vinyls and tapes produce analogue, or continuously changing, signals, (see page 154). Digital signals produce much clearer recordings, with less accidental 'noise', that happens during the recording of a tape or vinyl.

Computer memories

In 1975 computer programs were stored on punched cards or tapes, which were read very slowly into the computer. By 1980 programs for home computers were stored on magnetic tape that you used in tape recorders. In 1985 most home computers were beginning to use floppy discs, and now home computers have hard discs.

Magnetic discs work in a similar way to tape. However, they are far faster to use because the read head on the disc drive can reach any part of the disc very quickly.

Computers also store information in electronic memories, which are of two types: ROMs and RAMs. RAM stands for Random Access Memory; this is a read and write or erasable memory.

These various forms of computer disk can hold differing amounts of data. ranging from less than 1 Mbyte for some of the floppy disks up to 650 Mbyte for CD-ROMs

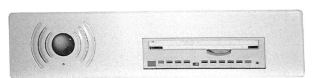

This magneto-optical drive is used to store computer data. It works through using a combination of a magnet and a laser. To write to the disk, both the magnet and the laser are used, while only the laser is required to read information from the disk. The drive uses a disk similar in size to a 3.5 inch floppy disk, but a magneto-optical disk can store much more data than a floppy disk

Questions

1 (a) (i) What are the similarities and differences between vinyl discs, tapes and CDs?
(ii) Give one advantage and one disadvantage of each.
(b) Explain how a video tape stores information.
(c) Digital audio disc-corders and digital audio tapes were new to the music market in 1992. Explain briefly what each of these new products can offer.
2 The microphone and recording head in a tape recorder are examples of transducers.
(a) Explain what is meant by a transducer.
(b) Explain how these two transducers work.
3 (a) Which of the following devices use erasable memories, ROMs or both? CD player, tape recorder, washing machine, digital watch, cash point, book, calculator, electronic till.
(b) A video recorder uses both sorts of memory. Explain how each is used.

SECTION K: QUESTIONS

1 With the components connected as shown in diagram (1) the LED glows.

(1)

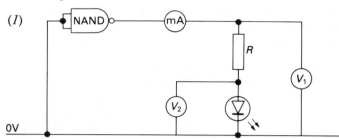

(a) What gate could be used instead of the NAND gate?
(b) What is the purpose of the resistor R?
   The voltmeters V_1 and V_2 have high resistances, so a negligible current passes through them. The readings on the voltmeters are: $V_1 = 3.8\,V$, $V_2 = 2.0\,V$. The milliammeter reads 10 mA.
(c) What is the resistance of R?
(d) The diode is now removed from the circuit and replaced by a light bulb labelled 3 V, 0.15 A. Explain why the bulb does not light.
(e) (i) The bulb does light when a transistor is used as shown in diagram (2). Explain the action of this circuit.
(ii) What value would you choose for R?

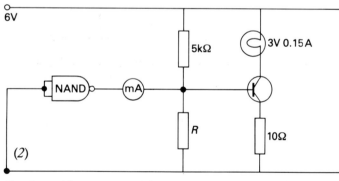

(2)

(f) A transistor cannot drive an electric motor by itself. Explain how you would modify diagram (2) so that an electric motor could be turned on and off, by switching the inputs of the NAND gate from a high to a low voltage.

2 The graph shows how the resistance of a light dependent resistor (LDR) depends on the intensity of light falling on it.

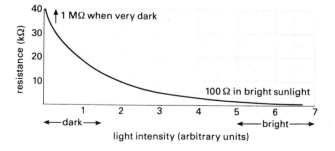

(a) Describe the action of the LDR.
(b) The milliammeter in the circuit below reads 1 mA.

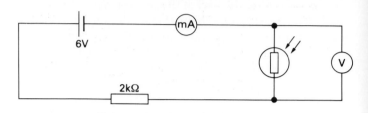

(i) Calculate the resistance of the LDR.
(ii) Use the graph to calculate the light intensity.
(c) The voltmeter across the LDR has a very high resistance. Estimate the reading on this voltmeter: (i) in bright sunlight, (ii) in very dark conditions.
(d) The truth table describes the operation of the diagram below it. Copy and complete the table.

LDR 1	LDR 2	Voltage states			Buzzer
		A	B	C	
light	light	1			OFF
light	dark				
dark	light				
dark	dark				

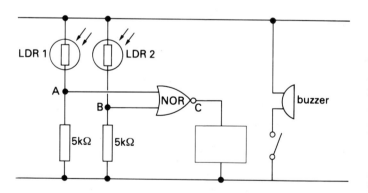

(e) Why is a relay needed to operate the buzzer?
(f) (i) Explain how this system could be used in a biscuit factory to reject any biscuits that are too long to fit into the packet.
(ii) How would you adapt the circuit to detect biscuits that are too short? **MEG**

3 (a) Explain how you would make a bistable using two NAND gates.
(b) The diagram shows a circuit with 5 NAND gates. The input receives a series of 1 Hz pulses, which switch it from high to low. Describe the behaviour of the three lamps, X, Y, Z when the 'hold' lead is (i) kept low, (ii) kept high.

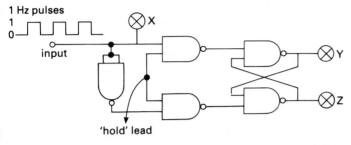

'hold' lead

(c) Explain how this circuit can be used as a memory; it is called a **data latch**. What advantage does this circuit have over an ordinary bistable memory?

4 (a) Copy and complete the truth table for the circuit below.

A	B	C	D	E	F	G
0	0	1				
0	1	1				
1	0	1				
1	1	0				

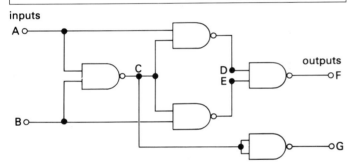

(b) What arithmetic function does this circuit carry out?

5 This circuit shows part of a car security lock system. The lock opens when the relay is switched on.

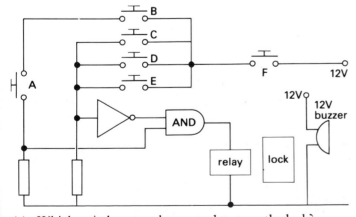

(a) Which switches must be pressed to open the lock?
(b) If the wrong switches are pressed the alarm sounds and *stays on*. Copy the diagram and add extra circuit components to make the alarm sound.

6 The circuit below controls the air conditioning in a factory. When it becomes too hot in the factory the fan switches on. The graph shows how the resistance of the thermistor changes with temperature. Explain why the fan switches on when the temperature rises to 25°C.

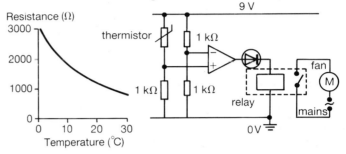

7 What is the output voltage in these circuits?

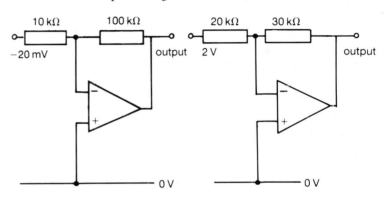

8 This diagram shows how an op–amp can be used as a non–inverting amplifier.

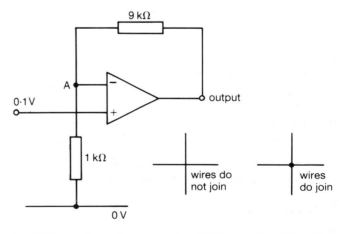

(a) What is the voltage at point A? (Remember V+ = V−).
(b) Calculate the current flowing through the 1 kΩ resistor.
(c) What current goes through the 9 kΩ resistor? (Remember no current goes into the op–amp.)
(d) Now calculate the output voltage.
(e) What is the voltage gain of this amplifier?

9 Mr Corrigan owns an amusement arcade. He is having some trouble with Tommy who is very rough with the pinball machines. So Mr Corrigan decides to fit anti-tilt devices to his pinball machines. When Tommy tips the machine, the power supply turns off. The circuit is shown below.

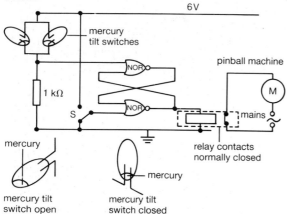

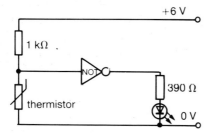

mercury tilt switch open   mercury tilt switch closed

(a) Explain how the mercury tilt switches work.
(b) Why are there two mercury tilt switches?
(c) What do the two NOR gates do?
(d) Why are the relay contacts normally closed?
(e) Explain how the machine can be started again by putting another coin in.

10 (a) For the circuit below, calculate the input voltage to the NOT gate (the voltage across the thermistor) when the thermistor has a resistance of 500 Ω.

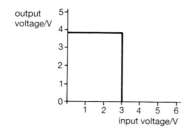

This graph shows the input/output voltage characteristics of the NOT gate used in this circuit.

output voltage/V

input voltage/V

(b) What is the output voltage of the NOT gate when the thermistor has a resistance of 500 Ω?
(c) When the light-emitting diode is on there is a voltage of 1.8 V across it. Calculate the current through the 390 Ω resistor.
(d) Write down what the answers to parts (a) and (c) will be if the thermistor and the 1 kΩ resistor are interchanged.

MEG

11 The circuit in the diagram below controls a motor.

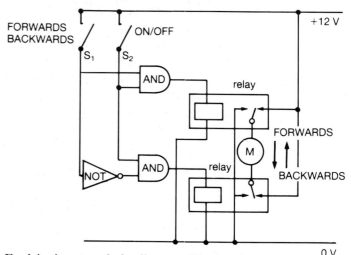

Explain, in as much detail as possible, how it works.

12 This question is about making a time control switch. When the switch is pressed and released, the capacitor gets charged up. For a while the voltage at B is high. The transistor is on and the bulb lights. However the capacitor loses its charge through R. After a while the voltage at B drops. This makes the bulb go out.
(a) Explain where you might use this idea in your house.
(b) What changes could you make to the circuit to make the bulb light for longer?

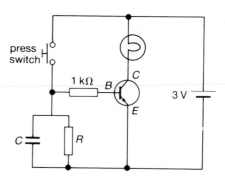

SECTION L
Radioactivity

This is the Conqueror, a nuclear powered submarine. The small nuclear reactor on board means that the Conqueror can remain submerged for many months without any need to refuel. It only surfaces to replenish food and other supplies, or for maintenance

1 Introducing Nuclear Physics

A nuclear war could completely destroy the world we live in. Many people near the explosion would be killed, and many more would die from the radiation emitted. In addition, the exploding bombs would drive smoke and dust high into the atmosphere, which could block out the Sun for a long time, resulting in yet more people dying, of cold, in the following years. This is called a 'nuclear winter'

At the Moscow Summit in 1988, the 'superpowers' (the USA and USSR) agreed to reduce their stores of nuclear missiles. The picture shows Ronald Reagan and Mikhail Gorbachev shortly after signing the agreement

For 30 years after the end of the Second World War, relationships between the West (USA and Europe) and the former Soviet Union were unfriendly. Quite a lot of people were afraid that a disagreement between the two sides might lead to a nuclear war. This situation improved after the Moscow Summit of 1988 – an agreement was made to reduce the number of nuclear weapons in the world. A nuclear war now seems unlikely. However, there is still the fear that an irresponsible government, or perhaps a group of terrorists for example, might develop and use a 'nuclear bomb'.

In the national news we frequently hear about the use of nuclear energy to produce electrical energy. On television we may see people arguing angrily about this issue. On one side we will hear scientists claim that the use of nuclear power is quite safe, and yet we will hear others claim that the use of nuclear power has caused more people than usual to die from cancer.

Which side is right? This is a difficult question to answer, but it is your job as a responsible citizen to read about these matters and to decide for yourself. You should understand why radioactive materials can be harmful to us, and also why they are useful to us.

However, before you can understand these issues you need to learn about nuclear physics and radioactivity.

The nuclear model of the atom

You will already know about atoms. You have learned how Brownian motion helped us to discover them. We think of atoms as hard bouncy balls that exert a pressure by hitting the walls of their container. You now need to know something about the *insides* of atoms.

In 1909 Geiger and Marsden discovered a way of exploring the insides of atoms. They directed a beam of **alpha particles** at a thin sheet of gold foil. Alpha particles were known to be positively charged helium ions, He^{2+}, which were travelling very quickly. They had expected all of these energetic particles to pass straight through the thin foil. Much to their surprise they discovered that a very small number of them bounced back, although most of them travelled through the foil without any noticeable change of direction.

Sellafield has been chosen as the site to reprocess nuclear fuel. Local residents are concerned about the level of radioactive contamination to the area

Rutherford produced a theory to explain these results; this is illustrated in Figure 1. He suggested that the atom is made up of a very small positively-charged **nucleus**, which is surrounded by **electrons** which are negatively-charged. His idea was that the electrons orbit around the nucleus in the same way that planets orbit around the sun. The gap between the nucleus and electrons is large, in fact the diameter of the atom is about 100 000 times larger than the diameter of the nucleus itself. Because so much of the atom is empty space most of the alpha particles could pass through it without getting close to the nucleus. Some particles passed close to the nucleus and so the positive charges of the alpha particle and the nucleus repelled each other causing a small deflection. A small number of particles met the nucleus head on, these were turned back the way they came. The fact that only a very tiny fraction of the alpha particles bounced backwards tell us that the nucleus is very small indeed. Rutherford proposed that the positive charge in the nucleus was carried by **protons**. Hydrogen has one proton, and this positive charge is balanced by the negative charge of one electron; helium has two protons whose charge is balanced by two electrons which orbit the nucleus.

Nuclear fuels are used to generate electricity at Dungeness B power station in Kent

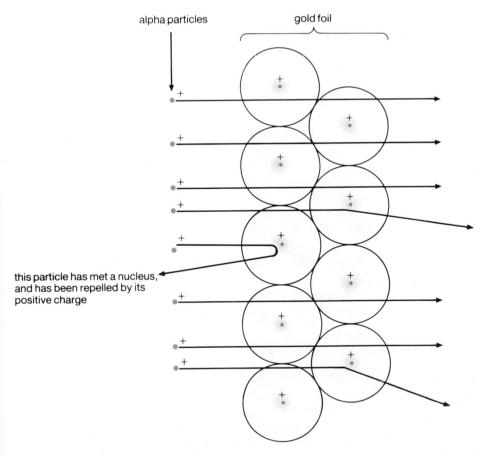

Figure 1
Most of the alpha particles pass straight through the gold foil or are deflected slightly, but a very small number bounce back

Questions

1 Explain carefully why Geiger and Marsden's experiment led us to believe that an atom has a nucleus.

2 In Geiger and Marsden's original experiment about one out of every 10 000 alpha particles 'bounced back'. What effect would the following changes have had on that number?
(a) using a thicker gold foil.
(b) using alpha particles that travelled more slowly.
(c) using a copper foil of the same thickness.

2 Atomic Structure

Part of the CERN super proton synchrotron, near Geneva. Here, protons hurtle around a 7 km long tube which is buried underground. They travel so fast that it takes them only 0.000 02 seconds to go round once. The synchroton helps nuclear physicists to understand more about the structure of the nucleus

Particle	Mass*	Charge*
proton	1	1
neutron	1	0
electron	$\dfrac{1}{1840}$	−1

*by comparison with a proton's
Table 1

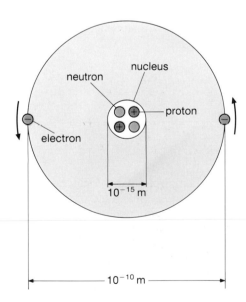

Figure 1
The helium atom; this is not drawn to scale – the diameter of the nucleus is about 100 000 times smaller than that of the atom itself

Neutrons, protons and electrons

Rutherford's model of the atom that you met in the last section is essentially the one that we accept today, except that it was not quite as simple as he thought. There is another particle in the nucleus called the **neutron**. The neutron was discovered after people realised that the nucleus of most atoms contained some extra mass. For example, the mass of a helium atom is four times that of a hydrogen atom, but it has only two protons in the nucleus. The neutron has no charge but it has the same mass as a proton. In comparison with the proton or neutron, the electron has virtually no mass but it carries a negative charge the same size as a proton's charge (see Table 1).

A hydrogen atom has 1 proton and 1 electron; it is electrically neutral because the charges of the electron and proton cancel each other. A helium atom has 2 protons and 2 neutrons in its nucleus, and 2 electrons outside that. The helium atom is also neutral because it has the same number of electrons as it has protons; it has 4 times the mass of a hydrogen atom because it has 4 particles in the nucleus (see Figure 1 and Table 2). You should remember that nearly all of an atom's mass is in the nucleus.

element	hydrogen, H	helium, He	lithium, Li
number of electrons	1	2	3
number of protons	1	2	3
number of neutrons	0	2	4
number of particles in nucleus	1	4	7
mass relative to hydrogen	1	4	7

Table 2

Ions

Atoms are electrically neutral since the number of protons balances exactly the number of electrons. However, it is possible either to add extra electrons to an atom, or to take them away. When an electron is added to an atom a **negative ion** is formed; when an electron is removed a **positive ion** is formed. Some examples are given in Table 3. The name ion is also used to describe charged molecules.

element	number of protons	number of electrons	total charge	ion
helium, He	2	1	+1	He^+
magnesium, Mg	12	10	+2	Mg^{2+}
chlorine, Cl	17	18	−1	Cl^-

Table 3

Proton and nucleon numbers

The number of protons in the nucleus of an atom determines what element it is. Hydrogen atoms have 1 proton, helium atoms 2 protons, uranium atoms 92 protons. The number of protons in the nucleus decides the number of electrons that are to be found surrounding the nucleus. The number of electrons determines the chemical properties of an atom. The number of protons in the nucleus is called the **proton (or atomic) number of the atom** (symbol **Z**). So the proton number of hydrogen is one, $Z = 1$.

The mass of an atom is decided by the number of neutrons and protons added together. Scientists call this number the **nucleon (or mass) number of an atom**. The name nucleon refers to either a proton or a neutron.

> Proton number = number of protons
>
> Nucleon number = number of protons and neutrons

For example, an atom of carbon has 6 protons and 6 neutrons. So its proton number is 6, and its nucleon number is 12. To save time in describing carbon we can write it as $^{12}_{6}C$; the nucleon number appears on the left and above the symbol C, for carbon, and the proton number on the left and below.

Nucleon number 12
Proton number 6 **C**

You should remember that the symbols $^{12}_{6}C$ describe only the **nucleus** of a carbon atom.

Isotopes

Not all the atoms of a particular element have the same mass. For example, two carbon atoms might have nucleon numbers of 12 and 14. The nucleus of each atom has the same number of protons, 6, but one atom has 6 neutrons and the other 8 neutrons. Atoms of the same element (carbon in this case) which have different masses are called **isotopes**. These two isotopes of carbon can be written as carbon-12, $^{12}_{6}C$ and carbon-14, $^{14}_{6}C$.

Questions

1 (a) An oxygen atom has 8 protons, 8 neutrons, and 8 electrons. What is its:
(i) its proton number? (ii) nucleon number?
(b) Why is the oxygen atom electrically neutral?
2 How many protons, neutrons and electrons are there in each of the following atoms?
(a) $^{17}_{8}O$ (b) $^{238}_{92}U$
(c) $^{235}_{92}U$ (d) $^{40}_{19}K$
3 Write an essay to describe the structure of an atom. In your essay, you ought to mention terms such as proton number, nucleon number, ions and isotopes.

3 Radioactivity

The nucleus of an atom is usually very stable; the atoms that we are made of have been around for thousands of millions of years. Atoms may lose or gain a few electrons during chemical reactions, but the nucleus does not change during such processes.

However, there are some atoms which have unstable nuclei which throw out particles to make the nucleus more stable. The first element discovered which emitted these particles was radium, and the name **radioactivity** was given to this process. There are three types of particle which can be released:

- **Alpha particles** are the nuclei of helium atoms, so they have a nucleon number of 4 and a proton number of 2. They have 2 positive charges since they are helium atoms stripped of their 2 electrons. When an alpha particle is emitted from a nucleus it causes it to change into another nucleus with a nucleon number 4 less and a proton number 2 less than the original one. It is usually only very heavy elements that emit alpha particles, for example:

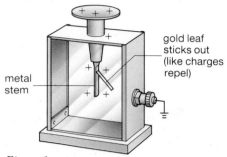

Figure 1
(a) Positively-charged electroscope

$$\underset{\text{uranium}}{\overset{238}{_{92}}\text{U}} \rightarrow \underset{\text{thorium}}{\overset{234}{_{90}}\text{Th}} + \underset{\substack{\text{alpha particle} \\ \text{(helium nucleus)}}}{\overset{4}{_{2}}\text{He}}$$

nucleus   nucleus

This is called
alpha decay

- **Beta particles** are electrons. In a nucleus there are only protons and neutrons but a beta particle can be created and thrown out of a nucleus when a neutron turns into a proton and an electron. Since an electron has a very small mass, when it leaves a nucleus it does not alter the nucleon number of that nucleus. However, the electron carries away a negative charge so the removal of an electron increases the proton number of a nucleus by 1. For example, carbon-14 decays into nitrogen by emitting a beta particle.

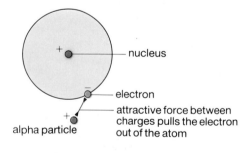

(b) Negative ions neutralise the electroscope

$$\underset{\text{carbon}}{\overset{14}{_{6}}\text{C}} \rightarrow \underset{\text{nitrogen}}{\overset{14}{_{7}}\text{N}} + \underset{\text{beta}}{\overset{0}{_{-1}}\text{e}}$$

nucleus   nucleus   particle

This is called
beta decay

- When some nuclei decay, by sending out an alpha or beta particle, they also give out a **gamma ray**. Gamma rays are electromagnetic waves, like radio waves or light. They carry away from the nucleus a lot of energy, so that the nucleus is left in a more stable state. Gamma rays have no mass or charge, so when one is emitted, there is no change to the nucleon or proton number of a nucleus (see Table 1).

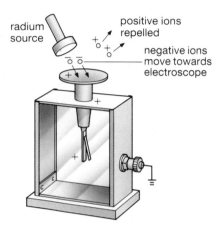

Figure 2

	Particle lost from nucleus	Change in nucleon number	Change in proton number
alpha decay α	helium nucleus $^{4}_{2}\text{He}$	−4	−2
beta decay β	electron $^{0}_{-1}\text{e}$	0	+1
gamma decay γ	electromagnetic waves	0	0

⚠ Pupils under 16 may not handle radioactive sources.

Table 1

Ionization

All three types of radiation (alpha, beta and gamma) cause **ionization** and this is why we must be careful when we handle radioactive materials. The radiation makes ions in our bodies and these ions can then damage our body tissues (see page 272).

Your teacher can show the ionizing effect of radium by holding some close to a charged gold leaf electroscope (Figure 1). The electroscope is initially charged positively so that the gold leaf is repelled from the metal stem. When a radium source is brought close to the electroscope, the leaf falls, showing that the electroscope has been discharged. The reason for this is that the alpha particles from the radium create ions in the air above the electroscope as the charges on these particles pull some electrons out of air molecules (Figure 2). Both negative and positive ions are made; the positive ones are repelled from the electroscope, but the negative ones are attracted so that the charge on the electroscope is neutralised. It is important that you understand that it is not the charge of the alpha particles that discharges the electroscope, but the ions that they produce.

Background radiation

There are a lot of rocks in the Earth that contain radioactive uranium, thorium and potassium, and so we are always exposed to some ionizing particles. In addition the Sun emits lots of protons which can also create ions in our atmosphere. These two sources make up **background radiation**. Fortunately the level of background radiation is quite low and in most places it does not cause a serious health risk.

In some jobs, people are at a greater risk. X-rays used in hospitals also cause ionization. Radiographers make sure that their exposure to X-rays is as small as possible. In nuclear power stations neutrons are produced in **nuclear reactors**. The damage caused by neutrons is a source of danger for workers in that industry.

The background radiation in Britain is highest in Cornwall. This is due to radioactive elements in the granite rocks

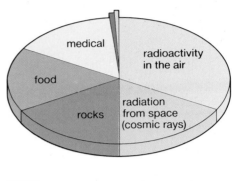

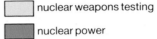

nuclear weapons testing

nuclear power

Figure 3
Sources of radiation in Britain

Questions

1 Fill in the gaps in the following radioactive decay equations:

(a) $^3H \rightarrow \, _2He + \, ^0e$

(b) $^{229}_{90}Th \rightarrow Ra + \, ^4_2He$

(c) $^{14}_6C \rightarrow ? + \, ^0_{-1}e$

(d) $^{209}_{82}Pb \rightarrow \, _{83}Bi + ?$

(e) $^{225}_{89}Ac \rightarrow \, _{87}Fr + ?$

2 $^{238}_{92}U$ decays by emitting an alpha particle and two beta particles; what element is produced after those three decays?

3 Explain what effect losing a gamma ray has on a nucleus.

4 Explain carefully how a radioactive source that is emitting only alpha particles can discharge a negatively charged electroscope.

5 (a) What is background radiation and where does it come from?

(b) Use the pie chart (Figure 3) to discuss whether the nuclear power industry in the UK, is likely to cause a serious health hazard.

4 More on α, β and γ Radiation

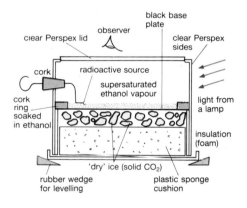

A false colour photograph of cloud chamber tracks. The green tracks are caused by alpha particles. One of them (yellow) collides with a proton (red) in the hydrogen gas that fills the chamber

Detecting particles

We make use of the ionizing properties of radiations to detect them. This is done using a **Geiger-Muller (GM) tube**. Figure 1 shows how such a tube works. A metal tube is filled with argon under low pressure; inside the tube there is a thin wire anode. A potential difference of about 450 V is applied between the inside and outside of the tube. When alpha, beta or gamma radiation enters the tube the argon atoms inside are ionized. These ions are then attracted to the electrodes in the tube so a small current flows. This current is then amplified and a counter can be used to count the number of particles entering the tube.

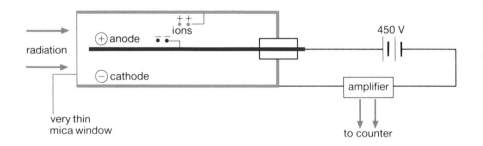

Figure 1
A Geiger-Muller tube

Cloud chambers

Another way of detecting radiation is to use a **cloud chamber** (Figure 2). The bottom of the cloud chamber is kept cold by placing some solid carbon dioxide ('dry ice') underneath the metal base plate. The inside of the chamber is filled with alcohol vapour. When a radioactive source is placed inside the cloud chamber, tracks are formed in the dense alcohol vapour. These white tracks can be seen clearly against the black bottom of the chamber. The alcohol molecules condense to leave a vapour trail in the region where ions have been produced by the passage of a particle. The tracks left by alpha particles are straight and thick; these show us that alpha particles are very strongly ionizing. Beta particles do not ionize air so strongly, so their tracks are much thinner. Gamma rays leave nearly no track at all because they produce few ions in a given distance.

⚠ The source in a cloud chamber is so weak that pupils may use it under supervision.

Properties of radiation

- Alpha particles will travel about 5 cm through air and they will be stopped by a sheet of paper (Figure 3). They ionize air very strongly. Alpha particles travel at speeds of about 10^7 m/s. This is more slowly than beta or gamma rays travel. They can be deflected by a very strong magnetic field, but the deflection is very small indeed because alpha particles are so massive.
- Beta particles can travel several metres through air and they will be stopped by a sheet of aluminium a few millimetres thick (Figure 3). They do not ionize air as strongly as alpha particles. Beta particles travel at speeds just less than the speed of light (3×10^8 m/s). Beta particles can be deflected quite easily by a magnetic field, because they are such light particles (see Figure 4).

Figure 2
A cloud chamber

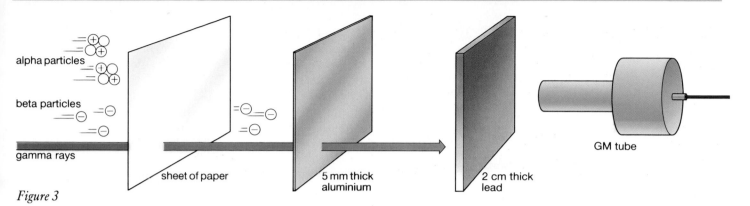

Figure 3

- Gamma rays can only effectively be stopped by a very thick piece of lead (Figure 3). They are electromagnetic waves, so they travel at the speed of light. Gamma rays only ionize air very weakly, and they cannot be deflected by a magnetic field because they carry no charge.

Alpha particles will cause the most damage to your bodies if they get inside you; this could happen if you were to breathe in a radioactive gas such as radon. A school alpha source is less dangerous because the particles will not reach you. You must keep well away from gamma sources since these rays can get right into the middle of your body and cause damage there. In all cases you will see your teacher take great care with sources, and handle them with tongs or special holders.

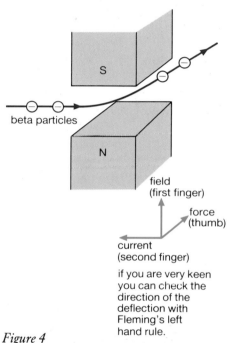

field
(first finger)

force
(thumb)

current
(second finger)

if you are very keen you can check the direction of the deflection with Fleming's left hand rule.

Figure 4
The deflection of beta particles by a magnetic field. Note that the deflection is not towards the poles of the magnet, but at right angles to them

Radiation	Nature	Speed	Ionizing power	Penetrating power	Deflection magnetic field
alpha α	helium nucleus	10^7m/s	very strong	stopped by paper	very small indeed
beta β	electron	just less than $3 \times 10^8 \text{m/s}$	medium	stopped by aluminium	large
gamma γ	electro-magnetic waves	$3 \times 10^8 \text{m/s}$	weak	stopped by thick lead	none

Table 1. A summary of radiation properties.

Questions

1 Explain carefully how you could use a Geiger-Muller tube and some pieces of paper, aluminium and lead to show that radium emits alpha, beta and gamma radiations.
2 Which type of radiation is most dangerous to us?
3 A beta particle can be deflected by a magnetic field. Why can an alpha particle only be deflected a little and a gamma ray not at all?

4 Why do gamma rays leave only very faint tracks in a cloud chamber?
5 This question is about testing the thickness of a metal sheet. The metal sheet moves past a β-source and Geiger counter at a speed of 0.2 m/s.

Count rate (s⁻¹)	75	80	77	73	76	75	63	57	50	55	67	75	77
Time (s)	0	10	20	30	40	50	60	70	80	90	100	110	120

(a) Plot a graph from the data below. Explain why the count rate changes over the first 50 s.
(b) Explain why the count rate drops and then rises again, after 50 s.

5 Radioactive Decay

Throw	Number of coins left
0	1000
1	500
2	250
3	125
4	62
5	31
6	16
7	8
8	4
9	2
10	1

Table 1 Coin tossing experiment

Time (hour)	Number of nuclei left
0	1 000 000
1	500 000
2	250 000
3	125 000
4	62 500
5	31 250
6	15 620
7	7 810
8	3 900
9	1 950
10	980

Table 2 The number of nuclei left in a sample: half life 1 hour

We know that the atoms of some radioactive materials decay by emitting alpha or beta particles from their nuclei. But it is not possible to predict when the nucleus of one particular atom will decay. It could be in the next second, or sometime next week, or not for a million years.

The radioactive decay of an atom is rather like tossing a coin. You cannot say with certainty that the next time you toss a coin, that it will fall head up. However, if you throw a lot of coins you can start to predict how many of them will fall heads up. You can use this idea to help you understand how radioactive decay happens. You start off with a thousand coins; if any coin turns up head up then it has 'decayed' and you must take it out of the game. Table 1 shows the likely result (on average). Every time you throw a lot of coins about half of them will turn up heads.

Radioactive materials decay in the same way. If we start off with a million atoms then we find that after a period of time (say one hour), half of them have decayed. In the next hour we find that half of the remaining atoms have decayed, leaving us with a quarter of the original number (Table 2). The period of time taken for half the number of atoms to decay in a radioactive sample is called the **half life**, and it is given the symbol $t_{1/2}$. You can see that after 10 tosses of the coins, nearly all the coins have fallen heads up, and that after 10 half lives, nearly all of the nuclei have decayed.

Measurement of half life

If you look at Table 2 you can see that the number of nuclei that decayed in the first hour was 500 000, then in the next hour 250 000 and in the third hour 125 000. So as time passes not only does the number of nuclei left get smaller but so does the rate at which the nuclei decay. So by measuring the decay rate of a radioactive sample we can determine its half life.

Figure 1 shows how we can measure the half life of radon, which is a gas. (You are not allowed to use radioactive materials until you are over 16, so you cannot do this experiment yourself.) The gas is produced in a plastic bottle; we can give the bottle a squeeze and force some gas into a chamber which is fixed on to the end of a GM tube. The GM tube is attached to a ratemeter, which tells us the rate at which radon is decaying in the chamber. We measure the rate of decay on the ratemeter every 10 seconds and plot a graph of the count rate against time, Figure 2. We can see that the count rate halves every 50 seconds, so that is the half life of radon.

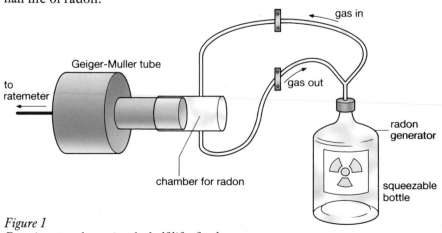

Figure 1
Experiment to determine the half life of radon

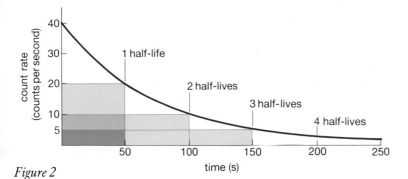

Figure 2

Dating archaeological remains

Carbon-14, $^{14}_{6}C$, is a radioactive isotope; it decays to nitrogen with a half life of about 5500 years. All living things (including you) have a lot of carbon in them, and a small fraction of this will be carbon-14. When a tree dies, for example, the radioactive carbon will begin to decay, and after 5500 years the fraction of carbon-14 in the dead tree will be half as much as you would find in a living tree. So by measuring the amount of carbon-14 in ancient relics, scientists can calculate their age. With this technique it was possible to date the Turin Shroud, which was thought to be the burial shroud of Christ.

Disposal of radioactive waste

Nuclear power stations produce radioactive waste materials, some of which have half lives of hundreds of years. These waste products are packaged up in concrete and steel containers and are buried deep underground or are dropped to the bottom of the sea. This is a controversial issue; some scientists tell us that radioactive wastes produce only a very low level of radiation, and that the storage containers will remain intact for a very long time. Others worry that these products will contaminate our environment and believe it is wrong to leave radioactive materials that could harm future generations.

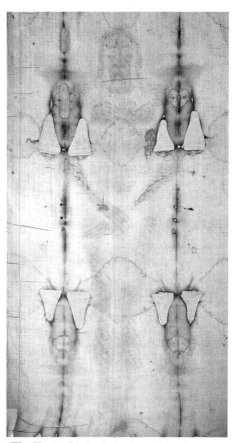

The Turin Shroud. Radiocarbon dating has shown that the shroud was made in the 13th century

Questions

1 A GM tube is placed near to a radioactive source with a long half life. In three 10 second periods the following number of counts were recorded: 150, 157, 145. Why were the three counts different?

2 A radioactive material has a half life of 2 minutes. What does that mean? How much of the material will be left after 8 minutes?

3 The following results for the count rate of a radioactive source were recorded every minute. Plot a graph of the count rate (*y*-axis) against time (*x*-axis), and use the graph to work out the half life of the source.

Counts per second	time (minute)
100	0
59	1
34	2
20	3
12	4
7	5

4 Why does radioactive waste worry some people?

5 When doing an experiment to measure the half life of radon you will also detect some background radiation. How can you correct for this?

Containers for radioactive waste materials must be strong enough to prevent leakage. Here, a technician uses a geiger counter to check a drum containing radioactive waste at Hinkley Point nuclear power station

6 *Nuclear Fission*

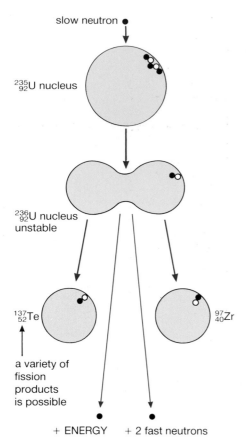

slow neutron •

$^{235}_{92}$U nucleus

$^{236}_{92}$U nucleus
unstable

$^{137}_{52}$Te $^{97}_{40}$Zr

a variety of
fission
products
is possible

+ ENERGY   + 2 fast neutrons

Figure 1
The fission of a uranium-235 nucleus

Fission

You read earlier (page 264) that the nuclei of some large atoms were unstable and that to become more stable they lost an alpha or a beta particle. Some heavy nuclei, ^{235}U for example, may also increase their stability by **fission**. Figure 1 shows how this works. Unlike alpha or beta decay, which happens at random, the fission of a nucleus is usually caused by a neutron hitting it. The uranium nucleus absorbs this neutron and turns into a ^{236}U nucleus, which is so very unstable that it splits into two smaller nuclei. The nuclei that are left are not always identical; two or three energetic neutrons are also emitted. The remaining nuclei are usually radioactive and will decay by the emission of beta particles to form more stable nuclei.

The fission process releases a tremendous amount of energy. The fission of a nucleus provides about 40 times more energy than the release of an alpha particle from a nucleus. Fission is important because we can control the rate at which it happens, so that we can use the energy released to create electrical energy.

Once a nucleus has divided by fission, the neutrons that are emitted can strike other neighbouring nuclei and cause them to split as well. This chain reaction is shown in Figure 2. Depending on how we control this chain reaction we have two completely different uses for it. In a controlled chain reaction, on average only one neutron from each fission will strike another nucleus and cause it to divide. This is what we want to happen in a power station. In an uncontrolled chain reaction all the neutrons from each fission strike other nuclei. This is how nuclear bombs are made. It is a frightening thought that a piece of pure uranium-235 the size of a tennis ball has enough stored energy to flatten a town.

Nuclear power stations

Figure 3 shows a gas-cooled **nuclear reactor**. The energy released by the fission processes in the uranium fuel rods produces a lot of heat. This heat is carried away by carbon dioxide gas which is pumped around the reactor. The hot gas then boils water to produce steam, which can be used to work the electrical generators.

- The **fuel rods** are made of uranium-238, 'enriched' with about 3 % uranium-235. ^{238}U is the most common isotope of uranium, but it is only ^{235}U that will produce energy by fission.
- The fuel rods are embedded in graphite, which is called a **moderator**. The purpose of a moderator is to slow down neutrons that are produced in fission. A nucleus is split more easily by a slow-moving neutron. The fuel rods are long and thin so that neutrons can escape. Neutrons leave one rod and cause another nucleus to split in a neighbouring rod.
- The rate of production of energy in the reactor is carefully regulated by the **boron control rods**. Boron absorbs neutrons very well, so by lowering them the reaction can be slowed down. In the event of an emergency they are pushed right into the core of the reactor and the chain reaction stops completely.

Figure 1 shows a possible fission of a uranium nucleus, described by:

$$^{1}_{0}n + ^{235}_{92}U \rightarrow ^{137}_{52}Te + ^{97}_{40}Zr + 2^{1}_{0}n$$

The nucleon and proton numbers on each side balance. However, very accurate measurement shows that the mass on the left hand side of the equation is slightly more than the mass on the right hand side.

The table opposite shows the masses of the nuclei in atomic mass units, u. $1\ u = 1.66 \times 10^{-27}$ kg, which is 1/12 of the mass of a carbon atom.

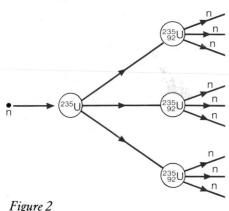

Figure 2
A chain reaction in uranium-235

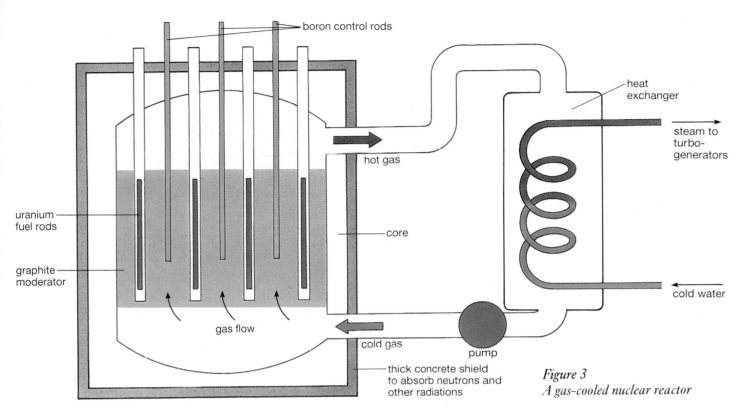

Figure 3
A gas-cooled nuclear reactor

Check that when ^{235}U undergoes this fission it loses a mass of 0.216 u. According to Einstein's Theory of Special Relativity, if mass disappears energy is created. The lost mass is turned into the kinetic energy of the fission fragments. The energy can be calculated using the equation $E = mc^2$, where E is the energy produced, m the lost mass, and c is the speed of light; $c = 3 \times 10^8$ m/s. The energy produced in this fission is:

$$E = mc^2$$
$$= 0.216 \times 1.66 \times 10^{-27} \text{ kg} \times (3 \times 10^8 \text{ m/s})^2 \quad = 3.2 \times 10^{-11} \text{ J}$$

Isotope	Mass in u
$^{235}_{92}U$	235.048
$^{137}_{52}Te$	136.918
$^{97}_{40}Zr$	96.906
$^{1}_{0}n$	1.008

Questions

1 Explain what is meant by nuclear fission. In what way is fission (i) similar to (ii) different from radioactive decay?

2 What is a chain reaction? Explain how the chain reaction works in a nuclear bomb and in a nuclear power station.

3 The following questions are about the nuclear reactor shown in Figure 3.
(a) What is the purpose of the concrete shield surrounding the reactor?
(b) Why is carbon dioxide gas pumped through the reactor?
(c) Which isotope of uranium produces the energy in the fuel rods?
(d) Will the fuel rods last for ever?
(e) What is the purpose of the graphite moderator?
(f) What would you do if the reactor core suddenly got too hot?

4 This question is about producing energy inside the nuclear reactor core, shown in Figure 3. This core has 1700 uranium fuel rods, each 1 m long with a diameter of 30 mm. Use the data provided to answer the following questions.
(a) How much ^{235}U is there in the core?

(b) What is the total amount of heat energy that this amount of ^{235}U can release?
(c) How long will this amount of nuclear fuel last for, if the core produces power continuously?

- Mass of one fuel rod is 14 kg
- 3% of the fuel is ^{235}U
- Power produced in the reactor core is 2400 MW
- 1 kg of ^{235}U produces 10^{14} J of heat energy
- There are 3×10^7 s in 1 year

7 The Hazards of Radiation

Measuring radiation

When scientists try to work out the effect on our bodies of a dose of radiation, they need to know how much energy each part of the body has absorbed. After all, the damage done to us will depend on the amount of energy that each kilogram of body tissue absorbs. The unit used to measure a **radiation dose** is the **gray**, (symbol Gy). A dose of 1 Gy means that each kilogram of flesh absorbs 1 joule of energy.

$$1\,\text{Gy} = 1\,\text{J/kg}$$

Some radiations are more damaging than others, so scientists prefer to talk in terms of a **dose equivalent**, which is measured in **sieverts** (symbol Sv).

$$\text{dose equivalent (Sv)} = Q \times \text{dose (Gy)}$$

Q is a number that depends on the radiation, as shown in Table 1. Alpha particles are very strongly ionizing and cause far more damage than a dose of beta or gamma radiation that carries the same energy. Usually the amounts of radiation that we are exposed to are very small. Most people receive about 1/1000 sievert each year, this is a **millisievert** (symbol 1 mSv).

Q	Type of radiation
1	beta particles/gamma rays
10	protons and neutrons
20	alpha particles

Table 1

The Chernobyl nuclear reactor near Kiev in Ukraine. During an unauthorised experiment in 1986, the core of the number 4 reactor unit melted, causing an explosion which released a large amount of radioactive material. This will be a very serious health hazard for many years to those living close to the reactor. Winds carried some of the radioactive material to many countries in Europe

Risk estimates

On 26 April 1986 there was an explosion in the Ukrainian nuclear reactor at Chernobyl, causing a large leakage of radiation. During May the **background count** in Britain increased causing us all to be exposed (on average) to an extra dose equivalent to 0.1 mSv. This is a small dose and no worse than going on holiday in Cornwall, where granite rock areas produce low amounts of radiation. However, estimates have been made to suggest that over the next 30 years, extra people will get cancer as a result of the Chernobyl disaster.

Research suggests that for a population of 1000, about 12 fatal cancers will be caused by a dose equivalent of 1 Sv. In Britain, the population is about 50 million, so the number of deaths expected by a dose for all of us of 1 Sv would be:

$$\frac{12}{1000} \times 50\,000\,000 = 600\,000$$

However, the dose from Chernobyl was only 0.0001 Sv, so the estimated number of deaths from the Chernobyl disaster, in Britain over the next 30 years, is about $600\,000 \times 0.0001 = 60$.

How dangerous is radiation?

Radiation affects materials by ionizing atoms and molecules. When an atom is ionized, electrons are removed or added to it. This means a chemical change has occurred. In our bodies such a chemical change could cause the production of a strong acid which will attack and destroy cells.

- **High doses** of radiation will kill you. There is only a 50 per cent chance of surviving a dose equivalent to 4 Sv. A 10 Sv dose equivalent would give you no chance of survival. Such high doses kill too many cells in the gut and bone marrow for your body to be able to work normally. You could be exposed to such doses in a nuclear war, and people certainly died in Hiroshima and Nagasaki as a result of such doses.

- **Moderate doses** of radiation below 1 Sv will not kill you. Damage will be done to cells in your body, but not enough to be fatal. The body will be able to replace the dead cells and the chances are that you would then recover totally. However, a study of the survivors from Hiroshima and Nagasaki shows that there is an increased chance of dying from cancer some years after the radiation dose. Even so, you would only have a chance of about 1 in 100 of getting cancer from such levels of radiation.

- **Low doses** of radiation, below 10 mSv, are thought to have little effect on us. However, some people think that any exposure to radiation will increase your chances of getting cancer.

There can be no doubt that radiation doses can cause cancer or leukaemia (see Table 2). Uranium miners are exposed to radon gas, and girls who painted luminous watch dials were exposed to radium.

Workers in the nuclear power industry are exposed to more radiation than the rest of the population. Special monitoring and remote control are used to keep this extra radiation exposure as small as possible

Questions

1 In Table 1, the value of Q for alpha radiation is 20. Why is it so high?

2 What does a sievert measure?

3 Summarise the effects of high, moderate and low doses of radiation on us.

4 Many modern watches do not have luminous dials. Instead they have small lights that turn on at the press of a switch. Explain why lights are safer.

5 Use the data in Table 2 to show that exposure to alpha radiation is more likely to cause cancer than exposure to gamma radiation or X-rays.

6 In a nuclear reactor disaster about 200 workers are exposed to a radiation dose equivalent to 2 Sv. Use the data in the text to estimate the number of them likely to die from cancer some time after the accident.

Source of radiation	Type of radiation	Number of people studied	Extra number of cancer deaths caused by radiation
uranium miners	alpha	3400	60
radium luminisers	alpha	800	50
medical treatment	alpha	4500	60
medical treatment	X-rays	14 000	25
Hiroshima bomb	gamma rays and neutrons	15 000	100
Nagasaki	gamma rays	7000	20

Table 2 This table illustrates the connection between radiation and the increased chance of cancer

8 Nuclear Power: The Future?

The outside (left) and inside (right) of a 'torus' (a hollowed-out doughnut shape) at a laboratory in Oxfordshire. A strong magnetic field traps ionised hydrogen inside the torus. Researchers hope to generate a sustained fusion reaction in such a vessel

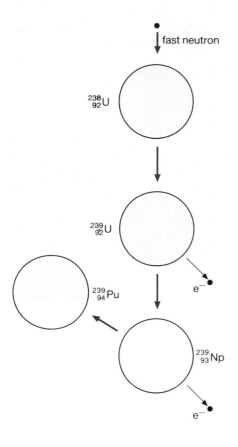

Figure 1
Production of plutonium from uranium

At the moment about 15% of our electricity is generated using nuclear power. Nearly all the rest is produced in coal-burning power stations. In 1987 the government announced plans to build more nuclear power stations, and it now seems likely that early next century more of our electricity will come from nuclear power.

The reason for the increased use of nuclear fuels is that our fossil fuels (coal, gas and oil) are running out; once they are gone we cannot replace them. In Britain we have saved up about 20 000 tonnes of uranium – how long will that last? At present we do not use our uranium very efficiently. Only 0.7 per cent of uranium is the fissionable ^{235}U, which produces energy in most reactors. However, in 1976 at Dounreay (Scotland) a new type of power station opened. This was a **fast breeder reactor**, which used a new fuel, **plutonium**.

Plutonium is an element that does not occur naturally, but it is produced from uranium, as shown in Figure 1. Plutonium nuclei will release energy by fission, but the process is triggered by the nucleus absorbing a fast neutron (hence the name 'fast reactor'). By producing plutonium from uranium our stocks of nuclear fuel will last for a few hundred years.

Nuclear fusion

The energy that is produced inside our Sun comes from the fusing together of hydrogen nuclei. **Fusing** means melting together, which is a good description of the process. At the centre of the Sun the temperature is about 15 000 000 K; at these temperatures the nucei of atoms are stripped of all their surrounding electrons, and they are moving very quickly indeed. Fusion involves two small nuclei colliding and sticking together to form a larger nucleus. As in the fission of a large nucleus, the fusion of two small nuclei releases a lot of energy (Figure 2).

At Culham in Oxfordshire, attempts are being made to get energy from nuclear fusion. This is an extremely difficult project; as you can imagine producing conditions similar to the inside of a star is no easy matter! If this experiment is successful it might solve the problem of producing electricity for a long time to come.

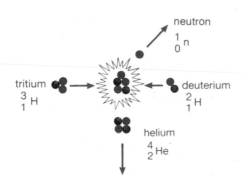

Figure 2
The fusion of deuterium and tritium

The nuclear debate

Is nuclear power the way into the next century, or should we be looking to produce energy from natural sources such as wind and the tides? A lot of people are pressing for more nuclear power stations and others are strongly against them. This is an important issue and you should have your own ideas about it. Below are listed some points that someone in favour of nuclear power might make, and also some points that could be made by someone against it.

Views of someone in favour of nuclear power stations.
- Fossils fuels are running out, so nuclear power provides a convenient way of producing electricity.
- Nuclear power stations produce a very small level of radiation. The extra radiation is very little in comparison with the background radiation, and is not a health hazard.
- Coal-fired power stations put out more radiation into the atmosphere than nuclear power stations, because coal is naturally radioactive. Burning coal also produces acid rain.
- Radioactive waste can be safely stored.
- The chances of a large nuclear accident in this country are very small. Our technology is far better than that of the Ukranians, so an accident like Chernobyl could not happen here.
- Accidents happen anyway; nobody seems to worry about the number of deaths caused in road accidents. Is anybody suggesting banning cars?
- People do not understand radiation. That's why they are afraid of it.

The prototype fast reactor at Dounreay power station

A lot of people are worried by the dumping of radioactive waste

Views of someone against nuclear power stations.
- Fossil fuels are running out, so we should be looking to conserve energy. Research should be done to use wind and wave power.
- Nuclear power stations produce dangerous quantities of radioactive waste. The government has ordered Sellafield to stop discharging waste into the Irish Sea. Statistics show that children are more likely to die of leukaemia near Sellafield.
- Coal-fired power stations cause acid rain and produce radiation. They should be closed down as well.
- It is irresponsible to store radioactive wastes with long half-lives; it pollutes the environment for our grandchildren.
- The fact remains that a power station blew up in 1986, belching radioactive stuff all over Europe. We may have escaped lightly, but a lot of people in Ukraine will die as a result.
- Cars have got nothing to do with it.
- We *do* understand radiation. That's why we're afraid of it.

Questions

1 Why does the production of plutonium allow us to get more energy than we would if we used uranium as a nuclear fuel?

2 Fast reactors do not have moderators. Explain why.

3 What is nuclear fusion? Why is it much more difficult to control nuclear fusion than nuclear fission?

4 You have now read a lot about nuclear energy. Write a short essay to explain whether it is the answer to our energy problems. After thinking about the arguments above, who do you agree with?

Radioactive materials have a great number of uses in medicine, industry and agriculture. People who work with radioactive materials must wear radiation badges which record the amount of radiation to which they are exposed.

Medicine

- Radioactive **tracers** help doctors to examine the insides of our bodies. Iodine-131 is used to see if our thyroid glands are working properly. The thyroid is an important gland in the throat which controls the rate at which our bodies function. The thyroid gland absorbs iodine, so a dose of radioactive iodine (the tracer) is given to a patient. Doctors can then detect the radioactivity of the patient's throat, to see how well the patient's thyroid is working.
- Cobalt-60 emits very energetic gamma rays. These rays can damage our body cells, but they can also kill bacteria. Nowadays nearly all medical equipment such as syringes, dressings and surgeons' instruments is first packed into sealed plastic bags, and then they are exposed to intense gamma radiation. In this way all the bacteria are killed and so the equipment is sterilised.
- The same material, cobalt-60, is used in the treatment of cancers. Doctors direct a strong beam of radiation on to the cancerous tissue to kill the cancer cells. The treatment is very unpleasant and causes the patient to be very ill, but it is often successful in slowing down the growth or completely curing the cancer.

To check the amount of radiation that workers in a nuclear power station are exposed to, they wear special radiation-sensitive badges, like the ones in this photograph. At the end of each month the sensitive film in the badges is developed. It can then be seen how much radiation the wearer has been exposed to

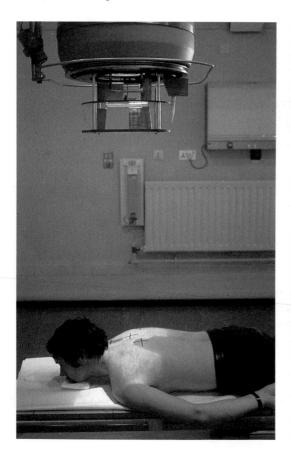

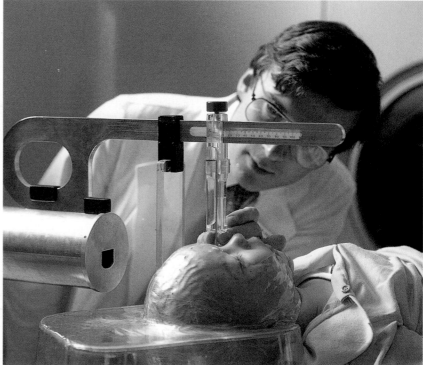

Gamma rays can be used to destroy cancer cells. Here, a child with a cancerous growth in the eye is being treated with gamma rays from cobalt-60. The red light is used to direct the equipment at the cancerous tissue, before the gamma rays are released. The photograph on the left shows a patient with Hodgkins disease (cancer of the lymph nodes) being treated by radiotheraphy

Industry

- Radioactive tracers may be used to detect leaks in underground pipes (Figure 1). The idea is very simple; the radioactive tracer is fed into the pipe and then a GM tube can be used above ground to detect an increase in radiation level and hence the leak. This saves time and money because the whole length of the pipe does not have to be dug up to find the leak.
- A radioisotope of iron is used in industry to estimate the wear on moving parts of machinery. For example, car companies want to know how long their piston rings last for. A piston ring which has radioactive iron in it is put into an engine and run for several days. At the end of the trial, the oil from the engine can be collected, and from the radioactivity of the oil the engineers can calculate how much of the piston ring has worn away.

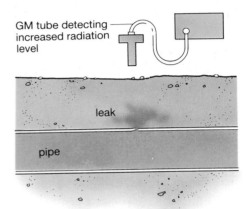

Figure 1
How to find leaks in pipelines without digging

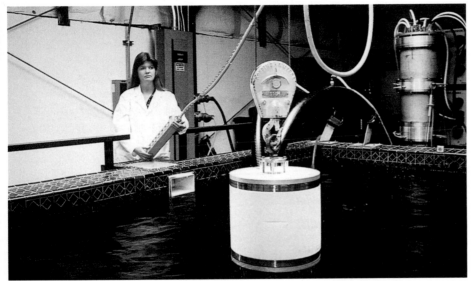

A technician lowers a casket of fruit into a pool of water used to study irradiated fruit. The casket is irradiated by a source of X-rays in the pool. This technique can be used to preserve foods that would otherwise become rotten very quickly

Agriculture

- Tracers are used in agriculture too. Phosphates are vital to the growth of plants and are an important component of fertilisers. Radioactive phosphorous-32 is used as a tracer to show how well plants are absorbing phosphates.
- Gamma radiation is used to prolong the shelf-life of food. Gamma rays are very penetrating so this process can be used on pre-packaged or frozen foods. Gamma rays kill the bacteria in the food and so can eliminate the chance of food poisoning. However, the gamma rays will also kill some cells in the food itself and therefore can alter the taste considerably. This process is allowed in Britain but most of our irradiated food is imported.
- Gamma rays are also used to help produce new types of crops. Large doses of gamma rays will kill cells, but smaller doses can cause mutations to the cells, which will change the nature of the crop. The seeds of crops are exposed to gamma radiation to encourage mutations. The new crops may show desirable qualities, like being stronger or producing a greater yield. These successful mutations can be kept and used in the fields.

Questions

1 Explain carefully why radioactive tracers are useful.

2 Do radioactive tracers show different chemical properties to other isotopes of the same element?

3 What uses does gamma radiation have in medicine and agriculture?

4 In the text you will have read that radioisotopes are useful in industry. Clearly they have important economic results; it saves a lot of money to find a leak with the help of a radioisotope. However, the people who find the leak and the general public will be exposed to radiation in the process and therefore there will be a health hazard. How can we balance the possible loss of human life with the saving of money? What is an acceptable risk?

1 The brain can suffer from a particularly nasty cancer called a glioblastoma. This cancer penetrates the brain and cannot be cured by surgery. Instead the neurosurgeon gives the patient an injection which contains some boron. The boron is absorbed by the glioblastoma. Then the patient is irradiated with neutrons. The following nuclear reaction occurs:

$$^{10}_{5}B + ^{1}_{0}n \rightarrow ^{7}_{3}Li + ^{x}_{y}He$$

(a) Copy the equation and fill in the missing number x and y.
(b) The boron nucleus splits up to form lithium and helium. What is this process called? Explain why the lithium and helium nucleus move away from each other very quickly.
(c) Explain how this process can kill the glioblastoma.
(d) Why is this process dangerous for healthy patients?

2 An experiment to measure the half life of caesium-140 is done in a region where there is a high background count. The results of the experiment are shown in the table. Determine the half life of caesium-140 by plotting a suitable graph.

Count rate (counts/s)	Time (s)
68	0
52	30
40	60
32	90
24	120
20	150
16	180
14	210
12	240
8	270
10	300

3 The age of archaeological remains can be found using carbon dating. All living things contain small amounts of carbon-14; this is a radioactive isotope. The concentration of carbon-14 is the same for all living things. But when the creature (or plant) dies the carbon-14 decays. Its half life is about 5700 years. So after that time the concentration of the carbon-14 has halved.

Cro-Magnon man is one of our ancestors. Five adult skeletons were found near Les Eyzies in France. A 1 g sample of charcoal from this site produced a radioactive count of 0.5 counts per minute. A modern sample of charcoal of the same mass produces a count rate of 32 counts per minute. Both counts were corrected for background radiation.

How long ago did Cro-Magnon man live?

4 Plutonium-241 is unstable and it decays by giving out an alpha particle. This is the start of a long decay series. By the emission of more alpha and beta particles, eventually a stable isotope of bismuth is made.

The table shows the decay series. Copy it and fill in the gaps.

Element	Symbol	Radioactive emission
Plutonium	$^{241}_{94}Pu$	α
Uranium	$^{237}_{92}U$	β
Neptunium	$^{?}_{?}Np$	α
Protactinium	$^{?}_{91}Pa$	β
Uranium	$^{233}_{?}U$	?
Thorium	$^{229}_{90}Th$	α
Radium	$^{?}_{?}Ra$	β
Actinium	$^{?}_{?}Ac$	?
Francium	$^{221}_{87}Fr$	?
Astatine	$^{217}_{85}At$	α
Bismuth	$^{?}_{?}Bi$	?
Polonium	$^{213}_{84}Po$	α
Lead	$^{?}_{?}Pb$	β
Bismuth	$^{209}_{83}Bi$	stable

5 A silver atom has a proton number of 47 and a nucleon number of 107. The atom is neutral.
(a) The atom contains protons and neutrons. How many of each does the atom have? Where are the protons and neutrons?
(b) What other particle does the atom have? How many of these particles are there?
(c) Why is the atom neutral?
(d) Silver-108 is another isotope of silver. Explain, with as much detail as possible, what this means.

6 This diagram shows a method which is used in factories to check the thickness of polythene being produced. In this case a long radioactive source is placed below the whole width of the polythene and a long Geiger-Müller tube is placed above it.

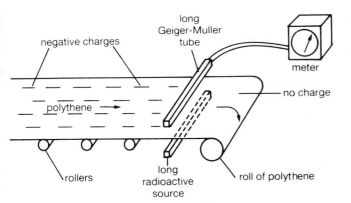

A Geiger-Müller tube is used for detecting radiation. The output pulses of current from the tube may go into a counter, an amplifier connected to a loudspeaker or a meter.

In this application, the reading on the meter can then be used as a measure of thickness of the polythene; the thicker the polythene, the lower the meter reading.

When the polythene passes over the rollers, effects of friction cause it to become negatively charged. The presence of the radioactive source enables the polythene to become discharged.
(a) It is suggested that because radioactive decay is random this method for checking thickness gives better results when the polythene is going through slowly.
(i) What is meant by *random*?
(ii) Why is the result likely to be more reliable when the polythene is going through slowly?
(b) Radiation from the radioactive source ionises the air. This produces many positive and negative ions from atoms in the air. Discuss how these ions are affected by the negative charge on the polythene and hence explain how the polythene becomes discharged.
(c) If sources of similar activity giving either α or β radiation were available, which one would be better for:
(i) measuring the thickness of the polythene,
(ii) discharging the roll? Explain your choice in each case.
(d) In view of the presence of the radioactive material state any *two* suitable precautions which should be observed for the safety of workers in the factory.
(e) A buyer of polythene visits the factory and is alarmed by the use of radioactive sources in the method shown. He is concerned that the polythene may become radioactive.
(i) Why has he no real cause for concern?
(ii) Explain briefly an experimental check which you could make to show that there is no cause for concern.

ULEAC

7 The equation below shows part of the reaction in a nuclear reactor.

$$^{235}_{92}U + ^{1}_{0}n \rightarrow ^{236}_{92}U$$

(a) Explain the significance of the numbers 235 and 92.
(b) $^{236}_{92}U$ atoms are unstable and disintegrate spontaneously into fragments approximately equal in size, together with two or three fast-moving neutrons and a large amount of energy. What is this process called and what is the source of the energy?
(c) $^{235}_{92}U$ is much more likely to absorb slow-moving (thermal) neutrons than fast-moving neutrons. Describe how neutrons may be slowed down in the reactor core.
(d) Explain what is meant by a *chain reaction*.
(e) How is the rate of energy production in the reactor core controlled?
(f) The energy is produced in the form of heat in the reactor core. How is the heat removed from the core and how is it converted into electricity?
(g) When the fuel rods are withdrawn from the reactor core they are *radioactive*, containing *isotopes* with long *half lives*. Explain the terms in italics.
(h) Outline three precautions which must be taken to ensure safe operation of the reactor.

NEAB

8 Iodine-131 is a *radioisotope*. This isotope is used as a *tracer* to investigate the working of patients' thyroid glands. The thyroid absorbs iodine and doctors can discover how well it is working by detecting *radioactive emissions*. Iodine–131 decays by emitting β and γ rays with a *half life* of 8 days.
(a) Explain the meaning of the phrases in italics.
(b) The table lists the proton numbers of some elements. Which element does $^{131}_{53}I$ decay to? Explain why this daughter element is safe for the patient.

Element	Proton number
Antimony (Sb)	51
Tellurium (Te)	52
Iodine (I)	53
Xenon (Xe)	54
Caesium (Cs)	55

(c) A patient receives a dose of 10 mg of iodine-131. After 24 days, what is the maximum mass of the isotope that can be left in her body?

9 The table shows some of the isotopes of the element uranium, the type of radiation they give out, their half lives and their abundances.

Proton number	Nucleon number	Type of radiation	Half life	Abundance of isotope
92	233	α	2×10^5 years	–
92	234	β	2×10^5 years	–
92	235	β	2×10^8 years	0.7%
92	236	β	2×10^7 years	–
92	237	β	7 days	–
92	238	α	4.5×10^9 years	99.3%

(a) Explain what is meant by the 'abundance' of an isotope. Why does the table show the abundance of only two isotopes?

(b) In a sample of naturally occurring uranium, explain why there are approximately 142 uranium–238 atoms for each uranium–235 atom.

(c) The Earth is approximately 4.5×10^9 years old.

(i) Work out the ratio:

$$\frac{\text{age of Earth}}{\text{half life of } {}^{235}_{92}\text{U}}$$

Express your answer to the nearest whole number.

(ii) Work out the relative abundances of ^{235}U and ^{238}U, just after the Earth was formed.

(d) Which will be more radioactive: 1 g of uranium–233 or 1 g of uranium–236?
Explain your answer.

(e) Write down the proton and neutron numbers of the nucleus after: (i) $^{233}_{92}$U, (ii) $^{237}_{92}$U, have decayed.

10 Melanie wishes to investigate the nature of the radiations emitted by three different radioactive sources (see diagram). She placed each source 20 mm from the window of a Geiger-Müller tube. The table shows the count rate when different absorbers were placed between the window and the source.

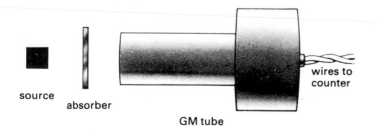

source · absorber · GM tube · wires to counter

Absorber	Count rate (counts per minute)		
	Source A	Source B	Source C
air	3127	890	4281
paper	3142	873	3752
1 mm of aluminium	3072	381	1215
10 mm of lead	1890	23	437
no source present	24		

(a) Which source is the most active?

(b) What effect did the paper have on the radiation from: (i) source A, (ii) source C?

(c) What effect did the lead have on the radiation from: (i) source B, (ii) source C?

(d) Use the data to reach a conclusion about the nature of the radiation from each of the sources.

11 Imagine that you are the Minister of State for the Environment, Countryside and Planning. The following report, produced by *Friends of the Earth* (FoE), has landed on your desk. Read it carefully; use the data in it to check the claims made and decide what action you are going to take. You may find the information about risk estimates in Section L, Unit 7, helpful.

Unacceptable radiation levels on the banks of the River Esk, West Cumbria

Scientists, working on behalf of the FoE, have identified radioactive contamination on the banks of the River Esk. The contamination is due to radionucleides carried in the discharges from the nearby Sellafield reprocessing plant. Sellafield discharges low level radioactive waste through pipelines into the Irish Sea. Although these pipes extend 2.1 km into the sea, waste is washed back onto the shore.

Dose rates on the banks of the River Esk have been measured as 0.4 μSv per hour. Thus a person working on or near the river for 40 hours a week is exposed to an unacceptably high dose. The Nuclear Radiation Protection Board's annual site specific limit is 0.5 mSv per year.

We make these recommendations:

● Further detailed radiological surveys should be carried out to assess the public's exposure to radiation in the area.

● The Minister should explain his failure to inform the public of the extent of the contamination.

● All discharges from British Nuclear fuels Limited at Sellafield should be stopped immediately.

Location of the River Esk in relation to the Sellafield reprocessing plant

R. Bleng — Sellafield Works — Seascale — R. Irt — R. Mite — Irish Sea — R. Esk — Ravenglass — Esk estuary — Eskmeals viaduct — 0 1 2 3 km

(Article reproduced with permission from *Friends of the Earth*.)

SECTION M
Earth and Atmosphere

The Earth and its atmosphere, a delicate balance of physical processes

1 Weather and Atmosphere

yellow background for degrees above zero

blue background for temperatures below freezing

sunshine and expected temperatures, 25°C or over

wind speed and direction, e.g. 40 mph from south-west

fog

fair weather clouds

dull weather clouds

sunny intervals

rain

showers and sunny intervals

snow

hail

sleet

thunderstorm

Figure 1
TV weather symbols

The Earth is a small, rocky planet. It is the only planet in our solar system which supports life. 4600 million years ago, the Earth was molten but it slowly cooled. Gases including water vapour were given off from the hot surface, forming an atmosphere. Eventually, the Earth cooled enough for water to collect on its surface.

Weather and climate

We all take an interest in the weather, particularly in the UK where the weather sometimes changes from one hour to the next. **Weather** is the state of the atmosphere at any one time or place. It includes the temperature, rainfall, windspeed and wind direction. **Climate,** on the other hand, is the average of these weather records over a long period of time.

The weather affects our daily lives. Many people try to watch the weather report on television every day. This is Ian McCaskill from the Meteorological Office giving a weather report. He is using standard symbols to explain the weather

Information about the weather is collected from weather stations on land and sea, from weather balloons and from satellites which view the Earth from space. They provide the vital basis for weather forecasts. There are various ways to describe the weather. The symbols used on television weather maps are shown in Figure 1. You can see some of these in the photograph, too.

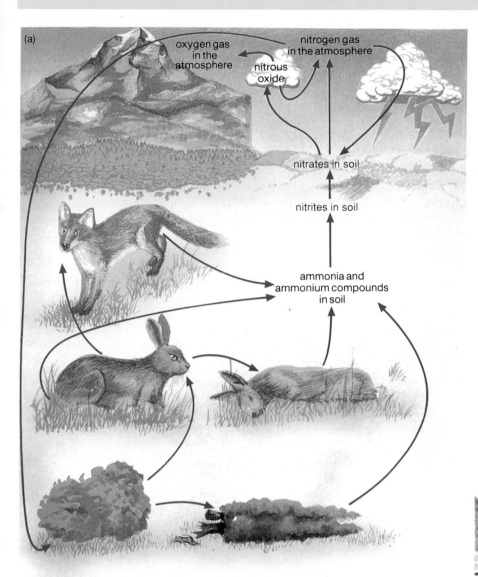

Figure 2
Gaseous cycles in the atmosphere
(a) Circulation of nitrogen and ammonia
(b) Circulation of carbon and methane

The evolution of our atmosphere

4600 million years ago, the Earth was still forming. At first, the atmosphere was rich in hydrogen and helium. As the molten, volcanically active surface cooled, other gases were added. These included methane, carbon dioxide and water vapour. When simple, green plants appeared 3500 million years ago, oxygen was formed from carbon dioxide by photosynthesis. Gradually complex plants evolved, adding more oxygen to the atmosphere. In time, animals evolved and used the oxygen for respiration. This helped to maintain a balance between the production and use of both oxygen and carbon dioxide. The composition of the atmosphere has remained more or less in balance for the last 500 million years.

Today, our industries are altering this delicate balance. Burning fossil fuels in excess of the quantity plants can cope with is removing oxygen and adding more carbon dioxide to the atmosphere. The clearing of forests has also resulted in a decrease in oxygen production by photosynthesis. Figure 2 shows how air is renewed by living creatures. Nitrogen, oxygen and carbon dioxide are used and restored to the soil and atmosphere in their cycles.

Questions

1 What measurements are always taken at a weather station?
2 (a) Using the weather symbols in Figure 1, record your own observations of the weather for today. Write a paragraph to record these same conditions in the kind of language used in weather forecasts.
(b) Use the television symbols to plot your own weather map for the British Isles for a cold windy winter day.

2 The Water Cycle

The never-ending circulation of the Earth's water supply is called the water cycle, or **Hydrological Cycle**. The watery layer of ocean, lakes and rivers which covers most of the Earth's surface, called the **hydrosphere**, was originally formed thousands of millions of years ago, as the Earth cooled. Water was gradually released from the solidifying crust as vapour, which rose into the atmosphere, cooled and condensed, and fell back to the Earth's surface as rain. Eventually this rain water formed the Earth's oceans which today cover 60% of our planet.

Oceans hold by far the most water on the planet, and are the beginning and end of the water cycle. The atmosphere and the land hold water in the intermediate stages of this huge system, which is powered by heat energy from the Sun. The atmosphere forms the link in the cycle between the oceans (which contain 97% of the Earth's water), and the continents (which contain almost 3%). A diagram of the water cycle is shown in Figure 1, which shows how the whole system works.

Water evaporates from oceans and the land when energy from the Sun heats the surface. The vapour rises into the atmosphere, where it forms tiny droplets which are transported as moist air by the wind. As more droplets collect, and join up to form larger droplets, clouds eventually form. If the clouds move over the sea from the land, or rise up over a mountain, the drop in temperature causes the droplets to condense into larger drops. Because the drops are too heavy to stay in the atmosphere, they fall as rain, hail, sleet or snow. Most of this **precipitation** falls back into the oceans, and will begin the cycle again. A smaller amount falls on the continents, from where it gradually makes its way back to the oceans. Some precipitation which falls onto land sinks, or **infiltrates** into the soil, while some becomes **run-off**, which flows over the surface, joins the **drainage system** of streams and rivers and flows back to the ocean. Water which infiltrates into soil, and then into rock, becomes **groundwater**, which eventually joins streams and becomes part of run-off. Some groundwater flows directly into the sea.

Figure 1
The water cycle

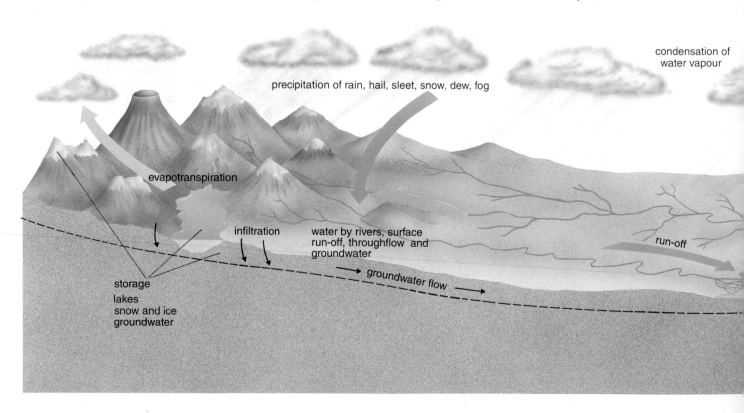

Snow or hail may stay frozen in cold continental areas. A great deal of fresh water is stored in snow fields and glaciers, though some is released as meltwater or vapour when temperatures get high enough. A lot of water which infiltrates or runs off the surface of the land is evaporated back to the atmosphere before it reaches the ocean, while some is taken up by plants and is lost during the transpiration. When water goes back into the atmosphere by evaporation and transpiration we call this **evapotranspiration**.

Oceans lose more water by evaporation into the atmosphere than they receive from precipitation. Continents receive more precipitation from the atmosphere than they lose by evaporation. It is the flow of run-off water and groundwater that balance the cycle, as shown in Figure 1.

Mist, fog, dew and frost

During the day, surface water evaporates from oceans, rivers and even puddles into the atmosphere. The water absorbs heat from the atmosphere as it evaporates, which results in a general cooling of the air. If the air temperature falls a little further, tiny droplets of water condense again, and a mist forms in the atmosphere. If the air temperature drops even further, more water droplets condense, the mist gets thicker and we call this fog. If the temperature of the ground falls at the same time, then water condenses at ground level as dew. As the vapour condenses, heat energy is released which warms the atmosphere.

Frost forms when the temperature of the ground and the atmosphere falls below 0°C. Particularly on starry nights with no cloud to prevent heat loss by radiation, the air quickly falls below 0°C and moisture forms needle-shaped ice crystals.

Water molecules turn to ice crystals if the temperature drops below freezing. The frost in this photo is an example. Water may become trapped in cracks in solid material such as rocks or building stone. If this water freezes, its expansion can cause breakages in the rock or stone

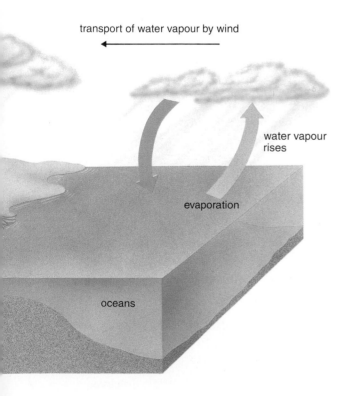

transport of water vapour by wind

water vapour rises

evaporation

oceans

Questions

1 Explain carefully the difference between water droplets and water vapour.

2 Describe and explain the conditions that are necessary for the formation of fog.

3 The highest rainfall in England is in Seathwaite, which lies about five kilometres east of the Scafell range of mountains; explain this.

4 Why is the annual average rainfall greater near the equator than it is in Britain?

3 Prevailing Winds

The Sun is the Earth's source of energy. As well as supporting life, solar energy controls our planet's climate and weather. As the Earth remains at a fairly steady temperature, there must be a balance between incoming energy (**solar radiation**) and outgoing heat loss (**terrestrial radiation**). 45% of the solar radiation reaches the Earth's surface. 55% of the energy is absorbed and re-emitted by the atmosphere.

The surface of the Earth is not warmed evenly; the equator receives far more radiation that the poles. However the equator only radiates a little more into space than the poles, because the temperature of space is close to absolute zero (Figure 1). This imbalance would cause the equator to become unbearably hot, if it were not for the fact that energy is carried away by convection currents. Most of the energy is carried by airstreams, though ocean currents (such as the Gulf Stream) also contribute to the transfer of energy towards the poles.

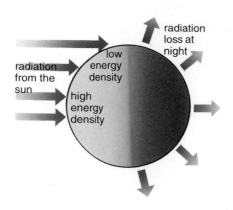

Figure 1
Heat budget for the Earth: an imbalance in radiation density creates a hot equator and cold poles

Global airstreams

The airstreams, or global winds, are convection currents on a very large scale. Warm, less dense air rises, and cold, denser air flows in to replace it. Powered by the Sun airstreams circulate constantly between the hot tropics and the freezing poles. Figure 2 shows how air might circulate on a slowly rotating Earth, but the Earth's rotation produces two major effects: first the air flow is broken up into three **cells**, and secondly the airstreams are deflected away from a northwards or southwards flow.

Figure 3 shows the global airstreams over one quarter of the globe. Hot air rising from the equator creates the **doldrums**, an area of low pressure and light, variable winds. As the air rises it cools and spreads outwards, then becomes denser and sinks. Belts of high pressure and relative calm, which we call **horse latitudes**, result from this movement of air. The dense, high pressure air at the horse latitudes flows back towards the equator. This circular flow of a large mass of air is known as a cell. Some air from the tropical cells flows towards the temperate latitudes, creating **mid latitude cells**, which end where **polar cells** of very cold air flow away from the poles, displacing the warm air. The cold polar air warms up in the temperate latitudes, rises and flows back to the poles where it cools and sinks again. Wind blows from regions of high pressure to regions of low pressure; for example, winds tend to blow from the high pressure areas in the horse latitudes to the lower pressure areas in the temperate latitudes.

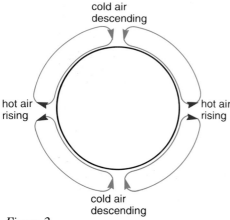

Figure 2
If the Earth rotated very slowly, heat would be carried away from the equator by simple convection currents such as those shown here

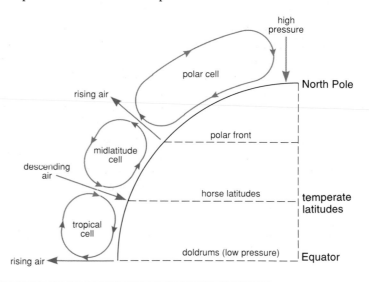

Figure 3
Global airstreams (over one-quarter of the globe)

Coriolis effect

The Earth rotates as a solid body. This means that the equator is moving faster than the surface of the Earth at higher latitudes. As a result of this airstreams are deflected east or west as they move over the Earth's surface, this is called the **Coriolis effect**. Hot air which rises from the equator has an eastwards velocity of about 1700 km/h, but air at a latitude of 30°N, only has an eastwards velocity of about 1500 km/h.

Therefore the hot air overtakes the air at higher latitudes and is deflected eastwards. On the other hand, cold air travelling southwards is deflected towards the west, since it is travelling more slowly than the more southerly air. The winds in the northern tropical cell travel towards the south west. Winds are named after the direction from which they blow, so these are the north-east trade winds. In the horse latitudes, cold air travelling northwards is deflected towards the east to produce the prevailing south westerlies.

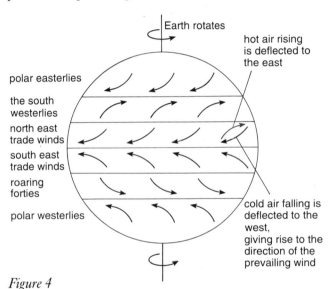

Figure 4
The Earth's prevailing winds

Jupiter

Questions

1 The Moon's surface reaches a temperature of 150°C, when sunlight falls on it, yet during the lunar night the Moon's surface temperature drops to –150°C. The Moon and Earth are the same distance away from the Sun, but the hottest and coldest recorded temperatures on Earth are 50°C and –90°C. Account for the differences between the lunar and terrestrial temperatures.
2 (a) Explain how convective cells arise on the Earth.
(b) Why is there more than one convective cell in each hemisphere?
3 In the northern hemisphere the trade winds blow from the north east, but in the southern hemisphere they blow from the south east. Explain why.
4 Look at the photograph of Jupiter. The dark bands are called belts; these are regions of gas low down in the atmosphere. The lighter bands are called zones; these are higher in the atmosphere than the belts and are colder. Jupiter has a diameter 11 times that of the Earth, and it completes one revolution on its axis in 10 hours.

(a) Explain how these belts and bands might be formed.
(b) Why do you think that there are more belts on Jupiter than there are convective cells on the Earth?
(c) Atmospheric features near the equator of Jupiter have been observed moving at several hundreds of kilometres per hour. Account for these very large Jovian wind speeds.
(d) Such large wind speeds give rise to turbulence. What features in Jupiter's atmosphere could be due to turbulence?

4 Highs and Lows

Air masses affecting the British Isles

Weather around the world is controlled by a number of air masses, which originate from source regions. Such regions are found in areas dominated by large semi-permanent high pressure systems. These are often associated with the interiors of large land masses, for example: northern Canada, Siberia and the Arctic Basin in winter. In summer, desert areas such as the Sahara, central Asia and central America, produce large masses of warm dry air.

Figure 1 shows the routes followed by the five main air masses which affect the British Isles. We live on the eastern edge of the Atlantic Ocean, in a region where prevailing westerly winds have collected moisture during their long passage over the sea. Our weather is dominated by two maritime air masses, polar and tropical.

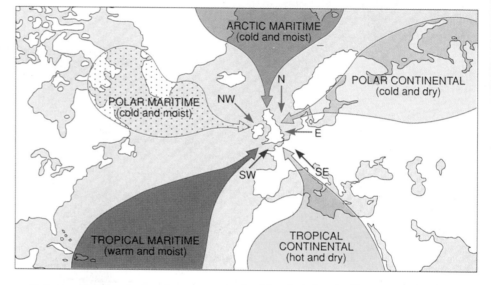

Figure 1
The air masses that affect the British Isles

- Polar maritime air, has its source in the North Atlantic Ocean south of Greenland. This air mass occupies Britain for nearly half the year in all seasons. This air brings with it cool and very moist air, which is associated with short-lived but heavy showers.
- Tropical maritime air, coming from the high pressure system near the Azores, controls our weather for about 15% of the year in all seasons. When the air leaves its source region it is warm and moist, and the skies are clear. Often the air mass can bring the same pleasant weather with it to the British Isles, but sometimes the air is cooled in its passage northwards and heavy thunderstorms can be the result in the summer.
- Polar continental air is less common over the British Isles, only affecting us as a rule during the winter months. Cold, dry air originates over Siberia and it brings with it clear skies, which allow a lot of heat to be radiated away into space overnight. This weather it typified by bitter cold and severe frosts.
- Arctic air is rare over Britain. It starts as Arctic continental air but after its long journey south it picks up some moisture, although cold air cannot hold as much water vapour as warmer air. These air streams can bring heavy snow showers in the north and on high ground.
- Tropical continental air is also rare; this reaches us from North Africa. These air masses are very stable and can lead to long heat waves, such as the summer we experienced in 1976.

Anticyclones

In the mid-latitudes where we live there are two types of **anticyclone** or high pressure system. One type is an extension of the polar continental or tropical continental high pressure cells. The second type is interspersed with low pressure systems moving in with maritime air masses. Atmospheric pressure increases towards the centre of an anticyclone. This causes the air to fall outwards in a descending spiral, rather like water going down the bath plug. As the air falls it warms up, which suppresses the formation of clouds. Therefore high pressure systems bring with them clear skies. The pressure gradient in the air is small, so the isobars are widely spaces and the winds are light; they blow in a clockwise direction around the centre of the system, in the northern hemisphere. You can see these features in the weather map shown in Figure 2.

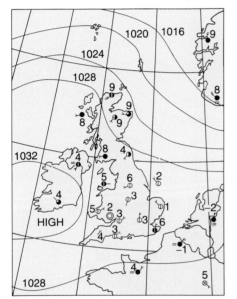

weather symbols		cloud coverage	
•	rain	○	0 oktas
⌐	drizzle	◐	1 okta
✳	snow	◕	2 oktas
∞	haze	◕	3 oktas
✱	sleet	◑	4 oktas
△	hail	◑	5 oktas
=	mist	◐	6 oktas
≡	fog	◑	7 oktas
⊤	thunderstorm	●	8 oktas
▽	rain showers	⊗	sky obscured
✳▽	sleet showers		(1 okta = 1/8 of the sky covered by cloud)
✳▽	snow showers		
△▽	hail showers		

wind speed

⦿	calm
—○	1 – 2 knots
⌐○	3 – 7 knots
⟍○	8–12 knots
⟍○	13–17 knots
⟍○	18–22 knots
⟍○	23–27 knots
⟍○	28–32 knots
⟍○	33–37 knots
⟍○	38–42 knots
⟍○	43–47 knots
▶○	48–52 knots

type of front

⌒⌒⌒	warm front
⌒⌒⌒	warm front above surface
▲▲▲	cold front
▲▲▲	cold front above surface
⌒▲⌒▲	occlusion

Figure 2
High pressure region over Britain. This anticyclone has pushed northwards from north Africa, bringing with it clear skies and hot, calm weather. The pressure on our weather maps is measured in millibars: 1 millibar = 100 Nm⁻². This is a synoptic chart for the British Isles on 11 January 1992 at 1200 hours

Figure 3
'Synoptic' symbols used for Met Office weather maps. Why do you think TV weather maps used different symbols?

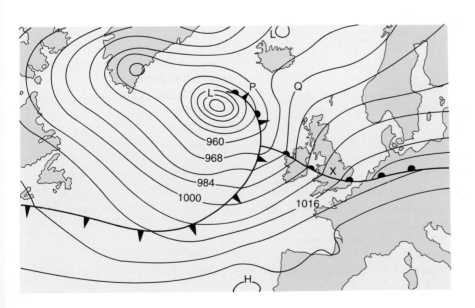

Figure 4
A frontal system moves over Britain

Depressions

In Britain we are all familiar with a **depression**, **cyclone** or low pressure system; usually these have frontal systems associated with them. The series of diagrams in Figure 5 shows how such a system arises.

- In diagram (a) a warm south westerly air stream meets a cold polar air mass travelling from the north east.
- In diagram (b) the warm air has begun to bulge northwards into the colder air. Distinct warm and cold fronts have now been formed, which begin to rotate around a centre of low pressure.
- By diagram (c) a deep central depression has been formed; low pressure air spirals upwards cooling as it rises. Cloud formation occurs as rising water vapour in the atmosphere cools and condenses. Heavy rain is associated with frontal systems. There is a strong pressure gradient and strong winds circulate in an anticlockwise direction (in the northern hemisphere) around the centre of the depression. The system moves westwards towards the British Isles.
- In diagram (d) of the sequence the warm air has been squeezed upwards to leave an **occluded front**. It is quite hard to visualise the nature of frontal systems using just views from above. Figure 6 shows two cross-sections through fronts in Figure 5(d).

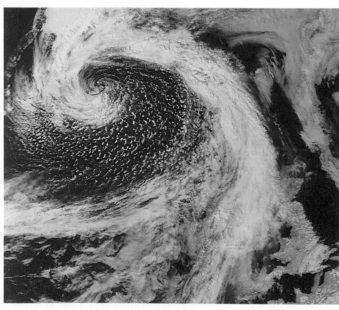

A false colour satellite image of a frontal system over the North Atlantic Ocean. The UK and France may be seen at the bottom right. At the top centre is the ice cap over Greenland. The frontal system is the swirling mass of clouds to the left of centre, with a low pressure area at the centre of the spiral. Low-level clouds are shown as yellow, high-level clouds as white. The pale yellow area at the top right is a large bank of sea fog.

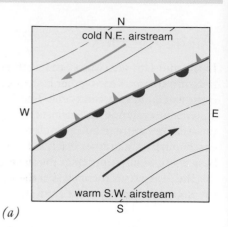

(a)

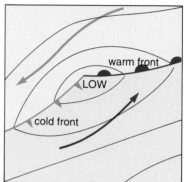

(b)

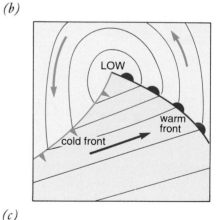

(c)

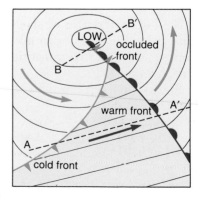

(d)

Figure 5 (right)
(a) Polar continental and tropical maritime air masses meet over the Atlantic Ocean
(b) The air masses begin to curl around, forming a cold front travelling south and a warm front moving north
(c) The air masses spiral around a central depression
(d) Warm air is forced upwards, and an occluded front is formed

Weather recording

Detailed information about the weather is collected at special weather stations on land and at sea, from weather balloons and from satellites which view the Earth from space. The information is recorded on weather maps called 'synoptic charts'. These are revised as new data comes in. Temperature, pressure, cloud cover, present weather, wind direction and wind speed are all recorded and plotted on to complex maps, which are analysed to forecast the coming weather conditions. Synoptic charts are the basis for all weather forecasts (Figure 2).

The daily recordings made at the weather stations include the maximum and minimum temperature, the amount of precipitation and the hours of sunshine.

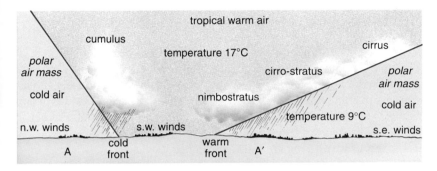

Figure 6
(a) Cross-section through a frontal system. Cold polar air pushes warm tropical air into an upwards spiral. Rain is heaviest at the cold front. The letters AA' connect this cross-section to Figure 3 (d)

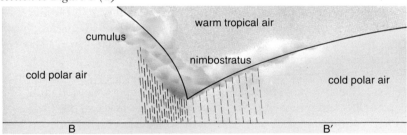

Figure 6
(b) This cross-section shows an occluded front. The warm tropical air has been lifted clear of the ground by the cold polar air masses

Questions

1 Explain what weather we should expect when the following air masses approach Britain: (i) polar maritime (ii) polar continental.

2 Explain the terms: warm front, cold front, occluded front.

3 (a) In the text it is stated that stronger wind blows where the pressure gradient is highest, explain why this is so.

(b) Explain what is meant by an isobar.

(c) Winds blow approximately along the lines of the isobars. Compare the directions and strengths of the wind at the points P and Q marked on the map in Figure 4.

4 The weather map shows a frontal system approaching Britain. Such a system might take 48 hours to pass over the country travelling west to east.

(a) Why do frontal systems arrive from the west?

(b) Imagine you are staying with a friend at point X on the map. Use the information in Figures 4 and 6 to describe how the weather changes over the period that the warm and then cold fronts pass over. In your answer you should mention: clouds, their height and type; the temperature; the wind direction and the rainfall.

(c) Look at the satellite photograph of the cold front. Describe carefully the atmospheric conditions immediately after the front has passed overhead.

5 Weathering and Erosion

Our landscape is slowly and constantly changing. Several processes act upon it, wearing down old rocks and creating new ones. These processes involve **weathering, erosion, transport** and **deposition**. They have operated constantly since the Earth was first formed. They have lowered mountains to plains, creating sediments which may later be built up into mountains.

Weathering and erosion

The first stage in the cycle which wears down the landscape is **weathering**. Weathering involves the breakdown, fracturing and decomposition of rocks on or near the surface of the Earth. There are many different kinds of weathering. Some occur on a large scale, others at molecular level. They take place in normal atmospheric conditions and are helped along by changes in temperature and the presence of water.

Hard crystalline rocks are not necessarily more difficult to weather than the 'softer' sediments. Granite is very hard, yet one of its main components, feldspar, is particularly vulnerable to reaction with water (**hydrolysis**). In hydrolysis, the feldspar weathers to china clay (kaolin). On the other hand clay rocks are very soft, but they weather very little because clay minerals are the end product of weathering (they cannot be broken down further).

The main stages of weathering are shown in Figure 1.

An agent of erosion is something which can pick up particles and carry them along. So moving water in rivers and waves, winds (especially in deserts) and glaciers are the important agents of erosion. You can see examples of erosion in the photographs.

For wind and water, a critical speed of flow is needed before particles of a certain size can be picked up and carried along. If the speed of flow falls below this critical level, heavier particles will be deposited. The same is true for glaciers, like the one in the photograph, although the slow-moving ice of glaciers works on a totally different time scale to wind and rivers.

The rock particles carried in the agent wear away the landscape, like a scouring pad. Rivers wear away their channels, deepening and widening valleys. Coastal waves undercut cliffs and scrape out caves. Glaciers carve slowly and deeply like giant files. Desert winds sculpture and polish bare rock surfaces over wide areas. Little by little, these agents erode and alter the landscape.

Figure 1 (diagram descriptions)

Stage 1 — UNALTERED ROCK e.g. granite

The rock gradually cracks as large scale weathering processes act upon it

Stage 2 — CRACKED ROCK

Rock is broken into smaller parts by the fracturing which occurs when water turns to ice in thin cracks, for instance.

Stage 3

Smaller rock fragments are more easily weathered by further molecular weathering processes

Stage 4 — DIFFERENT MINERALS separate

Chemical processes involving water, carbonic acid, rain and oxygen act on different minerals

RESIDUAL MATERIAL

NEW MINERALS e.g. kaolin

ROCK MATERIAL is now chemically changed

Figure 1
Sequence of weathering processes

Water is an agent of erosion. Weathered material is transported by rivers to the sea where it is carried by water currents and wave action. The Scarisdale River, shown here, is able to transport fine material, as well as large boulders when it is in flood. The load it carries helps to erode its valley

Transport

Processes which operate in the transporting agents of water, wind and, to a much lesser degree, ice, have the effect of altering the shape and size of weathered particles which are picked up and carried.

Freshly weathered rock particles are always angular but their sharp edges are gradually removed once they are picked up and transported. Particles carried by water or wind knock against each other, corners and sharp angles are broken off and the particles gradually become rounded. This is the process of attrition. Abrasion is the transport process where water and wind transported particles knock against the bedrock and erode it, gradually wearing away the landscape. This process of abrasion also helps to round the grains.

Deposition

When winds, rivers, waves and ice can no longer carry weathered material, it is deposited as sediment. With the transport agents of water and wind, the heaviest material is always deposited first and the lightest last. Down the course of a river, in shallow seas and in deserts particles are deposited in gradation from coarse gravels (the heaviest) to sands, silts and clays (the lightest). Water and wind deposit sorted sediments. When glaciers retreat or melt they deposit an unsorted load of all grain sizes from boulders to rock flour and there is no gradation in size.

Material transported by slow moving glaciers erodes the sides and floor of the valley

Sedimentation and soil formation

Most sediment is deposited on the shallow continental shelves surrounding land masses. Here they form horizontal beds or strata. As the beds become buried under younger layers of sediment the weight of the overlying beds compacts the grains together and squeezes out water from between them. Mineral fluids fill in the small spaces between the grains and cements them together. The once soft, unconsolidated sediments are converted to hard sedimentary rocks.

Soils gradually form on weathered material, deposited sediment or bedrock which is exposed on the land surface. Weathering processes disintegrate and decompose the rock material creating a regolith or stony layer upon which plants can grow. Plants add organic matter to the immature soil which can then support animals. Gradually, over hundreds or even thousands of years, a mature soil develops consisting of about 45% mineral matter 5% organic matter, and 25% water and 25% air. Poor soils tend to develop on sands and gravels, while we find better soils on silts and clays.

Questions

1 How does mechanical weathering make chemical weathering more effective?

2 Heat speeds up chemical reactions. So why does chemical weathering take place slowly in hot deserts?

3 Why is wind erosion relatively more important in arid (dry) regions of the world than in humid (wet) ones?

4 Describe two ways in which waves cause erosion. What typical landforms of erosion are seen along the coastline?

The 'towers' in Monument Valley, Arizona, have been formed by the eroding action of the wind

6 The Rock Cycle

Figure 1
Formation of igneous, sedimentary and metamorphic rock

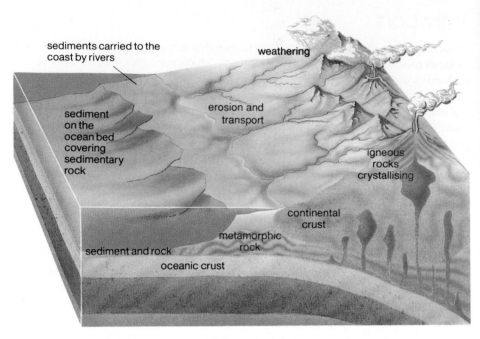

This is an example of an igneous rock. 'Igneous 'means formed by fire' because these rocks result from the cooling of molten magma on, or within, the Earth. The size of the crystals depend on how quickly the magma cools. This rock has cooled slowly so its crystals are large

This is an example of a sedimentary rock, formed from the breakdown products of older rocks

When the Earth first cooled, its molten crust solidified to form **igneous rock**. Igneous means 'formed by fire'. Initially there was no other type of rock. Over thousands and millions of years, two other types of rock were created. These were **sedimentary rocks and metamorphic rocks**. Figure 1 shows how these three types of rock are forming today.

Igneous rocks

Igneous rocks originate from the liquid rock, or magma, below the Earth's crust. **Lava** from volcanoes is magma pouring out from cracks in the Earth's crust. When magma or lava cools and crystallises, igneous rock is formed (Figure 1). The size of crystals in the igneous rock depends on the rate of cooling. Lava loses its heat very rapidly to the air, so above the Earth's surface it quickly becomes a solid. The lava crystallises in days or months depending on its thickness. This process produces rocks, such as **basalt,** with small crystals. Magma which is cooling *below* the Earth's surface retains its heat far longer. It may not cool to a solid for centuries so the crystals have a very long time in which to grow. This produces rocks like **granite** with large coarse-grained crystals.

Sedimentary rocks

When igneous rocks are exposed at the surface of the Earth, they undergo weathering and erosion. These processes form sediments which are carried from their original location and deposited elsewhere. Most sediment is deposited on the shallow **continental shelves** surrounding land masses (Figure 1). As the weathered material is brought down from the eroding land masses, successive beds, or **strata** form. In some areas, chalk sediments have formed from the calcium carbonate present in the shells of sea animals. As the beds become buried deeper, they are compacted under the weight of the layers above. This converts the soft sediments to harder sedimentary limestone rocks.

300

Metamorphic rocks

In some cases, sedimentary rock has been buried to great depths and been changed by enormous pressure and high temperatures. This has produced **metamorphic rocks (Figure 1)**. There are various types of metamorphic rock depending upon the type of sediment from which they originate. The minerals in the parent sediment are sometimes changed under the stresses of heat and pressure. In some cases, they are changed to chemically similar, but harder and more stable metamorphic rocks. This results in a series of physically different metamorphic rocks. Deep down inside the Earth, the base of the metamorphic rock may melt, creating magma. This will eventually solidify as igneous rock, beginning the rock cycle again.

Figure 2 shows the various stages of the rock cycle. The complete rock cycle, lasting hundreds of millions of years, does not always take place exactly as described. Sometimes, igneous rock, instead of weathering at the surface, may be subjected to heat and pressure and become metamorphic rock. Metamorphic and sedimentary rocks may also weather and erode to form new sedimentary rocks. Not all old rocks are turned into sediment and recycled. Some of them have been covered and protected by successive layers of younger rocks since the Earth was first formed 4600 million years ago.

This is an example of a metamorphic rock, formed from sedimentary rock under high temperature and pressure. The minerals in metamorphic rocks can help geologists to trace their origin

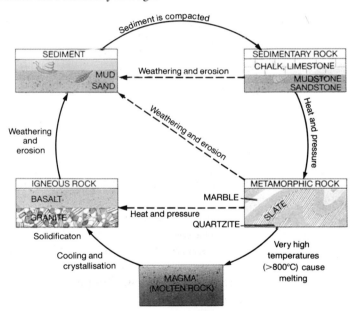

Figure 2
The rock cycle

Questions

1 Using the rock cycle in Figure 2, explain the following statement: 'one rock is the raw material for another'.

2 When the Earth first formed, there were only igneous rocks.
(a) Why was this?
(b) How did sedimentary and metamorphic rocks come into existence?

3 (a) Describe how sediments may form on land and in the sea.
(b) How are they turned into sedimentary rocks?

4 You are given three pieces of rock by a friend who wants to know more about them.

Rock A is made of rounded pebbles held together by a hard, sandy layer.

Rock B is white in colour and contains several small fossils which look like shells you have seen at the seaside.

Rock C is very hard and is made of large crystals which you can see quite clearly.

What could you tell your friend about the rocks?

7 The Structure of the Earth

Internal structure of the Earth

The Earth is shaped like an orange, spherical but slightly flattened at the poles. Its structure and surface is like a badly cracked egg. The 'cracked shell' is the very thin **crust**, the 'white' is the **mantle** and the 'yolk' is the **core** (Figure 1). These concentric layers increase in thickness, density and temperature towards the centre. Evidence for the Earth's structure comes from various sources. Information about the crust comes from studying mines, volcanoes and earthquakes. Evidence for the mantle and the core comes from the study of earthquakes, deep volcanoes, meteorites and the Earth's magnetic field.

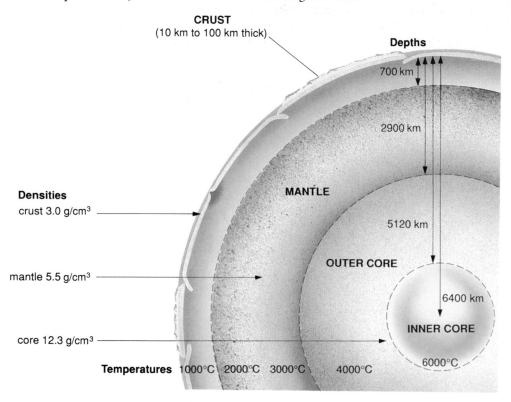

Figure 1
Cross-section of the Earth to show its internal structure

Figure 1 shows a cross-section of the Earth. The Earth's crust is composed of igneous, metamorphic and sedimentary rocks. The crust is very thin and is broken into large and small parts, called **plates**. These float on the denser mantle, parts of which are molten. Over long periods of geological time, changes occur in the crust. Immense internal forces cause the plates to move resulting in faults, rift valleys, volcanoes and earthquakes. In some parts, the crust is being stretched and is under tension. Deep cracks appear in the crust, rocks break and one side slips down filling the space created. This forms a **fault**. **Rift valleys** form where two normal faults lie alongside each other (Figure 2). Deep cracks in the Earth's surface can also lead to volcanoes (Unit 8). In other parts of the Earth, the crust is being pushed together and compressed (Figures 2(b) and 2(c)). This causes layers of the crust to ride over each other and fold, resulting in earthquakes (Figure 2(d)).

The mantle and the core are so deep in the Earth that they are normally inaccessible but, in The Lizard in Cornwall, mantle rock has been pushed to the surface and can be seen. Mantle rocks can also be studied from **xenoliths** ('strange rocks') brought up in volcanoes.

Evidence for the Earth's Structure

Much greater detail of the Earth's structure comes from the study of earthquakes and magnetic fields. During an earthquake, the crust ruptures. Energy generated at the focus of the earthquake creates a train of shock **(seismic)** waves. These extend outwards through the Earth and their pathways are shown in Figure 3. Earthquake energy creates three main types of waves: P or **primary waves; S, secondary or shear waves; L, longitudinal or surface waves.** L waves are slow moving and travel through the crust. P and S waves move faster and travel through the deeper layers of the Earth (Figure 3). P waves travel through liquids and solids. S waves travel through solids; they cannot pass through the Earth's liquid outer core. From the movement of these waves we can obtain evidence for the Earth's internal structure. Surface waves roll invisibly through the Earth's crust. Around the **epicentre** of an earthquake (right above the focus) these waves cause the most damage to buildings.

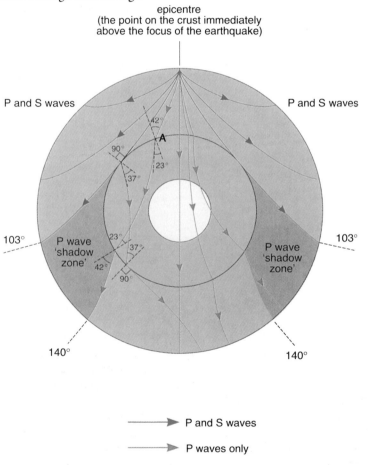

Figure 3
Cross-section of the Earth showing the paths of seismic waves from an earthquake. Both P and S waves spread out to angles up to 103° from the epicentre, but S waves cannot travel through the liquid core. P waves entering the liquid core slow down and are focused (like light waves through a lens) into the region at the bottom of the diagram (angles >140°). No P wave travelling through the liquid outer core reaches the 'P wave shadow zone'. However weak P waves are detected in that zone, which have been speeded up, and therefore refracted, by the solid inner core

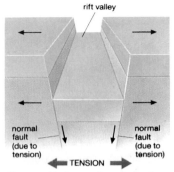

Two normal faults opposite each other form a rift valley (plates are drifting apart and the land in between drops down)

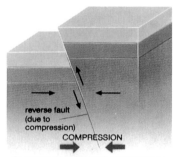

Rock strata ride over each other as plates are forced together; a reverse fault results

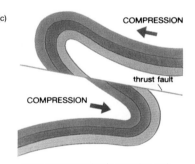

Folding occurs under compression, and cracking results in thrust faults

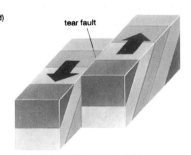

Sometimes plates are forced sideways past each other causing tear faults

Figure 2
(a) Normal faults and a rift-valley
(b) Reverse fault
(c) Thrust fault (d) Tearfault

Seismic waves

The velocities of seismic waves depend on the density and elasticity of the rock they are passing through. Rigid rock transmits waves faster than loose sediments or molten rocks, because it springs back more readily when compressed or distorted. The more dense a rock is the more slowly it transmits a wave. Both the density and elasticity of rocks increase with depth, but the elasticity rises faster than the density. Consequently, both P wave and S wave velocities increase as they go deeper into the Earth's mantle. However, when P waves enter the molten outer core they slow down considerably. P waves speed up again slightly on entering the solid inner core. Figure 4 shows the variation of wave velocities with depth.

S waves are transverse waves. The rock oscillates at right angles to the direction of the wave motion. S waves can only travel through solids, because only a rigid body can provide the sideways forces to restore the rock to its original position. Liquids flow so they cannot provide the sideways or shear forces. P waves are longitudinal waves. They are like sound or shock waves, which are well illustrated by a series of compressions and expansions passing along a slinky. See Figure 5.

S waves do not register on seismometers at an angle greater than 103° from the epicentre of the earthquake, because they cannot travel through the liquid outer core. Most P waves also cut out at angles greater than 103° but reappear at angles greater than 140°. Between 103° and 140° a **shadow zone** exists in which little earthquake information is received. Some weak P waves, however do enter the shadow zone, either by being refracted outwards by the solid inner core, or by being reflected back off the Earth's surface.

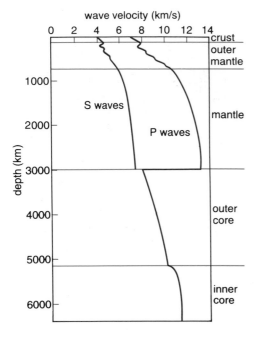

Figure 4
This graph shows the variation of P and S wave velocities with depth in the Earth

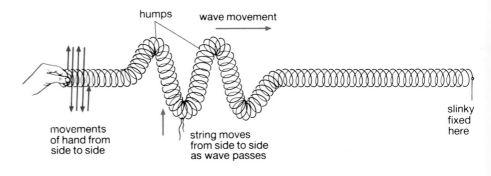

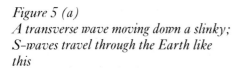

Figure 5 (a)
A transverse wave moving down a slinky; S-waves travel through the Earth like this

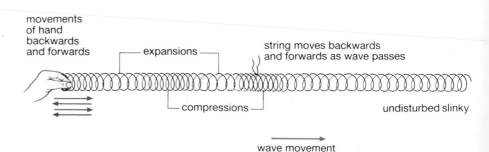

Figure 5 (b)
A longitudinal wave moving down a slinky; this is how P-waves move through the Earth

When light passes from air to glass it is slowed down and it is refracted towards the normal; when light leaves glass and enters air it refracts the other way (Figure 6). Seismic waves obey the same rules of refraction. In Figure 3 you can see how the P waves are refracted towards the normal as they enter the outer core. At such a boundary some waves will also be reflected; similarly when light strikes a surface some is transmitted and some is reflected.

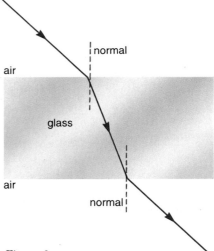

The Earth's magnetic field

The Earth has a strong magnetic field which is produced as the Earth spins on its axis. The fluid outer core allows the mantle and the crust to rotate faster than the solid inner core. This produces a magnetic field similar to an enormous bar magnet (Figure 7). The magnetism exists because the Earth has an iron-rich core.

Some rocks particularly Basalt, contain iron-rich minerals. When these rocks formed they were magnetised in the direction of the Earth's magnetic field, thus leaving a magnetic 'fossil' record. These records show that the Earth's magnetic field has reversed many times over geological history. Also putting together a jigsaw puzzle of magnetic rocks, shows that rocks which once lay side by side ages ago, have been moved around the Earth's surface. This drift of land masses is examined further in the next section.

Figure 6
Light rays refracted by a glass block; seismic waves obey the same rules of refraction

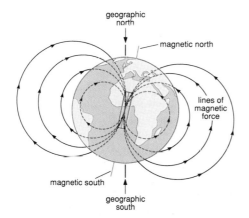

Figure 7
The Earth's magnetic field

Questions

1 In California, you can see orange groves with trees growing in perfect, straight lines. However, there are some orange groves in which the lines of trees are for some reason kinked. They were not planted like that.
(a) What is the cause?
(b) In what interesting geological area are these orange groves?
2 Although some very deep holes have been drilled into the Earth, none has ever reached the mantle. We have no direct evidence of the nature of the interior of the Earth. Our knowledge is based on indirect evidence. What is

this evidence and what has it told us about the Earth's interior?
3 This question is about seismic waves; you will find it useful to refer to Figures 3 and 4.
(a) When P waves enter the core at Point A they are refracted. Explain carefully using the data in Figure 4 the direction of this refraction.
(b) Explain fully why seismic waves travelling through the Earth follow curved paths.
(c) S waves can travel only to depths of about 3000 km. Why is this the case?

(d) How can you tell from Figure 4 that the outer mantle of the Earth has many separate layers or discontinuities in it?
(e) A seismic station is about 1000 km from the epicentre of an earthquake. P and S waves reach it travelling close to the Earth's surface. Approximately how long is the interval between the two types of waves reaching the station?

8 Plate Tectonics

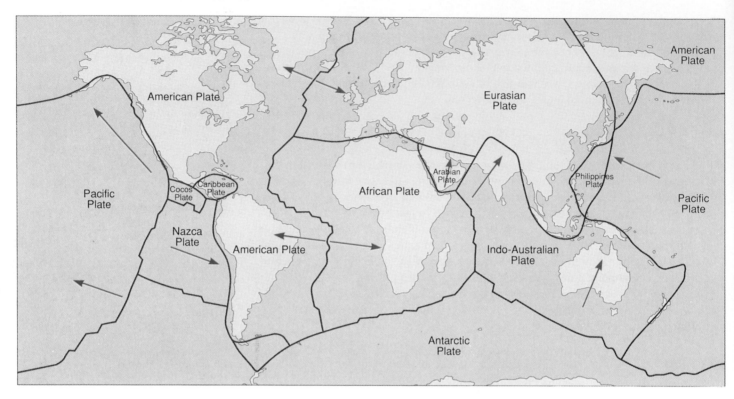

Figure 1
World map showing the main plates and their directions of movement

The structure and physical properties of the Earth are the key to understanding **plate tectonics.** The Earth's core is as hot as the surface of the Sun. This causes slow convection currents in the mantle which result in slow movements in the plates of the Earth's crust. This is the start of the **tectonic cycle.** Figure 1 shows how these plates are moving. Although we normally think of mantle rocks as solids, over millions of years they can show fluid properties.

Many years ago, Africa and Arabia used to be joined together. They are now growing apart as the Red Sea gradually widens. This sea may eventually become as large as the Atlantic Ocean

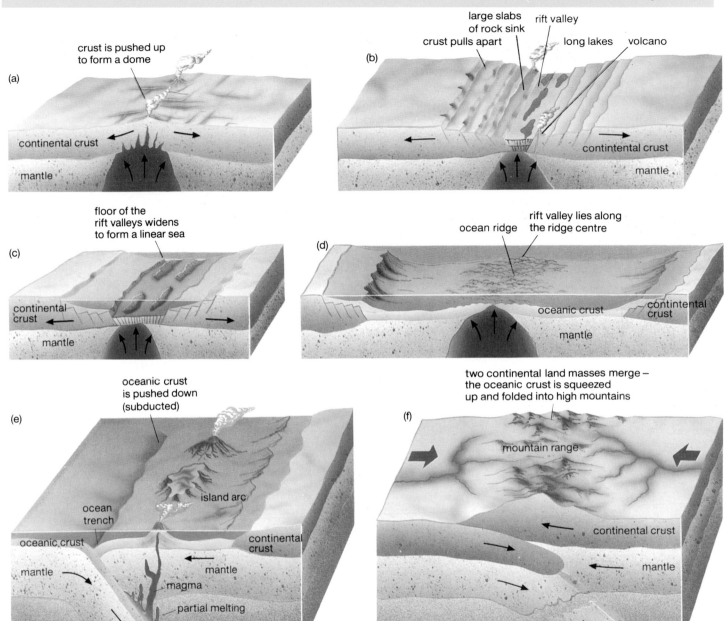

The convection currents circulating in the liquid mantle may take millions of years to rise to the surface. If currents of hot, molten rock rise under the thin oceanic crust they form '**hot spots**' of intense volcanic activity. The Hawaiian islands lie above one of these hot spots. On these islands, magma rises to the crust surface forming '**shield volcanoes**'; the 'runny' basalt lava flows quickly so shield volcanoes have gentle slopes. Under the thicker continental crust, rising convection currents push the crust up into a dome, causing tension and cracking (Figure 2(a)). As the crust is pulled apart, large slabs of rock sink and rift valleys form. Volcanoes appear where the magma escapes from cracks in the rift valley (Figure 2(b)). This is typical of the rift valleys in East Africa.

The Red Sea and Gulf of Aden illustrate the next stage of the tectonic cycle when the floor of the rift valley can widen to form a **linear sea** (Figure 2(c)).

Figure 2
(a) The continental crust is pushed up into a dome
(b) Rift valleys form with volcanoes
(c) A linear sea develops
(d) An ocean ridge forms and the ocean widens
(e) An ocean trench forms with island arcs; the oceanic crust is subducted (pushed down)
(f) As continents collide, the ocean closes and a larger land mass with high mountains is formed

This region will gradually form a new ocean, separating Africa from Arabia. The Atlantic is an example of an expanding **young ocean**. It is widening as fast as your fingernails grow. A **submerged ocean ridge** has formed down the centre of the Atlantic from the Arctic to the Antarctic (Figure 2(d)). This submerged ridge follows the boundary between the American Plate and the African and Eurasian Plates (Figure 1). A rift valley lying along the centre of this ridge, slowly generates new oceanic crust.

The closing of the Pacific Ocean

As the Atlantic Ocean widens, the Pacific Ocean is slowly closing. The Pacific is a much older ocean than the Atlantic. Over millions and millions of years, thick deposits of sediment have collected on the ocean floor. The layers of sediment are particularly thick near continents where rivers have carried silt into the ocean. As the oceanic crust is thin, the weight of sediment makes it sag. More sediment collects, and eventually the crust breaks. As new crust is being created at the ocean ridge, the old oceanic crust, near to the continent, is pushed down or **subducted** (Figure 2(e)).

The deep depression in the crust where subduction occurs is called an **ocean trench**. Subduction of the ocean crust and wet sediment into the mantle creates magma which rises up to the surface. Some of this escapes as thick, explosive lava, forming volcanic islands in arcs which fringe the trench (Figure 2(e)). These **island arcs** extend from the Aleutian Trench in the north-west Pacific to the Tonga Trench in the south-west Pacific. Some islands, with metamorphosed sediments, form larger island groups like Japan.

The creation of the Andes

As the Atlantic grows, the American continents are moving westwards. Along the western edge of South America, the Nazca Plate is being pushed against the American Plate (Figure 1) Here the ocean crust is being forced under the advancing landmass. This is pushing up marine sediments and creating the Andes mountains.

The end of a tectonic cycle

A tectonic cycle ends when two continental land masses converge and the ocean between them disappears. The layers of oceanic crust are squeezed into tight folds forming high mountains (Figure 2(f)). This is what happened when India moved north to collide with Asia. The ancient Tethys Ocean disappeared and the Himalayan mountains were formed. The crust here is so thick that volcanic activity has stopped, although earthquakes are common.

If the Earth was filmed from outer space using time lapse photography over a period of 1000 million years, you would be able to see the tectonic cycle clearly: continents moving apart, oceans waxing and waning, new continents joining and mountain chains forming.

These mountains were formed from horizontal beds of sediment at the bottom of an ancient ocean. As the continents moved together and the ocean closed up, this sediment was pushed up into a fold mountain range

Evidence for plate tectonics

A wealth of evidence supports the theory that continents are not fixed in their position, but drift apart over great lengths of time.

Continental shelves can be 'fitted' together like pieces of a jigsaw puzzle. African and South America illustrate this very well. Not only do they fit well together but the picture is completed by a matching of rock types and fossils on both sides of the South Atlantic Ocean. Mesosaurus, a small swimming reptile, not capable of swimming an ocean, is found on both continents. We believe that hundreds of millions of years ago Africa and South America were joined; as they drifted apart two populations of Mesosaurus were left on either side of the Atlantic Ocean.

Mountain chains can be traced from one continent to another. Mountains in Norway and Scotland fit perfectly with those of Greenland and North America. They were once one long chain which has been broken as the North Atlantic widened.

Coral reefs grow only in warm tropical waters, and coal can form only in equatorial swamp conditions, yet fossil coral reefs and thick coal seams are found in Britain and Northern Europe. This suggests that the British Isles were once near the tropics, but they have drifted northwards over periods of geological time.

In the 1950s, maps drawn to show the position of magnetic north appeared to show that it had gradually moved; this was called polar wander. We now think that the pole has stayed in the same position but the continents have moved.

Geologists also use space age technology to monitor the movements of the Earth's surface. A geologist can take bearings simultaneously on three orbiting satellites using radio signals; this enables him to mark his position exactly. Returning some years later to the same point allows him to measure the movement of the Earth's surface. Such measurements are made regularly in New Zealand, which lies along the boundary between two plates. The east and west parts of the islands that make up New Zealand are moving past each other in opposite directions, giving rise to volcanic and seismic activity.

As continents gradually move together and the oceans between them become smaller, the rocks of the Earth's crust on the ocean bed are pushed up into folds. These small folds at Lulworth in Dorset are an example

Questions

1 Using an atlas, find named examples of volcanoes for each of the following stages of the tectonic cycle (the areas in brackets will help you):
(a) continental rifting (East African Rift valley)
(b) linear sea spreading (Red Sea and Gulf of Aden)
(c) growth of ocean (islands on the Atlantic ridge)
(d) subduction of ocean floors far from continents, (island arcs near deep trenches)
(e) subduction of ocean floor near to continents (the Andes and Rockies).
(f) Are there any volcanoes in the final stage of the tectonic cycle, e.g. in the Himalayas? Where do we find volcanoes and why are volcanoes to be found there?

2 Explain clearly three pieces of evidence to support the theory of continental drift.

3 In the Americas we find jaguars and alligators and in Africa we find leopards and crocodiles. Explain how such similar species can exist on different sides of the Atlantic, and explain how small differences between similar species might have arisen.

4 How does the theory of plate tectonics depend on the fluid nature of part of the inner Earth?

5 Summarise the evidence which suggests that Britain has drifted slowly northwards over a period of millions of years.

6 What do we mean by the 'jigsaw fit' of continents?

SECTION M: QUESTIONS

1 What are the main sources of carbon dioxide in the atmosphere? What is the role of carbon dioxide in the atmosphere? What could happen if atmospheric levels of carbon dioxide changed significantly?

2 What is the significance of ozone in the atmosphere? What is causing the amount of ozone to change? What may be the long term consequences of increasing or decreasing the amount of ozone in the atmosphere?

3 The following diagram shows three samples of rock material. They are the same rock type and have an equal volume. Calculate the total surface area for each sample. Which sample would be most susceptible to weathering processes? How will their rates of weathering compare?

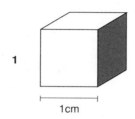

1

1cm

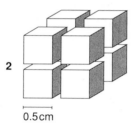

2

0.5cm

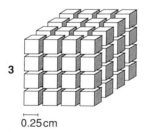

3

0.25cm

4 The diagram below shows an air mass moving across the sea from one land mass to another. Describe and explain the changes to the air mass as it moves from A to B to C.

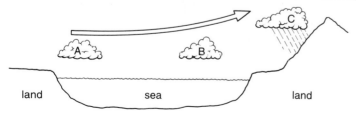

land   sea   land

(a) From A to B
(i) the change, if any, in the amount of water in the air mass
(ii) the change, if any, in the temperature of the air mass
(iii) the change, if any, in the pressure.
(b) From B to C
the change, if any, in the amount of water in the air mass.
(c) Explain why precipitation occurs at C.

ULEAC

5 (a) Explain what is meant by the following terms:
(i) igneous rock
(ii) sedimentary rock
(iii) metamorphic rock
(b) A student examined two samples of the same type of igneous rock using a microscope at the same magnification. She made the drawing below.

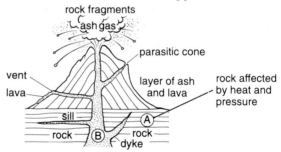

sample A   sample B

What difference in the origin of the two samples is likely to have caused this difference in appearance?

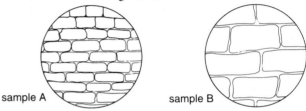

(c) The diagram represents a volcano erupting
(i) What type of rock would you expect to be found at A?
(ii) In time another type of rock will form at B.
Name the type of rock and explain how it is formed.

ULEAC

6 (a) In the study of earthquakes and Earth tremors, scientists can locate the **epicentre** of any disturbance from a study of the different waves produced. The epicentre is the point on the Earth's surface directly above the source of vibrations. The two main types of wave produced are the **P waves** and the **S waves**. These have different speeds through the Earth.

The behaviour of these waves from the time of the tremor is displayed in the graph on the next page.

Some scientists in a Cambridge laboratory detect a time lag of 10 seconds between the P and S waves for a particular tremor.

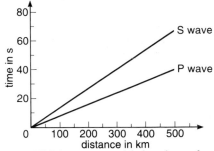

(i) Which wave would they have detected first? Why?
(ii) How far from Cambridge is the epicentre of this tremor?
(iii) Explain why it is not possible to determine the exact location of the epicentre from this single measurement.
(iv) What other measurements would be needed to work out the exact location of the epicentre?
(b) Variations in the velocities of P and S waves with depth below the surface of the Earth are shown on the graph below.

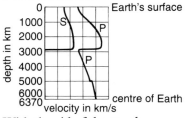

With the aid of the graph
(i) describe and explain what happens to P waves at the boundary between the mantle and outer core.
(ii) deduce the nature of the outer core from the behaviour of the S waves.
(c) Copy the diagram and mark the approximate depths of the mantle, outer core and inner core.

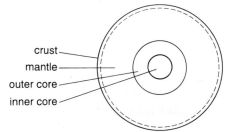

Explain how you deduced these values.
(d) (i) How does the theory of plate tectonics depend on the liquid nature of part of the inner Earth?
(ii) What happens when plates continuously move apart on the ocean bed?　　　　**ULEAC**

7 It is often said that the Sun is the source of energy which drives the weather. The energy supplied by the Sun can affect the movement of air and it can affect the amount of wet or dry weather.
(a) Copy the diagram at the top of the next column and indicate possible movements of air which might be due to the greater warming of the Earth near the equator.

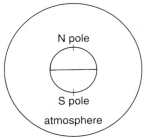

(b) In terms of the energy changes which take place, explain the following weather features:
(i) The difference between summer and winter temperatures in areas near the middle of continents is greater than in areas near the ocean shores.
(ii) Mountains near to a sea coast often have very high rainfall.
　　　　MEG

8 (a) The diagram below shows part of a weather map.

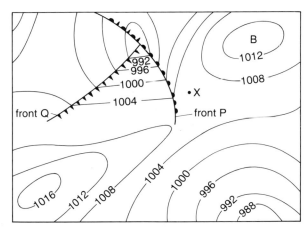

(i) What type of pressure centre is B?
(ii) What is a weather front?
(iii) What type of weather would be forecast for point X as the weather front P approaches and passes?

(b) During the Autumn the ground is often wet. Fog may form on a cloudless night. Explain how this occurs.
　　　　MEG

9 An earthquake is caused by the movement of rocks below the Earth's surface. The rock movement produces an earthquake wave.
The diagram shows a record of the shaking produced by an earthquake wave.

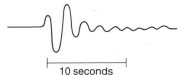

10 seconds

(a) What is the frequency of the earthquake wave recorded in the diagram? Show how you do your calculation.

(b) Earthquake waves travel at a speed of 8 km/s. What is the average wavelength of the earthquake wave recorded in the diagram? Show how you do your calculation.

An earthquake happened in a city where there were buildings of various heights. Only some of the buildings were destroyed. Shorter or taller buildings than those destroyed suffered less damage.

To try to understand this, Amy clamped two rulers to a table, as shown below. She made one project by 90 cm and the other by only 30 cm.

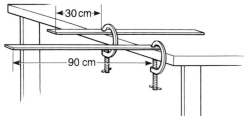

She set them both vibrating and found that the longer ruler vibrated with a longer period than the shorter one.

(c) Use Amy's observations to help you explain why the earthquake affected buildings of one particular height and not others.

(d) Give a brief plan of an investigations which could be used to test the explanation you have given in (c) above.

MEG

10 The table shows data obtained by school children doing a weathering project in a graveyard.

(a) Why do you think they chose a graveyard for this study?

(b) What do the results show?

Dates on gravestones	Type of rock		
	Sandstone	Marble	Igneous
1720–1770	extremely badly weathered	very badly weathered	moderately weathered
1771–1819	very badly weathered	badly weathered	moderately weathered
1820–1870	badly weathered	moderately weathered	slightly weathered
1871–1919	moderately weathered	slightly weathered	unweathered
1920–1970	slightly weathered	unweathered	unweathered
1971–1991	unweathered	unweathered	unweathered

11 The map below shows the location of three seismological stations which record earthquakes. The time that each station receives information on its equipment depends upon its distance from the epicentre of the earthquake. P waves arrive at stations faster than S waves; the time between the arrival of the two types of wave can help the stations to locate the epicentre. The table shows how far each station is from the epicentre.

Continent	Distance of station from epicentre/km	
	Quake 1	Quake 2
A	2500	1800
B	2000	1500
C	1100	3200

(a) Trace the map accurately and, using a compass, construct arcs to locate each epicentre.

(b) Why is it essential to know three distances?

(c) Why do stations receive little data for some earthquakes?

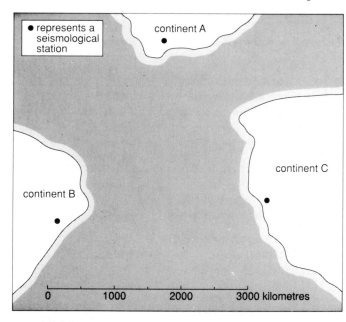

12 The map shows how geologists believe the world might look in eight million years time. What major changes have taken place? What evidence has led geologists to predict these changes?

Index

Acknowledgements

The following companies, institutions and individuals have given permission to reproduce photographs in this book:
Action Plus (42 top left), Adams IAL Limited (226, both); Airfotos (140); All–Action Photographic (177, top); J Allen Cash (86, top left; 157; 158, top); Allsport (4, both; 16, bottom; 26; 28; 31, middle left; 40, top; 86, top middle; 129); Alton Towers (78); Heather Angel (52); Associated Press (266, middle); Barnaby's Picture Library (75; 89; 266, top; 281); Black and Decker (223); Bob Thomas Sports Photography (40, bottom; 41, middle right); Martin Bond/Science Photo Library (156); Dr Tony Brain/Science Photo Library (260, top); Peter Brooker/Rex Features, London (251); CERN Photo (268); Channel 4 (208, top right); Chris Bonington (104); British Airways (2); British Antarctic Survey (114, bottom); British Rail (83); British Steel Strip Products; Bruce Coleman Limited (179); Central Electricity Generating Board (88; 89; 185; 192; 195; 232; 235; 267); Courier Newspapers (42); © Crown Copyright/MOD. Reproduuced with the permission of the Controller of HMSO (148); Dennis Di Cicco/Science Photo Library (56, top left); ECON Group Limited (102, bottom); The Electricity Council (186); © Fisons plc Scientific Equipment Division (258; 259); Ford (31, top); Nigel Forster (90, top left); Nick Fox (114, top); General Electric Company (219); Bob Gibbons (Natural Image) (24, top right; 96, top; 122); J H Golden (11, bottom); GreenGate Publishing Services (134; 145; 149; 150; 200; 260; 261); Robert Harding Picture Library (299, left); Hawaii Visitors Bureau (23); Holt Studios LTD (248); Kit Houghton (76); Hoverspeed (5); Independent Broadcasting Authority (17; 134, top); Joint European Torus (280, both); Rosalind Jones (309); © Stephen Krasemann/NHPA (299, top right); Frank Lane Picture Agency (193, top); David Lea Associates Limited (116); Vincent Martinelli (6, bottom; 27; 84, bottom; 200); Jerry Mason/Science Photo Library (58, right); Meteorological Office (163); Mercury Communications (146, bottom right); Ministry of Defence (Navy) (81, 225, 265); Larry Mulvehill/Science Photo Library (207); National Aeronautics and Space Administration (11; 39; 41; 58 left); Motorola LTD (252); NASA/Science Photo Library (50, top; 61; 64, top left; 66, top; 67, top and bottom right; 69, top right; 73; 177; 293); National Trust (100, bottom; 133; 271); Natural History Museum (300, both; 301); Natural History Picture Agency (1; 31, middle right; 160; 165); New Zealand Tourist Board (101); Mark Newcombe Photography (32, top); NOAO/Science Photo Library (56, bottom; 57; 63, middle right); National Portrait Gallery (60); NRSC Ltd/Science Photo Library (296); Pacific and Orient (32, bottom); Phillips (199); P & H S Architects (180); Pilkington (10); Redbreast Industrial Equipment (107); Roddy Paine, Photographer (153); Rolls-Royce (45); Ronald Royer/Science Photo Library (62); The Royal Institution (227); St. Bartholomews Hospital (282, bottom left and right); John Sandford/Science Photo Library (63, left); Barrie Schwartz (275, top); Science Photo Library (49, both; 50 bottom, 29; 44; 79; 95; 97, bottom; 115, both; 143, both; 147;, 166 both; 208 top left; 209, 213, 239, 272); Mr KED Shuttleworth/BUPA Hospitals Ltd for St Thomas' Lithotripter Centre (159); Sodel Photothetique (113); Syndication International (136); Colin Taylor Productions (86, top right); Tefal (105); Transport and Road Research Laboratorv (38, bottom); United Kingdom Atomic Energy Authority (275, bottom; 281; 2282, top); United Press International (188, top); US Geological Survey/Science Photo Library (66, bottom); Volvo(12; 38,top); PDWaghorn (126); WessexWater (102,top); ZEFA Photo Library (308).

We are grateful to the following examining bodies for permission to reproduce questions from specimen GCSE papers and from recent joint 16+ examinations in Physics:

The University of London Examinations and Assessment Council (ULEAC), The Midland Examining Group (MEG), The Southern Examining Group (SEG), The Welsh Joint Education Committee (WJEC), The Northern Examining Association (Associated Lancashire Schools Examining Board, Joint Matriculation Board, North Regional Examinations Board, North West Regional Examinations Board, Yorkshire and Humberside Regional Examinations Board (NEA), and the Northern Ireland Schools Education Committee (NISEC).

Orders: please contact Bookpoint Ltd, 39 Milton Park, Abingdon, Oxon OX14 4TD. Telephone: (44) 01235 400414, Fax: (44) 01235 400454. Lines are open from 9.00–6.00, Monday to Saturday, with a 24 hour message answering service. Email address: orders@bookpoint.co.uk

British Library Cataloguing in Publication Data
England, Nick
    Physics Matters. – 2Rev. ed
    I. Title
    530

ISBN 0 340-63935-0

First published 1989
Second edition 1995
Impression number 10 9 8 7 6 5 4 3
Year   2004 2003 2002 2001 2000 1999 1998

Artwork by Oxford Illustrators.
Text for second edition typeset in Ehrhardt by GreenGate Publishing Services, Tonbridge, Kent.
Original text typeset by Tradespools Ltd, Frome
Printed and bound in Hong Kong for Hodder & Stoughton Educational, a division of Hodder Headline Plc, 338 Euston Road, London NW1 3BH, by Colorcraft Ltd.